D1191312

POLYMER PROCESSING AND PROPERTIES

POLYMER PROCESSING AND PROPERTIES

Edited by

Gianni Astarita
and
Luigi Nicolais
University of Naples
Naples, Italy

PLENUM PRESS • NEW YORK AND LONDON

Library of Congress Cataloging in Publication Data

European Meeting on Polymer Processing and Properties

(1983: Capri, Italy)
Polymer processing and properties.

"Proceedings of the European Meeting on Polymer Processing and Properties, held June 13-16, 1983, in Capri, Italy." T.p. verso.
Includes bibliographical references and index.
1. Polymers and polymerization—Congresses. I. Astarita, Giovanni. II. Nicolais, Luigi. III. Title.
QD380.E88 1983 547.7 84-13265
ISBN 0-306-41728-6

Proceedings of the European meeting on Polymer Processing and Properties, held June 13-16, 1983, in Capri, Italy

A Division of Plenum Publishing Corporation
233 Spring Street, New York, N.Y. 10013

Printed in the United States of America

PREFACE

During the First Conference of European Rheologists, which was held in Graz, Austria, in April 1982, the Provisional Committee of European Delegates to the International Committee on Rheology held a meeting to discuss future European activities in the general area of rheology. It was agreed, among other things, that the organization of meetings in Europe on specific topics related to rheology would be done in cooperation, so as to avoid conflicts of dates and/or subject areas. Any such meeting, if approved by the Provisional Committee, would be named a European Meeting; the European Societies of Rheology would help the organizers with distribution of circulars, membership lists, and any required technical assistance. One of the very first meetings organized within this procedural scheme has been the European Meeting on Polymer Processing and Properties, which was held in Capri, Italy, on June 13-16, 1983. This book constitutes the Proceedings of that meeting.

The meeting was organized by the Italian Society of Rheology and the Polymer Engineering Laboratory of the Engineering School of the University of Naples. The basic idea of the meeting was to cover the whole area of polymer engineering with the exclusion of the actual production of the polymer: the whole area which is concerned with the processing of the polymeric material, and the properties of the finished product. Since the latter are very strongly influenced by the former, the core session of the meeting was the one dedicated to the relationships between processing and properties, part III of this book.

On the processing side, it was recognized that, while most of the scientific literature on the subject is concerned with operations the purpose of which is to obtain a finished product of assigned shape, in industrial practice there is a large number of important processing steps that have other purposes. Consequently, two sessions, corresponding to parts I and II of this book, were dedicated to the shaping and non-shaping operations.

On the properties side, it was recognized that there are two somewhat distinct areas to be considered. The first one is concerned with the actual identification, measurement, and possibly prediction of polymer properties. The second one is concerned with structural modeling, i.e., with the understanding of the relationship between macroscopic properties and microscopic structure of the polymer. The importance of the second area is related to the fact that different processing procedures will induce different structures, and hence different properties, in the finished product. The final two sessions of the meeting, corresponding to parts IV and V of this book, were dedicated to theses areas.

For each session, an internationally known expert was invited to give a broad-scope lecture; these are the first papers in each part. We wish to thank the five invited speakers, M. R. Kamal, C. D. Denson, H. Janeschitz-Kriegl, G. Menges, A. J. de Vries, for their authoritative contributions to the success of the meeting and to this book. We also wish to thank all other authors for their contributions.

During the organization of the meeting, a lot of the work was done by our friend and coworker, dr. P. Masi; we wish to thank him sincerely for all his efforts. We also wish to thank all our coworkers, students, and friends who have helped in a variety of ways. Finally, we wish to acknowledge support received from the following Institutions:

Azienda Autonoma di Soggiorno,Cura e Turismo, Capri
Ceast
Comune di Capri
Comune di Napoli
Consiglio Nazionale delle Ricerche
ENI
Istituto G. Donegani, SpA
Programma Finalizzato Chimica Fine e Secondaria, CNR
Regione Campania
Rheometrics, Inc.
Universita' di Napoli

Capri, Italy, June 1983
Gianni Astarita
Luigi Nicolais

CONTENTS

PART THREE - RELATIONSHIPS BETWEEN PROCESSING AND PROPERTIES

PART FIVE - STRUCTURAL MODELING OF POLYMER PROPERTIES

PART I

PROCESSING: SHAPING OPERATIONS

CURRENT TRENDS IN THE ANALYSIS OF POLYMER SHAPING OPERATIONS

Musa R. Kamal* and G.L. Bata**

*Chem. Eng. Dept., McGill University
Montreal, Canada H3A 2A7
**Industrial Materials Research Institute
Montreal, Canada

INTRODUCTION

Polymer processing, in its most general context, involves the transformation of a solid (sometimes liquid) polymeric resin, which is in a random form (e.g. powder, pellets, beads), to a solid plastics product of specified shape, dimensions, and properties. This is achieved by means of a transformation process: extrusion, molding, calendering, coating, thermoforming, etc. The process, as demonstrated in Figure 1, in order to achieve the above objective, usually involves the following operations: solid transport, compression, heating, melting, mixing, shaping, cooling, solidification, and finishing. Obviously, these operations do not necessarily occur in sequence, and many of them take place simultaneously.

Shaping is required in order to impart to the material the desired geometry and dimensions. It involves combinations of viscoelastic deformations and heat transfer, which are generally associated with solidification of the product from the melt.

Shaping includes: (a) two-dimensional operations, e.g. dieforming, calendering and coating, and (b) three-dimensional molding and forming operations. Two-dimensional processes are either of the continuous, steady state type (e.g. film and sheet extrusion, wire coating, paper and sheet coating, calendering, fiber spinning, pipe and profile extrusion, etc.) or intermittent as in the case of extrusions associated with intermittent extrusion blow molding. Generaly, molding operations are intermittent, and, thus, they tend to involve unsteady state conditions. Thermoforming, vacuum forming, and similar processes may be considered as secondary shaping opera-

SOLID RESIN

RANDOM CONFIGURATION
(PELLETS, BEADS, GRANULES
POWDER, ETC.)

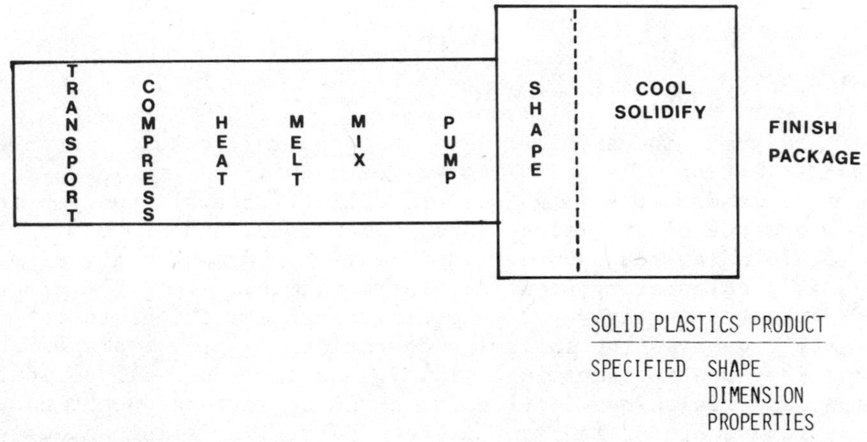

SOLID PLASTICS PRODUCT

SPECIFIED SHAPE
DIMENSION
PROPERTIES

Figure 1: Sketch of a Typical Plastics Processing Sequence

tions, since they usually involve the reshaping of an already shaped form. In some cases, like blow molding, the process involves primary shaping (parison formation) and secondary shaping (parison inflation).

Shaping operations involve simultaneous or staggered fluid flow and heat transfer. In two-dimensional processes, solidification usually follows the shaping process, whereas solidification and shaping tend to take place simultaneously inside the mold in three-dimensional processes. Flow regimes, depending on the nature of the material, the equipment, and the processing conditions, usually involve combinations of shear, extensional, and squeezing flows in conjunction with enclosed (contained) or free surface flows.

The thermo-mechanical history experienced by the polymer during flow and solidification results in the development of microstructure (morphology, crystallinity, and orientation distributions) in the manufactured article. The ultimate properties of the article are closely related to the microstructure. Therefore, the control of the process and product quality must be based on an understanding of the interactions between resin properties, equipment design, operating conditions, thermo-mechanical history, microstructure, and ultimate product properties. Mathematical modeling and computer simulation have been employed to obtain an understanding of these interactions. Such an approach has gained more importance in view of the expanding utilization of computer aided design/computer assisted manufacturing /computer aided engineering (CAD/CAM/CAE) systems in conjunction with plastics processing.

The following discussion will highlight some of the basic concepts involved in plastics shaping operations. It will emphasize recent developments relating to the analysis and simulation of some important commercial processes, with due consideration to elucidation of both thermo-mechanical history and microstructure development. More extensive reviews of the subject can be found in standard references on the topic (1,2,3,4,5,6).

GENERAL APPROACH TO PROCESS MODELLING

As mentioned above, shaping operations involve combinations of fluid flow and heat transfer, with phase change, of a visco-elastic polymer melt. Both steady and unsteady state processes are encountered. A scientific analysis of operations of this type requires solving the relevant equations of continuity, motion, and energy (i.e. conservation equations).

In general tensorial formulation, the conservation equations, in differential form, are as follows:

Continuity

$$\frac{D\rho}{Dt} = -\rho(\nabla.\vec{v})$$

Momentum

$$\rho \frac{D\vec{v}}{Dt} = -\nabla P - \left[\nabla.\vec{\vec{\tau}}\right] + \rho\vec{g}$$

Energy

$$\rho C_p \frac{DT}{Dt} = -(\nabla.\vec{q}) + \left(\frac{\partial \ln V}{\partial \ln T}\right) \frac{DP}{Dt} + (\vec{\tau}:\nabla\vec{v})$$

where ρ is the density, $\vec{v}$ is the velocity, P is the pressure, τ is the extra stress tensor, $\vec{g}$ is the gravitational acceleration, C_p is specific heat, T is temperature, $\vec{q}$ is the heat flux, V is the specific volume, and t is time.

Most work in the area of polymer processing has been limited to the treatment of one- or two-dimensional flow systems. Recently, more emphasis has been placed on the treatment of two-dimensional flow systems. Therefore, we shall present here the relevant equations for this case. The two dimensional equations, in Cartesian coordinates, take the following forms.

Continuity:

$$\frac{\partial \rho}{\partial t} = -\left[\frac{\partial}{\partial x}(\rho V_x) + \frac{\partial}{\partial y}(\rho V_y)\right]$$

Momentum:

$$\rho\left(\frac{\partial V_x}{\partial t} + V_x \frac{\partial V_x}{\partial x} + V_y \frac{\partial V_x}{\partial y}\right) = -\frac{\partial P}{\partial x} - \left(\frac{\partial \tau_{xx}}{\partial x} + \frac{\partial \tau_{xy}}{\partial y}\right) + \rho g_x$$

$$\rho\left(\frac{\partial V_y}{\partial t} + V_x \frac{\partial V_y}{\partial x} + V_y \frac{\partial V_y}{\partial y}\right) = -\frac{\partial P}{\partial y} - \left(\frac{\partial \tau_{xy}}{\partial x} + \frac{\partial \tau_{yy}}{\partial y}\right) + \rho g_y$$

Energy:

$$\rho C_p\left(\frac{\partial T}{\partial t} + V_x \frac{\partial T}{\partial x} + V_y \frac{\partial T}{\partial y}\right) = -\left(\frac{\partial q_x}{\partial x} + \frac{\partial q_y}{\partial y}\right)$$

$$-\left(\tau_{xx} \frac{\partial V_x}{\partial x} + \tau_{yy} \frac{\partial V_y}{\partial y}\right) - \tau_{xy}\left(\frac{\partial V_x}{\partial y} + \frac{\partial V_y}{\partial x}\right)$$

$$+\left(\frac{\partial \ln V}{\partial T}\right)_p \left(\frac{\partial P}{\partial t} + V_x \frac{\partial P}{\partial x} + V_y \frac{\partial P}{\partial y}\right)$$

Similar equations can be written in cylindrical or spherical coordinates.

In dealing with polymer processing operations, analysis of the shaping stage involves solving the above equations with appropriate initial and boundary conditions, in conjunction with realistic representation of material properties. Generally, none of these aspects is a simple task, and, therefore, many assumptions and approximations are employed.

Analytical solutions to the above equations exist only for the simplest cases of isothermal, incompressible, Newtonian, one-dimensional flow. In spite of this, first approximations are based on such assumptions, and the results obtained have been usually helpful in equipment and process design and optimization. Even when numerical methods are employed, solutions are not readily economically obtainable, and compromises are necessary. In many cases, convergence is not obtainable even at high cost. The current trend, however, is to analyze shaping operations as non-isothermal, visco-elastic two-dimensional flows, with compressible or incompressible materials, depending on the system. To achieve this, finite element techniques, which also have the advantage of adaptability to boundaries having complex shape, are becoming an important tool of analysis. However, limitations still exist, and a number of problems must be resolved before many important questions can be answered.

In view of the above complexities, no standard or general approaches have evolved yet for dealing with the shaping aspects of polymer processing. However, the following section outlines some of the methods most commonly used by polymer processing engineers.

(i) Equations of State. As mentioned above, the fluid is assumed to be incompressible in a large number of situations. However, in cases where density variations are important, as in the case of packing and solidification during injection molding, it is desirable to employ an appropriate equation of state to describe pressure-volume-temperature (P-V-T) variations. Three such equations have been found to be useful.

Spencer and Gilmore equation (7):

$$(P + \pi)\ (V - \omega) = RT$$

Breuer and Rehage equation (8):

$$V = V_o + \phi_o T - \frac{K_o}{a}\ (1 + bT)\ \ln(1 + aP)$$

Kamal and LeVan equation (9):

$$\rho = \rho_\infty + \left(\frac{\partial\rho}{\partial P}\right)_{P=o} T + (\alpha + \beta T)P + \tfrac{1}{2}(c + dt)P^2$$

In the above, $\pi, \omega, R, V_o, \phi_o, K_o, a, b, \rho_\infty, \alpha, \beta, c$, and d are constants.

(ii) Rheological Constitutive Equations. This represents one of the most complex issues involved in the analysis of polymer processing operations. The complexity is related to two aspects: (a) generally, equations that contain a description of the viscoelastic nature of polymer melts are quite complex mathematically, and (b) at the moment, there does not exist one rheological constitutive equation that can describe accurately all aspects of the viscoelastic behavior of polymer melts. In view of the above, polymer processing engineers have learned the art of rheological compromise.

The most commonly used rheological constitutive equations in conjunction with polymer melts are the Newtonian and Power Law models. Since most process models are based on one- or two-dimensional flow representations, these rheological equations take the following forms for two dimensional flow.

Newtonian.

$$\tau_{xy} = \eta\left(\frac{\partial V_y}{\partial x} + \frac{\partial V_x}{\partial y}\right)$$

$$\tau_{xx} = 2\eta \frac{\partial V_x}{\partial x}$$

$$\tau_{yy} = 2\eta \frac{\partial V_y}{\partial y}$$

Power Law.

$$\eta(\gamma) = m\left(\tfrac{1}{2} I_2\right)^{\frac{n-1}{2}} = m\left[\tfrac{1}{2}(\gamma_{xx}^2 + \gamma_{yy}^2 + 2\gamma_{xy}^2)\right]^{\frac{n-1}{2}}$$

where γ_{ij} refers to the rate of strain component.

There are many other more complex models that are commonly used in polymer processing analysis. Some of the subsequent discussion in this review will deal with the following three fluids: the convected Maxwell fluid, the White-Metzner Model and the Criminale-Ericksen-Filbey (CEF) fluid. The tensorial forms of these models are listed below, and the important parameters for two-dimensional flow are given in the Appendix.

Maxwell:

$$\overline{\overline{\tau}} + \lambda_o \frac{D\overline{\overline{\tau}}}{Dt} = -\eta_o \overline{\overline{\gamma}}$$

White-Metzner (10):

$$\overline{\overline{\tau}} + \frac{\eta}{G} \frac{D\overline{\overline{\tau}}}{Dt} = -\eta \overline{\overline{\gamma}}$$

CEF Model (11):

$$\overline{\overline{\tau}} = \eta \overline{\overline{\gamma}} + \tfrac{1}{2}(\Psi_1 + \Psi_2)\{\overline{\overline{\gamma}} \cdot \overline{\overline{\gamma}}\} - \tfrac{1}{2}\Psi_1 \frac{D\overline{\overline{\gamma}}}{Dt}$$

The Maxwell model has a constant viscosity and first normal stress coefficient and zero second normal stress coefficient, and its elongational viscosity becomes infinite at some finite elongation rate. In the White-Metzner modification of this fluid, the viscosity is a function of the rate of strain. For the CEF model, the viscosity and the first and second normal stress coefficients depend on the magnitude of the rate of strain tensor

$$\left|\gamma\right| = \left[\tfrac{1}{2} I_2\right]^{\frac{1}{2}} = \left[\sum_{ij}\sum \gamma_{ij} \cdot \gamma_{ji}\right]^{\frac{1}{2}}$$

In addition to the above constitutive equations expressed in the differential form, there are a number of integral constitutive equations that have been found useful and convenient in a variety of processing situations. Some of these equations are listed below.

Lodge's Rubberlike model (12):

$$\vec{\vec{\tau}} = + \int_{-\infty}^{t} M(t-t') \, \vec{\vec{\gamma}} \, (t') \, dt'$$

K-BKZ model (13)

$$\vec{\vec{\tau}} = + \int_{-\infty}^{t} \left[M_1(t-t',I,II)\vec{\vec{\gamma}}(t') + M_2(t-t',I,II)\{\vec{\vec{\gamma}}(t').\vec{\vec{\gamma}}(t')\} \right] dt'$$

(iii) Heat Transfer and Crystallization. Heat transfer is a critical aspect of plastics shaping, since the cooling or solidification time usually represents the longest proportion of the processing sequence and since the optimization of cooling rates is essential for controlling the rheological behavior during processing and the ultimate product properties. The treatment of heat transfer is based on the solution of the equation of energy, in conjunction with the equations of continuity and motion. In general, viscous dissipation is an important factor, especially at high processing speeds, due to the large viscosity of polymer melts. The boundary conditions usually reflect the cooling or heating conditions at the interfaces between the polymer and surrounding solid, liquid, or gaseous boundaries. Radiation can be an important factor as a heating source in some processes like thermoforming and blown film embossing, or as heat loss to the environment in many other situations. The treatment of convective heat transfer coefficients and radiation effects follows standard engineering procedures, and is well covered in standard polymer processing references (1-6).

One of the important areas involving heat transfer in polymer processing relates to heat generation or loss in the bulk of the material as a result of crystallization, melting, or chemical reaction as in the case of curing. In general, the above phenomena can be treated using approximately the same approach. In the following discussion, we shall concentrate on the analysis of heat transfer accompanied with crystallization in partially crystalline polymers. The treatment of melting is usually well covered in works dealing with melting during extrusion (14). Heat transfer in reactive thermosetting systems has been also discussed elsewhere (15). The approach discussed below incorporates the effects of both crystallization and the variation of the specific heat of each of the phases with temperature. Moreover, it permits the estimation of crystallinity development during processing (16).

A typical curve showing the dependence of the specific heat of a crystallizing polymer on temperature is shown in Figure 2. Assuming that the specific heats of the melt and the solid are independent of the cooling process, within the solidification range, the latent heat of solidification is

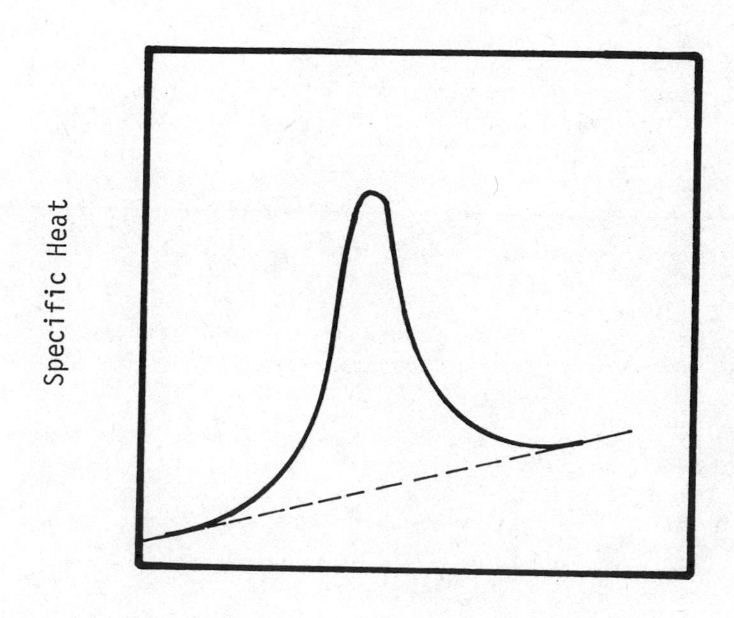

Figure 2: Typical DSC Scanning Curve for Specific Heat of a Crystallizing Polymer

included in the specific heat as follows:

$$C_p = C'_p + \frac{d\lambda c}{dT} = C'_p + \lambda_{pc}\frac{dx}{dt}$$

where C_p is the measured specific heat of the crystallizing melt, C'_p is the specific heat obtained by interpolation on the baseline, λ_c is the heat of crystallization released by the sample, λ_{pc} is the heat of fusion of the pure crystal, and X is the degree of crystallinity. Thus, it is important to follow the development of crystallinity in order to calculate the specific heat at any desired temperature during the process.

Nakamura et. al. (17) developed a procedure based on the Avrami equation, which was modified to apply to nonisothermal crystallization. The fundamental equation was written on the basis of isokinetic conditions:

$$X(t) = 1 - \exp\left[-\left(\int_o^t k(T)\,d\tau\right)^n\right]$$

where X(t): degree of phase transformation at time t.
n: Avrami index determined in isothermal experiments

$K(T) = \{k(T)\}^{\frac{1}{n}}$ where k(T) is the Avrami isothermal rate constant. From this model, nonisothermal crystallization can be analyzed in terms of data from isothermal crystallization experiments. The Avrami equation has the form:

$$X(t) = X(\infty)\left[1 - \exp\,(k(T)t^n)\right]$$

where $X(\infty)$ refers to the fact that the sample will reach a limiting crystallinity level. $X(\infty)$ will depend on temperature, and is below 1.0. It can be obtained from quenching experiments (16).

Figures 3 and 4 show comparisons between calculated and experimental temperature and crystallinity profiles obtained for the cooling of high density polyethylene is a static mold (16). The agreement is quite good. A similar approach, employing a somewhat different nonisothermal crystallization model, has been proposed by Dietz (18).

(iv) Numerical Methods. The formulation of polymer shaping problems, consisting of the continuity, momentum, and energy equations results in a system of multiple, simultaneous,

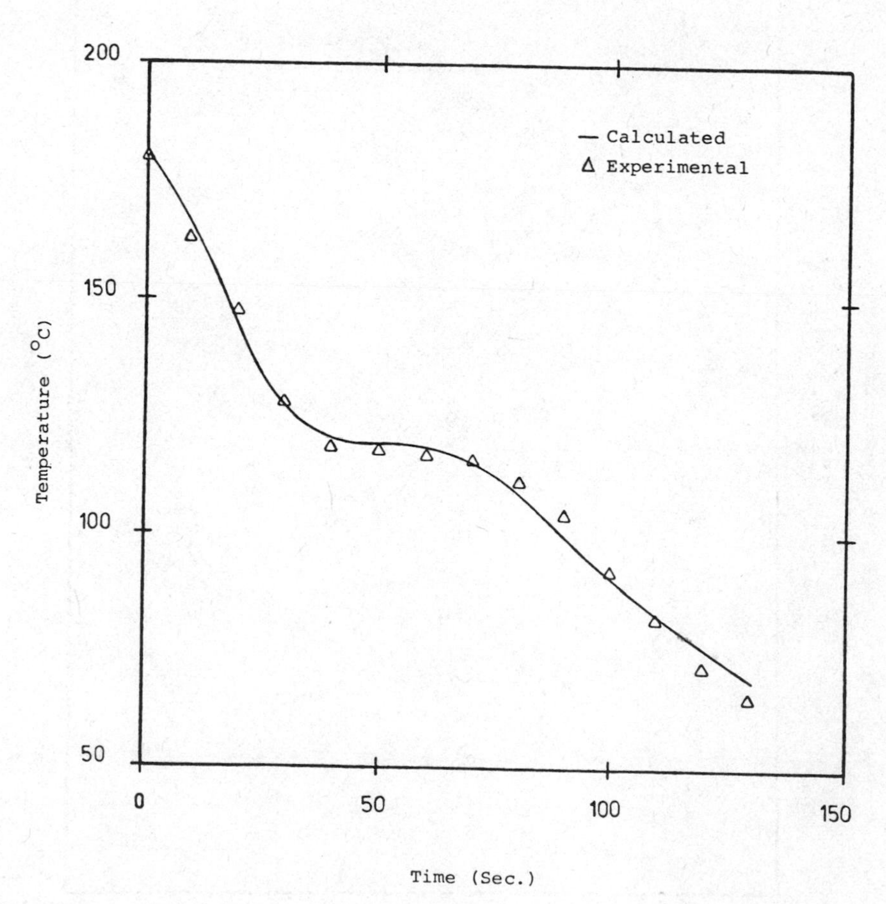

Figure 3: Temperature Profile During Static Cooling Experiment (16).

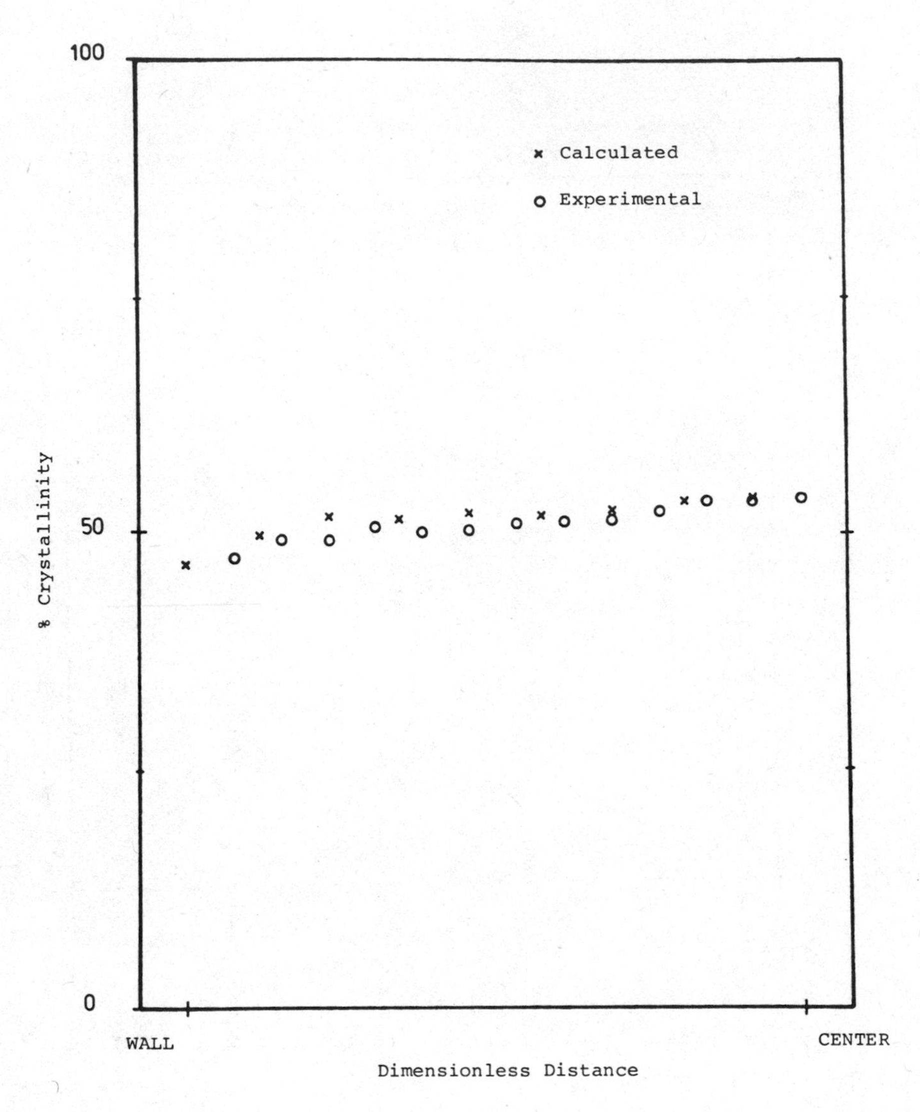

Figure 4: Crystallinity Profile During Static Cooling Experiment (16).

coupled, non-linear partial differential equations, which can only be solved by a numerical technique. Generally, the numerical techniques employed are based on finite difference or finite element methods.

When transient and inertial terms are neglected and when only simple, regular boundaries are involved, it is possible to employ an inexpensive explicit finite difference scheme. In some cases, involving simple shapes, a Marker-and-Cell technique may be used to treat free surface and transient flows (19). However, when the unsteady state terms of the momentum equations are included, it is necessary to use implicit finite difference equations because of the restriction on the time step due to the usually low Reynolds numbers. In this case, it might be more advisable to use finite element techniques.

Finite element techniques are finding wider applications, especially in systems involving free surface flows and complex geometries. For example, they have been used to study phenomena like die swell and entry and exit flows, in addition to molding situations involving complex molds and weld lines (20-26). One of the problems associated with these methods relates to the lack of convergence at Deborah Numbers ($\lambda\dot{\gamma}$ where λ is the characteristic time of the material and $\dot{\gamma}$ is the rate of deformation) exceeding unity. Recently, however, a number of workers have obtained convergence at Deborah numbers approaching 5.0 (23,27). Wang and Co-workers (26) have reported convergence for much higher Deborah numbers, employing a Leonov constitutive equation.

The use of finite element methods has also gained popularity because of the expansion in using CAD/CAM/CAE systems that incorporate three dimensional representation of objects having complex surface geometry.

EXAMPLES OF POLYMER SHAPING OPERATIONS

The remaining sections of this review outline the features of various models describing four typical plastics shaping operations. Two of the selected operations are basically continuous, steady state operations employed in generating two-dimensional (cross-section control) polymeric products: calendering for the production of film and sheet and wire coating for applying an annular coating over a cylindrical wire. The other two operations represent intermittent, unsteady state shaping operations for the manufacturing of three-dimensional objects: extrusion blow molding for the production of hollow containers and injection molding, which is the most important commercial process for the production of three-dimensional plastics articles. The discussion outlines the basic principles relating to the four operations and reviews recent developments regarding the analysis and modeling of the phenomena involved.

CALENDERING

Calendering is a continuous, steady state process involving a pair of corotating heated rolls. The system is usually fed with a thermoplastic melt which is squeezed into a thin film or sheet at the outlet end. A sketch of the process is shown in Figure 5. The diameters and temperatures of the two rolls are not always equal.

The relevant equations for flow between narrow calendars, when coupled with the lubrication approximation, reduce to the following forms:

Continuity:

$$Q = 2 \int_o^h V_x dy$$

Momentum:

$$\frac{\partial P}{\partial x} = \frac{\partial}{\partial y} (\tau_{yx})$$

Boundary Conditions

at the point of contact	, $y = H$	, $P = 0$	
at the leave-off point	, $y = H_1$	, $P = 0$	, $\frac{dP}{dx} = o$
at the upper cylinder	, $y = h^1$	, $u = U_1$	
at the lower cylinder	, $y = -h$	, $u = U_2^1$	

The early analysis of calendering was developed by Gaskel (28) and McKelvey (2) for Newtonian and Power Law fluids. The isothermal flow analysis for a Newtonian fluid yields the following pressure profile for rolls of equal diameter and velocity.

$$P = \frac{\mu U}{Ho} \left(\frac{9R}{32Ho}\right)^{\frac{1}{2}} \left[g(\rho,\lambda) + c \right]$$

$$g(\rho,\lambda) = \left[\frac{\rho^2 - 1 - 5\lambda^2 - 3\lambda\rho^2}{(1 + \rho^2)^2} \right] \rho + (1 - 3\lambda^2) \tan^{-1} \rho$$

$$C \approx 5\lambda^3$$

$$\rho = \frac{x}{\sqrt{2RH_o}} \quad ; \quad \lambda^2 = \frac{x_1{}^2}{2RH_o}$$

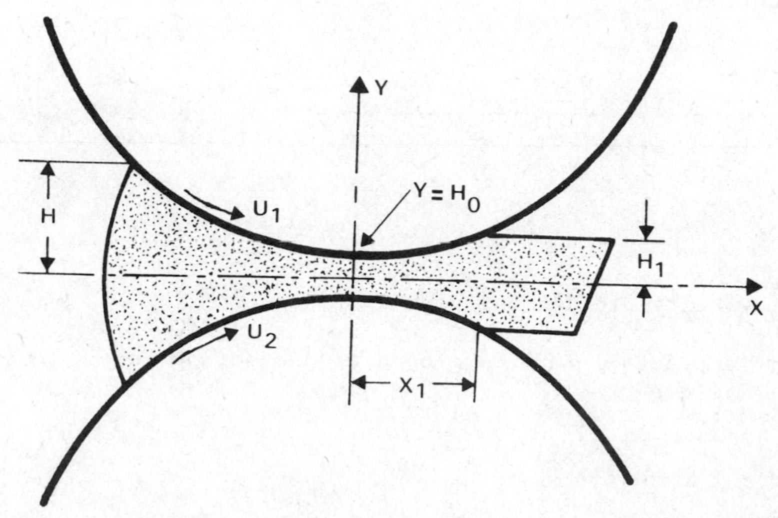

Figure 5: Notation for the Flow Analysis in the Gap between Two Rotating Calendar Rolls

In the above equations, P is the pressure, U is the velocity of the roll, and μ is the Newtonian viscosity of the fluid. The solution employs the lubrication assumption and assumes a no-slip condition. It yields a maximum pressure, P_{max}, at a point located at $-\lambda$ from the nip.

$$P_{max} = P(-\lambda) = \frac{5\mu U\lambda^3}{H_o} \left(\frac{9R}{Ho}\right)^{\frac{1}{2}}$$

and shows that the pressure at the nip is equal to half of the maximum pressure. Typical pressure profiles are shown in Figure 6 (29). The plots also show the effect of the Power Law index, n.

The velocity distribution for the system is given by the following equations:

$$\frac{V_x}{U} = \frac{2 + 3\lambda^2(1-\eta)^2 - \rho^2(1-3\eta^2)}{2(1 + \rho^2)}$$

where $\eta = \frac{y}{h}$

A typical velocity profile is shown in Figure 7 (2).

Alston and Astill (30) developed a calendering model for a "tanh" model fluid

$$\eta = A - B \tanh \left(\frac{\dot{\gamma}}{k}\right)^n$$

where A,B,k, and n are constants. Chung (31) carried out a similar analysis for a Bingham plastic fluid.

Recently, Vlachopoulos and Hrymak (32) extended the McKelvey and Gaskell analysis to allow for temperature rise due to viscous dissipation and for slip of the Power Law fluid at the rolls. They observed that the incorporation of slip yielded improved agreement between model predictions and experimental data for polyvinyl chloride. In more recent work, Kirpassidis and Vlachopoulos (29) and Mitsoulis, Vlachopoulos and Mirza (27) employed finite element methods to analyze the isothermal and non-isothermal two-dimensional calendering of Newtonian and Power Law fluids. The results showed that recirculating flow occurs in the entry, while flow becomes parallel in the vicinity of the minimum gap width. Figure 8 depicts the temperature development in a typical calender gap of a Power Law fluid (27). Mitsoulis, et. al. (27) have improved the

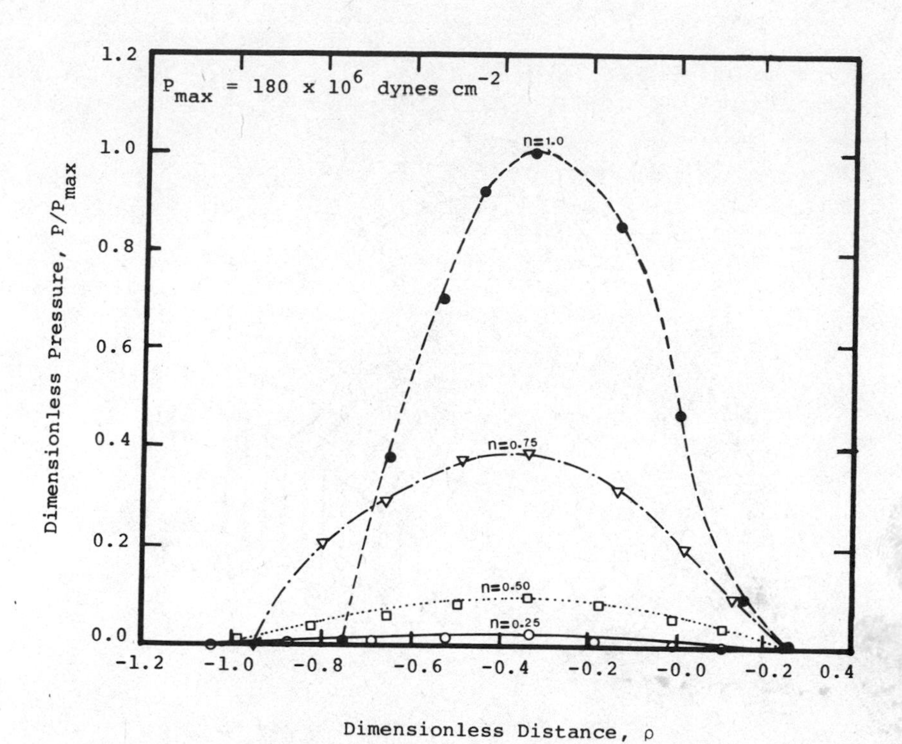

Figure 6: Effect of Power Law Index on Pressure Distribution in Calendering (29)

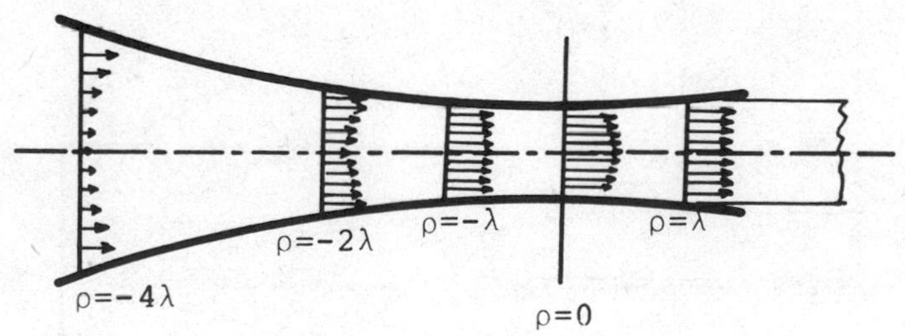

Figure 7: Velocity Profiles in Calendering (λ^2 = 0.10) for Isothermal Newtonian Flow (2)

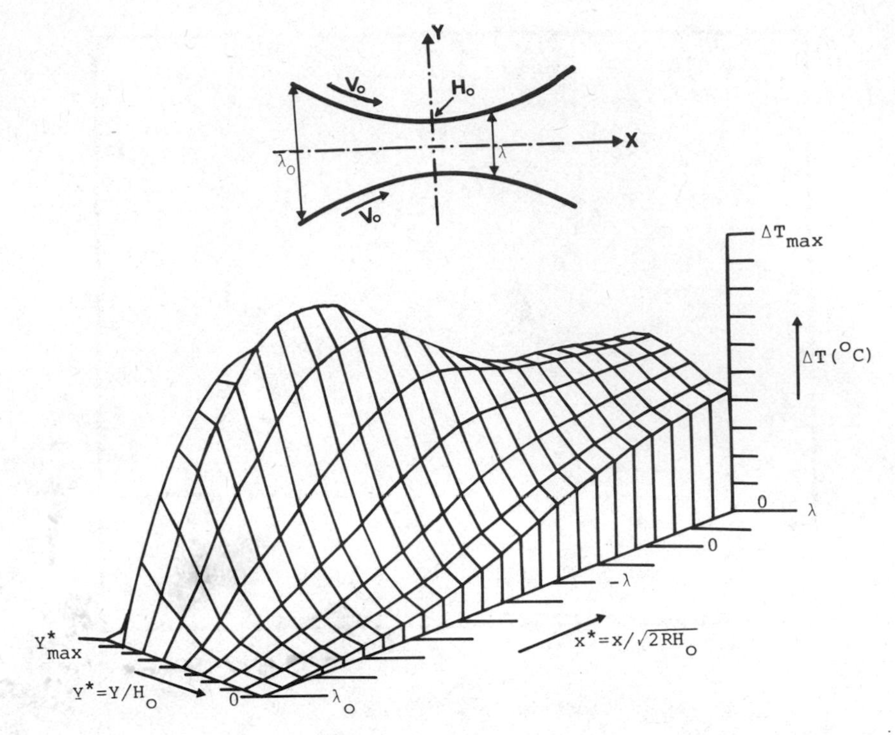

Figure 8: Temperature Distribution in the Calendering Gap (27).

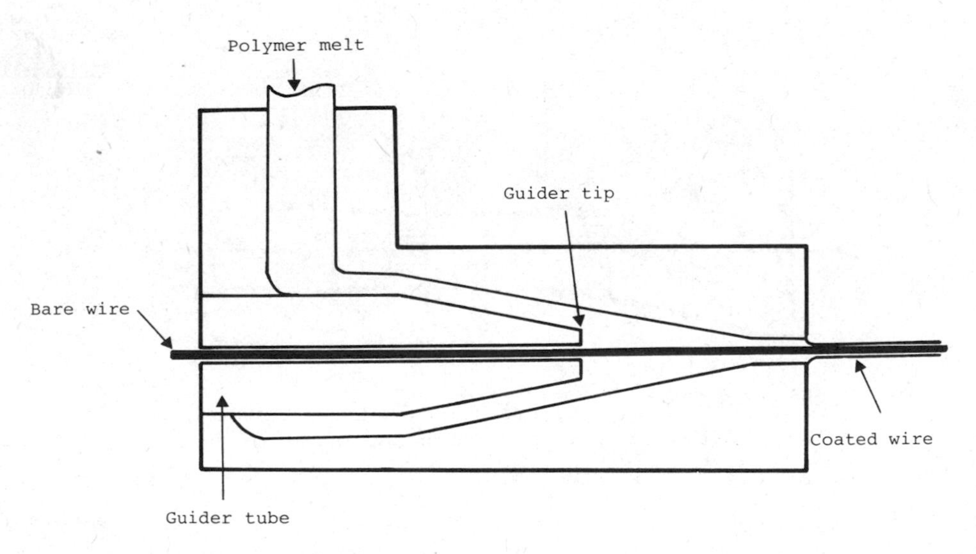

Figure 9: Pressure Type Wire Coating Die.

stability of the solution of the non-isothermal problem by employing an "upwind" scheme. Moreover, they extended the finite element treatment to the analysis of the flow of a CEF viscoelastic fluid between parallel plates and the entry zones.

WIRE COATING

The wire and cable coating process is an extrusion operation which involves the coating of wires of various diameters with polymers in order to provide protection, insulation, and other desirable properties. The wire is fed to a crosshead die which leads to a guider tube. As the wire exits through the guider tip and enters the die forming regions, the polymer melt coats the wire circumferentially. The polymer is supplied to the die in molten form through the use of an extruder.

There are two basic wire coating die designs. These are known industrially as "tube type" and "pressure type" dies. In the first type, the coating is extruded in the form of a cylindrical tube, which is later drawn onto the moving wire, either by vacuum or by the pull of the wire, after the wire and the tube emerge separately from the die. In pressure dies, the wire contacts the pressurized molten polymer within the die, as shown in Figure 9. Both the wire and the polymer emerge simultaneously as coated wire from the die.

Both of the above die configurations can be used effectively for low speed wire coating. In the case of high speeds, the pressure type die is preferred, since it takes advantage of the wire drag and, therefore, requires lower operating pressures. At high wire speeds (4,000-5,000 ft/min), surface roughness and melt fracture are encountered.

Carley (34) and McKelvey (2) proposed the treatment of an idealized isothermal wire coating process for Newtonian fluids by addition of the contributions of drag and pressure flow in cylindrical annuli. Under these conditions, the overall flow rate, Q, is given by

$$Q = \frac{V_w WH}{2} + \frac{WH^3 \Delta P}{12\mu L}$$

and

$$\Delta P = \frac{6\mu L V_w (2h-H)}{H^3}$$

where V_w is the wire velocity, W is the mean circumference of the annular space, H is the size of the annular gap, ΔP is the applied pressure, L is the contact length between the wire and melt inside the die, and μ is the Newtonian viscosity. Han and Rao (35) extended the above treatment to Power Law fluids. They divided the flow regime into two radial zones, one on each side of the point at which the maximum velocity occurs, $r/R_o = \lambda$, where R_o is the outer radius of the annulus. The expression for the volumetric flow rate, Q, becomes

$$Q = \pi R_o^2 (\lambda^2 - k^2) V_w + \pi R_o^3 \left(\frac{R}{2k} \frac{dP}{dz}\right)^s \int_k^1 \frac{(\lambda^2 - \xi^2)^{s+1}}{\xi^s} d\xi$$

where k is R_i/R_o with R_i being the inside radius of the annulus, s is 1/n where n is the Power Law index, ξ is the dimensionless radial position (r/R_o), and $\frac{dP}{dz}$ is the axial pressure gradient. The solution produces expressions for velocity distributions in the two radial zones, and it is necessary to employ a numerical technique to obtain λ by solving the equation for the identity of the velocity at the interface of the two zones. Typical results for the dependence of flow rate on wire speed and pressure gradient are shown in Figure 10 (5).

Carley, Endo, and Kranz (36) have obtained a finite difference solution for the two-dimensional (axial and radial) steady flow of an incompressible non-Newtonian fluid in both the tapered and die sections of the process. Their model employs a temperature and shear rate dependent viscosity and deals with both the isothermal and adiabatic situations. In order to avoid stability problems, they dropped the term $v_r \left(\frac{\partial T}{\partial r}\right)$ from the energy equation. Typical calculated temperature and velocity distributions in the tapered section of a commercial die, operating adiabatically with low density polyethylene at 4,000 ft/minute, are shown in Figures 11 and 12 (36). Similar solutions were obtained for a Power Law fluid by Vergnes (37). Fenner (38) and Fenner and Nadiri (39) employed finite element methods to solve a similar problem under isothermal conditions, employing the lubrication approximation. Basu (40) employed a finite difference scheme to solve coextrusion wire coating problems under non-isothermal conditions, but employing only one velocity component in the solution.

Although the lubrication approximation is helpful in cases involving small gap and taper angles, the procedure does not yield a clear and accurate description of stress distributions and irregularities that might lead to effects like melt fracture. Caswell and Tanner (41) have developed a finite element scheme to solve the relevant equations of change, without the lubrication approximation, in commercial wire coating dies. Since such dies usually include complex and changing geometries and involve abrupt changes in boundary conditions, the finite element method appears to be most

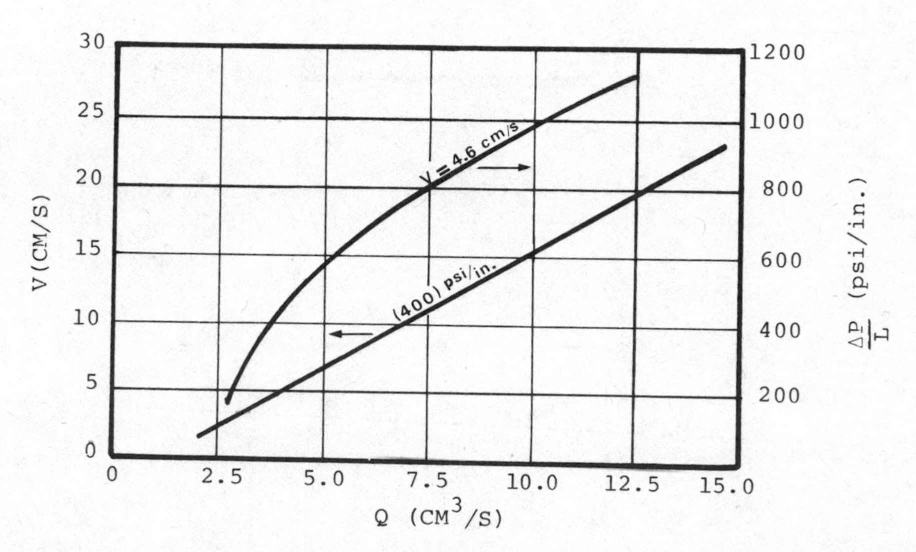

Figure 10: Calculated Effects of Wire Speed and Extrusion Pressure on Flow Rate of HDPE at 190°C. (5)

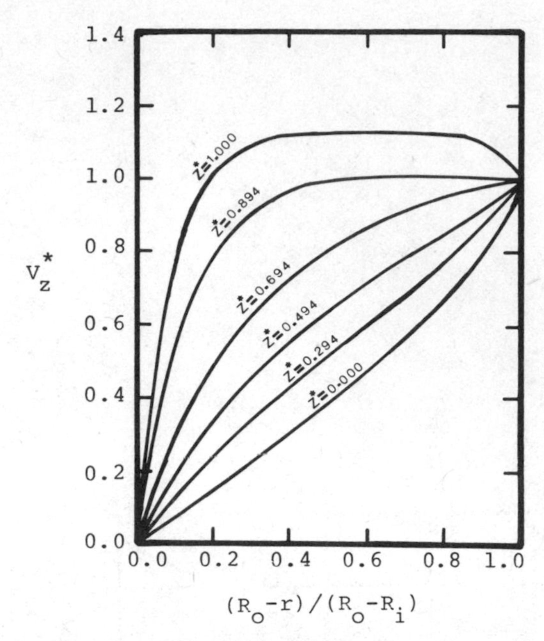

Figure 11: Velocity Profiles at Selected Axial Positions in LDPE Wire Coating Reference Velocity: 4000 ft/min (36).

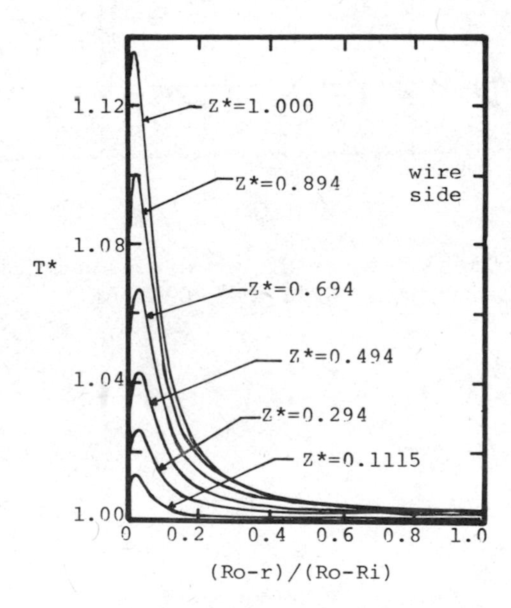

Figure 12: Temperature Profiles for LDPE at Wire Speed of 4000 ft/min. Reference temperature is 500°K (36).

suited for dealing with the wire coating problem. The application of this technique to isothermal Newtonian fluids shows circulation patterns in the area where die geometry changes, as indicated by the streamlines shown in Figure 13 (42). It has been possible to model isothermal pressure-type wire coating systems with this technique over a broad range of commercial operating speeds, for both Newtonian and Power Law fluids (42). However, instabilities in temperature predictions when nonisothermal solutions were attempted. The same approach was employed by Viriyaynthakorn and Deboo (43) for the analysis of tube-type wire coating dies, in conjunction with a generalized Newtonian fluid. An iterative procedure was employed to determine the shape of the unknown draw-down surface and the initial polymer-wire contact point. Typical stream lines for this situation are shown in Figure 14 (43).

BLOW MOLDING

Blow molding is the preferred commercial process for the production of hollow plastics containers. Extrusion blow molding consists of the following basic steps: (1) the extrusion of a molten plastic tube, "parison", from an annular die, (2) the clamping of the parison between the ends of a mold having the deisred shape, (3) the introduction of compressed air to inflate the parison so that it contacts the cold walls of the mold where it solidifies, and (4) the opening of the mold and ejection of the molding, when it has reached an acceptable level of rigidity. The cycle is repeated to produce identical articles. Blow molded articles cover a broad range of bottle shapes and sizes and include containers as large as 55 gallon drums.

Depending on the thermal stability of the material, the extrusion process can be continuous or intermittent. The former is the preferred technique with material having lower stability. Intermittent extrusion can employ either a reciprocating screw or an independent ram.

As a result of both parison swell and sag and the design characteristics of commercial containers, it is necessary to continually adjust the extrusion rate and/or parison thickness during extrusion. This is accomplished with the aid of parison programming, which involves movement of a mandrel to vary the size of the die opening or the control of the movement of the screw or the ram in intermittent blow molding. New machines employ electronic parison programming.

Injection blow molding offers the advantage of accurate dimensional control, elimination of scrap, and the molding of threads before blowing. A preform is injection molded over a rod, then the preform is introduced to the blow molding machine, where it is blown to the desired shape. Stretch blow molding is a process

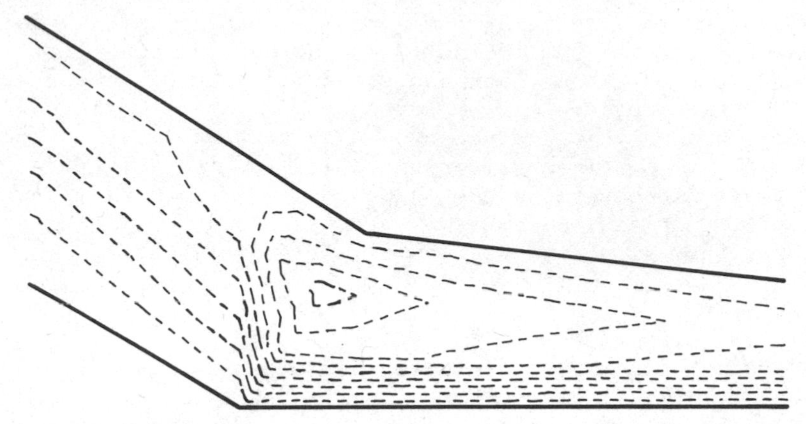

Figure 13: Calculated Streamlines and Circulation in Wire Coating Dies for A Newtonian Fluid ; Finite Element Method. (42)

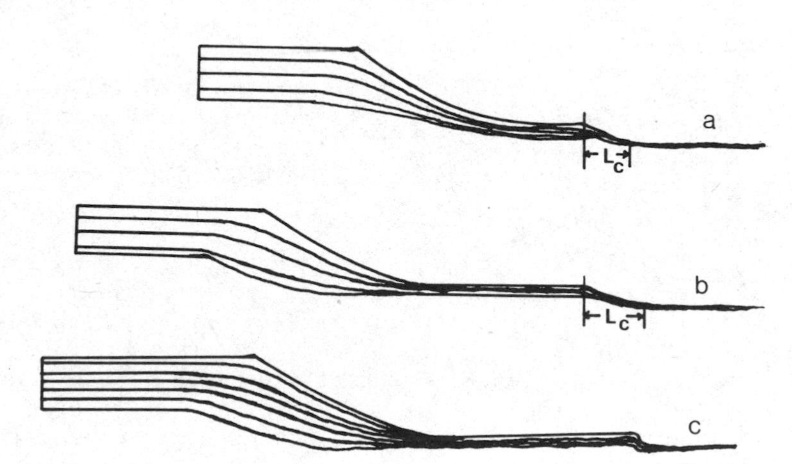

Figure 14: Streamlines Determined by Finite Element Analysis of Tube Type Coating Dies (43)

(a) Die A , no vacuum
(b) Die B , no vacuum
(c) Die C , high vacuum

whereby biaxial orientation (in the axial and radial directions) is optimized to yield desirable mechanical, optical, barrier and other physical-chemical properties. This process may be employed in conjunction with either extruded or injection molded parisons.

The optimization of the cooling system is an important aspect of blow molding, since it influences greatly production rates, production costs, and product quality. While high cooling rates (by employing refrigerated fluids) are desirable for high production rates, they could lead to frozen stresses and deficiencies in the product.

The shape and dimensions of the parison are the result of complex interactions between the molten polymer, which has been formed and shaped under the influence of thermal and mechanical conditions that exist in the die, and the thermo-mechanical conditions that influence the melt after it leaves the die. Parison swell is obtained as a result of exit phenomena and processes whereby the viscoelastic melt revovers from the shear and normal stresses which it experiences in the die. Sag or drawdown results from the action of gravity on the parison.

In a recent study, Orbey (44), evaluated the effects of die geometry on the isothermal swell and combined swell and sag for a variety of polyethylene blow molding resins. She discovered that diameter swell is highest for converging dies and lowest for diverging dies, while straight dies show intermediate swell between these extremes. The situation is not so clear for thickness swell, although it appears that converging dies exhibit somewhat higher thickness swell. The results indicate that the equilibrium diameter swell ratio, B_1, may be expressed in terms of the weight swell ratio, B_w, and the thickness swell ratio, B_2, according to the following relationship (45,46):

$$B_1 = AB_w^a = A(B_1B_2)^a$$

where A and a are constants depending on both the resin and geometry of the die. Generally, A is close to unity, while a is in the range 0.2 - 0.6. In the above:

$$B_1 = D_{o,e}/D_{o,d}$$

$$B_2 = G_e/G_d$$

$$B_w = B_1B_2$$

where $D_{o,e}$ is the outside diameter of the parison, $D_{o,d}$ is the outside diameter of the annular die, G_e is the thickness of the parison, and G_d is the annular gap at the die exit. It should be emphasized that the above relationship applies to equilibrium die swell, since swell is a transient phenomenon (45,47,48).

Conventionally, researchers have employed the parison pinch-off mold to determine the diameter swell characteristics of blow molding parisons (46,49,50). However, cinematographic studies have shown that this technique is not necessarily accurate (51).

Kamal, et. al. (52,53) have employed the approach proposed by Ajroldi (54) to estimate the dimensions of the parison under the combined effects of swell and sag. According to this approach, the area of a differential ith element of the parison (see Figure 15), A^i (z,t), is

$$A^i(z,t) = \pi \left[B_1(t) D_{o,d} - B_2(t) G \right] \left[B_2(t) G \right]$$

where, t, refers to time and the transient nature of swell, and z is the axial distance below the die. Under the influence of sag, the area becomes A_s^i (t), where

$$\ln A_s^i(t) = \ln A^i(t) - gJ(t)\{\Sigma W^i(t)\}/A_s^i(t)$$

where g is the gravitational constant, J(t) is the tensile compliance, and $W^i(t)$ is the weight of the ith pillow. The summation is taken to represent the weight of the elements below the its pillow. Kamal et. al. (52,53) obtained good agreement between calculated and measured parison diameters, as shown in Figure 16 (53). Similar agreement has been obtained for parison thickness and length.

Orbey (44) has proposed a somewhat similar model for the treatment of combined swell and sag in isothermal parisons. She bases her model on Lodge's rubber-like liquid model (12), which incorporates the relaxation spectra of the melt.

The shape of the parison at the end of extrusion tends to be circular. During the clamping stage, the parison is deformed both at the top and at the bottom. Fukase and coworkers (55) proposed equations for the estimation of the profile made by the outer surface of the parison during inflation. On the basis of this approach, it is possible to estimate the distribution of parison thickness in both the axial and circumferential directions (52,52).

More elaborate models describing the behavior of the parison formation stage (56) and the inflation stage (56-58). However,

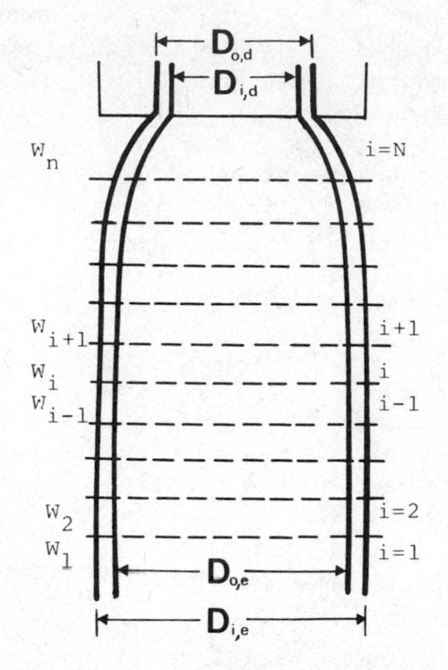

Figure 15: Schematic Diagram of Parison Segments During Extrusion

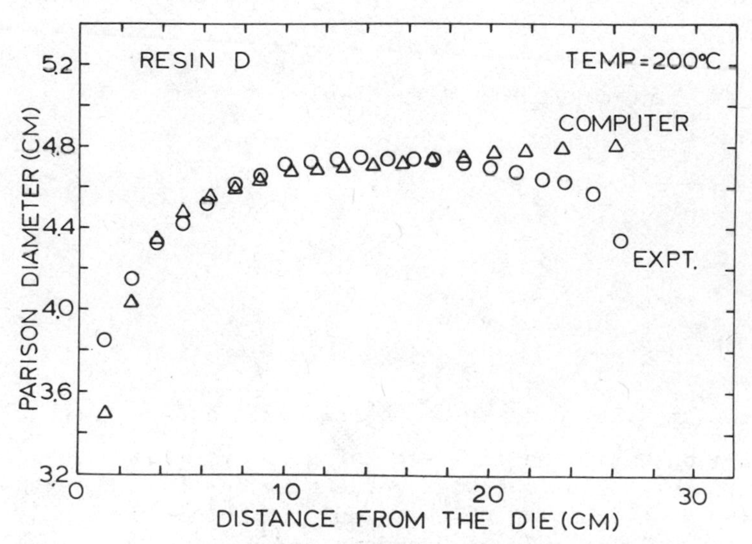

Figure 16: Comparison of Calculated and Experimental Parison Diameter Distributions for HDPE (53)

the predictions of these models have not been compared in sufficient detail to experimental data. In the following, we outline the model proposed by Ryan and Dutta (BM14).

According to the technique proposed by Ryan and Dutta (56), if the curvature of the surface is reasonably small (i.e. if $\partial h/\partial x \ll 1.0$), the relevant equation of motion for the parison during extrusion becomes

$$\frac{\partial V_x}{\partial t} + V_x \frac{\partial V_x}{\partial x} = \frac{3}{3\rho} \frac{\partial \tau_{xx}}{\partial x} + \frac{3\tau_{xx}}{2\rho h} \frac{\partial h}{\partial x} + g_x$$

where x is the extrusion direction and h is the parison thickness. The free surface is governed by the equation

$$\frac{\partial h}{\partial t} = V_y - V_x \frac{\partial h}{\partial x}$$

and the equation of continuity yields upon integration

$$\frac{\partial h}{\partial t} = V_y - V_x \frac{\partial h}{\partial x}$$

At the free surface, both normal and shear stresses vanish. Ryan and Dutta neglect the convective terms and employ a corotational Maxwell model with parameters λ_o and η_o to obtain the following relation:

$$U_x = V_E \exp(-\alpha\lambda_o \chi) + \frac{Lg}{\lambda_o} (1-\exp(-\alpha\lambda_o \chi)$$

$$- \frac{1}{\alpha\lambda_o^2} (\alpha\lambda_o \chi + \exp(-\alpha\lambda_o \chi))$$

where

$$\alpha = \rho g_x / 3\eta_o$$

V_E is the extrusion velocity at the die exit and Lg is the length of the parison under the influence of gravity. The actual length of the parison will be somewhat between Lg and Ls, where the latter represents the length that would be obtained if only swell took place in the absence of drawdown.

In order to determine the parison behavior during inflation, Ryan and Dutta write the equations of continuity and motion in the radial direction

$$\frac{\partial}{\partial r}(rV_r) = o$$

$$\rho\left(\frac{\partial V_r}{\partial t} + V_r \frac{\partial V_r}{\partial r}\right) = -\frac{\partial P}{\partial r} + \frac{\partial \tau_{rr}}{\partial r} + \frac{\tau_{rr} - \tau_{\theta\theta}}{r}$$

The boundary conditions relate the values of the pressure, P, at the inside and outside walls of the parison. The authors employed both a modified Zaremba-Fromm-Dewitt model recommended by Pearson and Middleman (57), as well as a modified White-Metzner fluid (10).

In most extrusion blow molding operations, especially those involving high production speeds, the melt may be considered to be isothermal throughout the parison formation and free inflation stages (59,60). Recently, a number of studies have been carried out relating to heat transfer during the cooling stage (56,59-63). Typical temperature distributions during the cooling stage are shown in Figure 17 (59). These results suggest a distribution of cooling rates in the material during solidification. This leads to a distribution of crystallinity between the outer and inner surfaces of molded bottles, which may be related to the cooling rate as shown in Figure 18 (59). The combined thermal and mechanical history of the material leads to the development of frozen stresses and orientations, in addition to the above morphological effects. Therefore, the mechanical and optical properties of blow molded articles are generally anisotropic (59,64,65).

INJECTION MOLDING

Injection molding is the most important commerical process for the manufacturing of three dimensional plastics articles. The process consists of three basic steps. During the filling stage, a viscoelastic, compressible polymer melt flows into the cavity. The walls of the cavity are maintained at a temperature significantly below the solidification temperature of the material. When the cavity is full, more melt is injected into the mold during the packing stage. This step, which takes advantage of the compressibility of the melt, causes a rapid rise of the pressure in the cavity and contributes to reducing the ultimate shrinkage of the material as a result of cooling and colidification. Cooling of the material, which starts at the beginning of the filling stage, continues throughout the cycle. In particular, after packing is complete, the cooling stage proceeds until the material has solidified to the point where it can be ejected from the cavity without damage. Subsequently, the above sequence of steps is repeated. Under normal commercial operating conditions, the filling stage represents 20-25% of the total time, packing 5%,

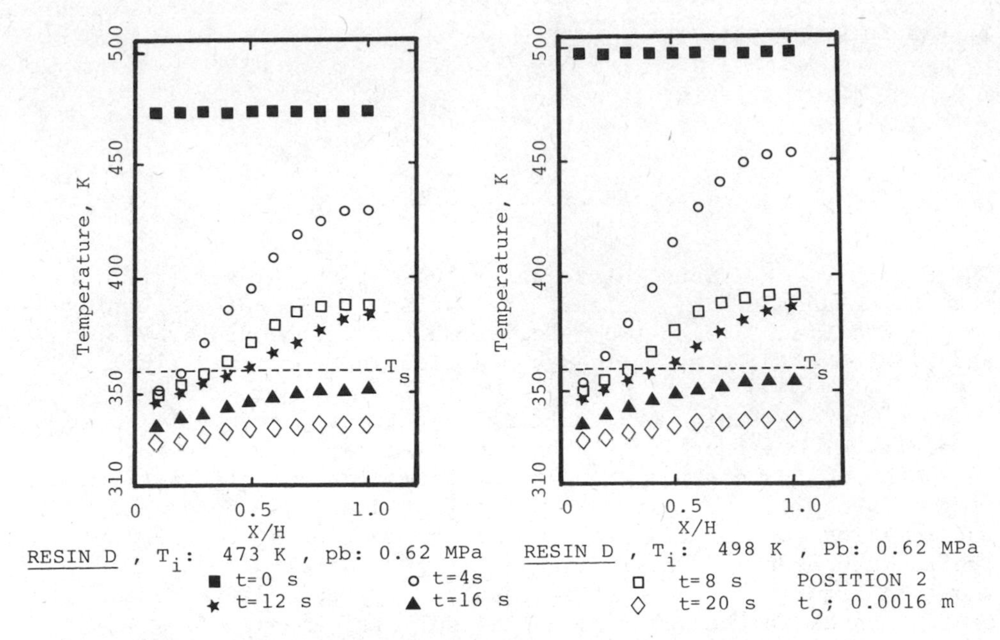

Figure 17: Temperature History During Cooling in Blow Molding of HDPE (59)

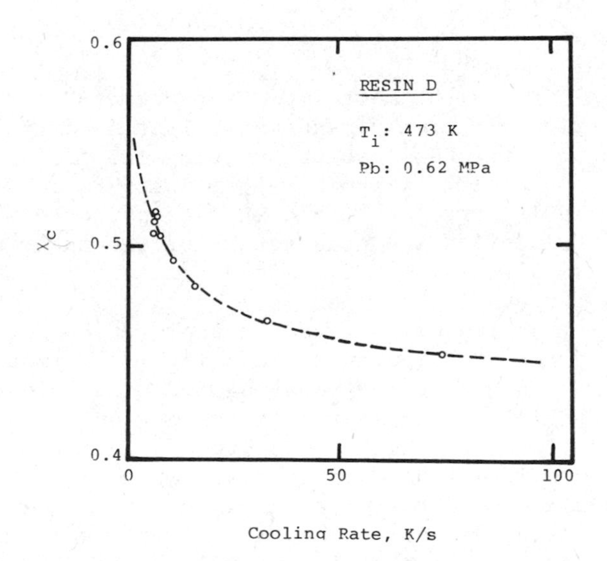

Figure 18: Crystallinity Versus Cooling Rate (59)

and cooling 70-75%. Thus, from the economic point of view, cooling is very important. However, the thermomechanical history experienced by the material during the filling and packing stages has a great influence on the ultimate behavior of the molded article. Therefore, a detailed understanding of all the steps involved in the injection molding process is essential for process and product optimization.

The early models of injection mold filling were based on simple concepts of isothermal flow of Newtonian and Power Law fluids in simple channels (e.g. circular tubes and between flat plates) (66,67,68). During the late sixties and early seventies, a number of models were developed on the basis of the solution of the transport equations for circular, semicircular, or rectangular cavities (69-82). These models usually involved Power Law fluids and one-dimensional, non-isothermal flow. In some cases, two-dimensional flow (76), elastic effects (80), latent heat effects (72) and heat of reaction (15,75) were considered. The results of these models have shown that isothermal treatment of the filling stage yields acceptable results for filling time and pressure distribution during the filling stage, in the case of thick cavities, which do not have extensive thin sections (69,70,81). Similarly, the incorporation of elastic effects does not seem to yield substantial improvements in the prediction of filling time and pressure distribution during the filling stage (80). In recent years, some commercial software packages for computer aided design/manufacturing/engineering (CAD/CAM/CAE) have been developed on the basis of analyses of the type mentioned above. These packages include CADMOLD (IKV, Aachen, West Germany) and MOLDFLOW (Colin Austin, Australia). The main utility of these models is for the prediction of filling times, pressure distributions, and short shots (7,8). They are also useful for balancing delivery channels in multi-cavity or multi-gated molding. Complex geometries may be treated as combinations of three primitive geometries: disc, circular tube, and rectangular channel.

During the last few years, more elaborate models have been proposed to deal with complex mold geometries and to elucidate, in more detail, the structure of the flow in the cavity. The model proposed by Wang et. al. (26) employs a constitutive equation developed by Leonov (83), in conjunction with both finite difference and finite element numerical schemes. The authors claim convergence of the finite element technique for quite high Deborah numbers. However, solutions have been reported for only simple flow between parallel plates.

Kamal and Lafleur (84,85) have proposed a model which incorporates the effects of fountain flow, elasticity, unsteady and inertial effects, and crystallization kinetics. The model, which employs a finite difference scheme combined with a Marker and Cell technique, yields very good agreement with experimental data regarding the progression of the melt front and pressure distribution (Figures 19 and 20). It also yields useful information regarding the structure of the flow, especially the transverse flow near the melt front due to the fountain effect (Figure 21). The model employs a modified White-Metzner (10) constitutive equation to describe the viscoelastic behavior of the Polymer melt.

An important aspect of the mold filling process relates to the balancing of the delivery channels (sprues, runners, and gates) in the case of multi-cavity or multi-gated molds to insure equal filling times and to avoid stress imbalances in the molding. The analysis of the flow in the delivery channels may be based on the application of the above models to the melt while it flows into these channels (86). Alternatively, the analysis may be based on the step-wise or lumped application of available equations for isothermal Newtonian and non-Newtonian flow in the relevant channel geometry (5,87). Other aspects include analysis of the behavior of the melt during a pressure jump (88) or change of flow area as in the case of a step change in the cross-section of the mold (89).

In general, the treatment of the packing stage has lagged significantly behind that of the filling stage. Spencer and Gilmore (IM-1) combined an equation of state and an empirical equation for the determination of the filling time, in order to estimate the maximum pressure during packing. Kamal and Coworkers (72-75, 90) proposed simplified numerical methods for dealing with the packing stage. More recently, Kamal and Lafleur (16,84,85), have proposed a treatment based on the analysis of the unsteady, non-isothermal, compressible flow of a visco-elastic melt (White-Metzner model) into the cavity. In addition to predicting velocity and pressure distributions throughout the injection molding cycle, the model also predicts the evolution of the distributions of temperature and shear and normal stresses. For example, Figure 22 shows predictions of the distribution of shear stress as a function of time and distance from the midplane of the cavity, halfway from the gate to the far end of the cavity. Note the radical change that occurs during the packing stage and the effect of relaxation after packing, prior to solidification. Also, Figure 23 shows the distribution of normal stress, at the end of the packing stage, at various nodes (node 2 near the gate and node 25 near the far end of the cavity) as a function of the distance from the midplane.

During the cooling stage, the material solidifies. Depending on the type of material, solidification may involve only vitrification at the glass transition temperature, Tg, or it may also involve crystallization, as in the case of polyethylene. The thermomechanical history experienced by the material will lead to a

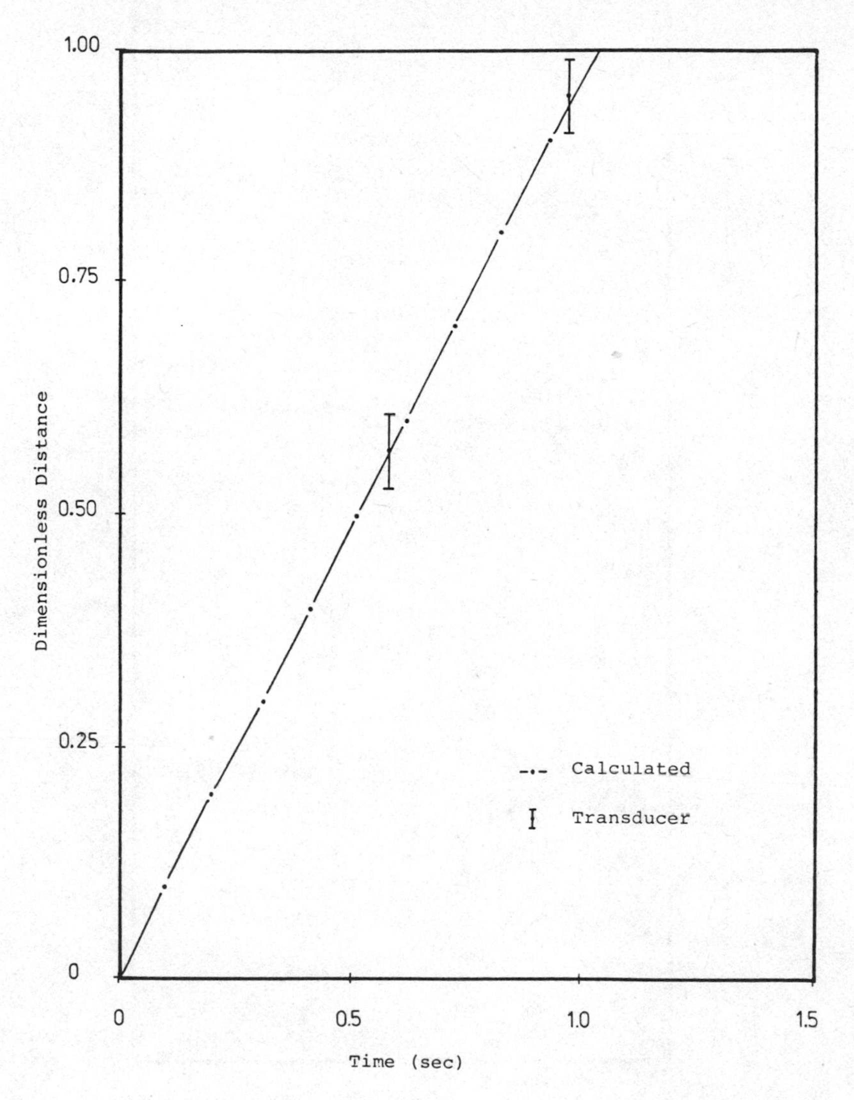

Figure 19: Progression of the melt front (85)

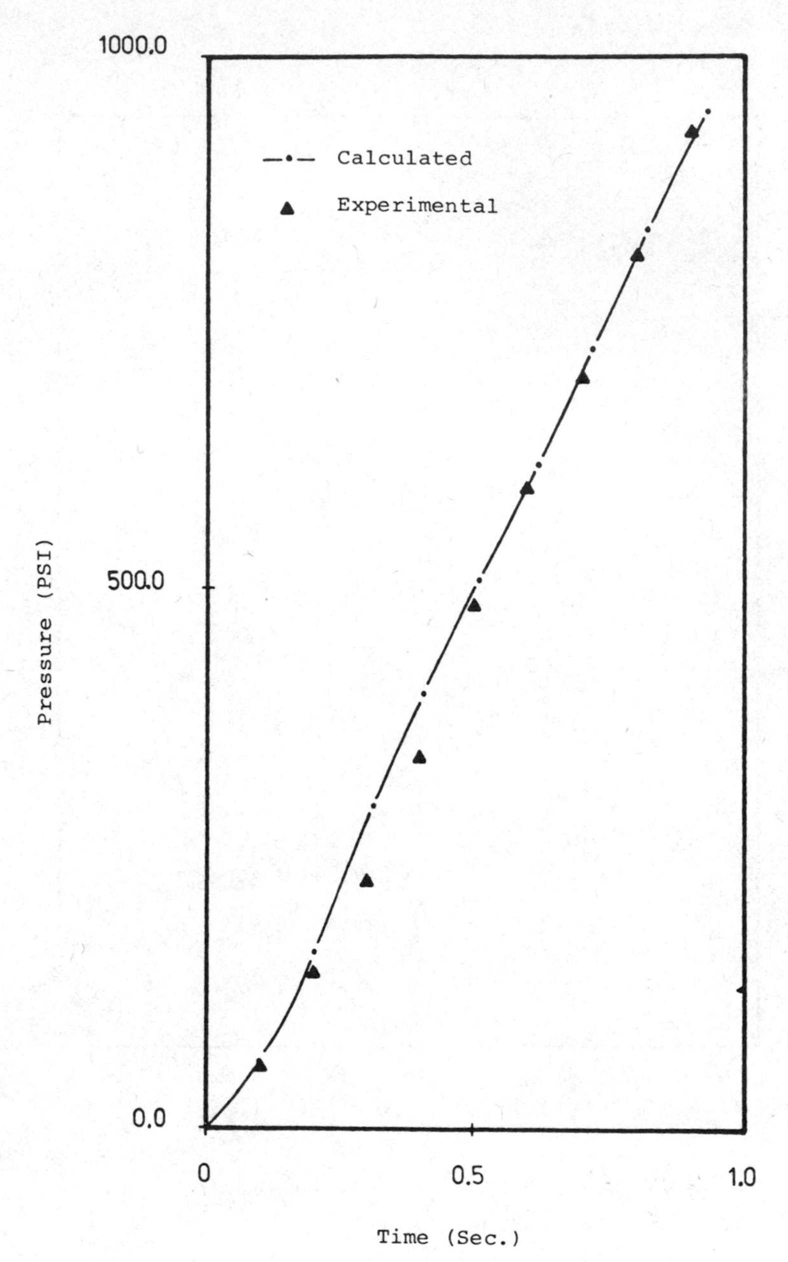

Figure 20: Pressure Variation with Time at the Entrance of the Mold (85)

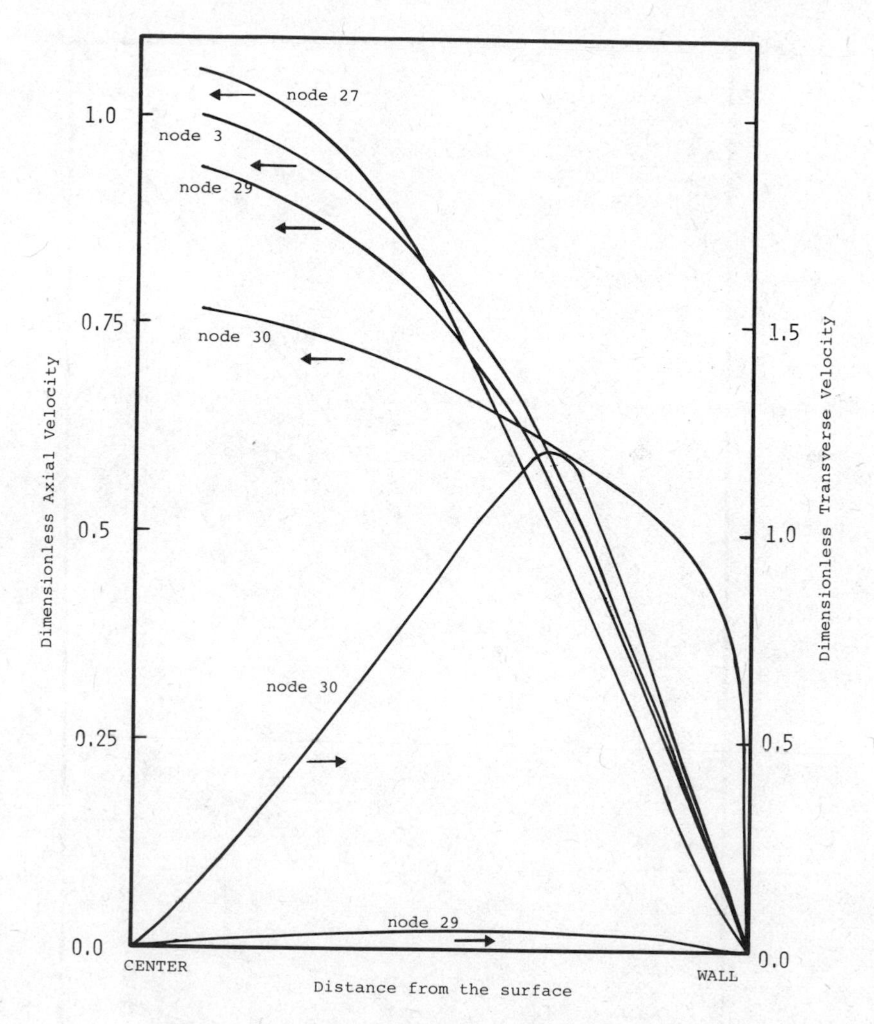

Figure 21: Axial and Transverse Velocity Distributions During Injection Mold Filling of HDPE (Fountain Flow) (85)

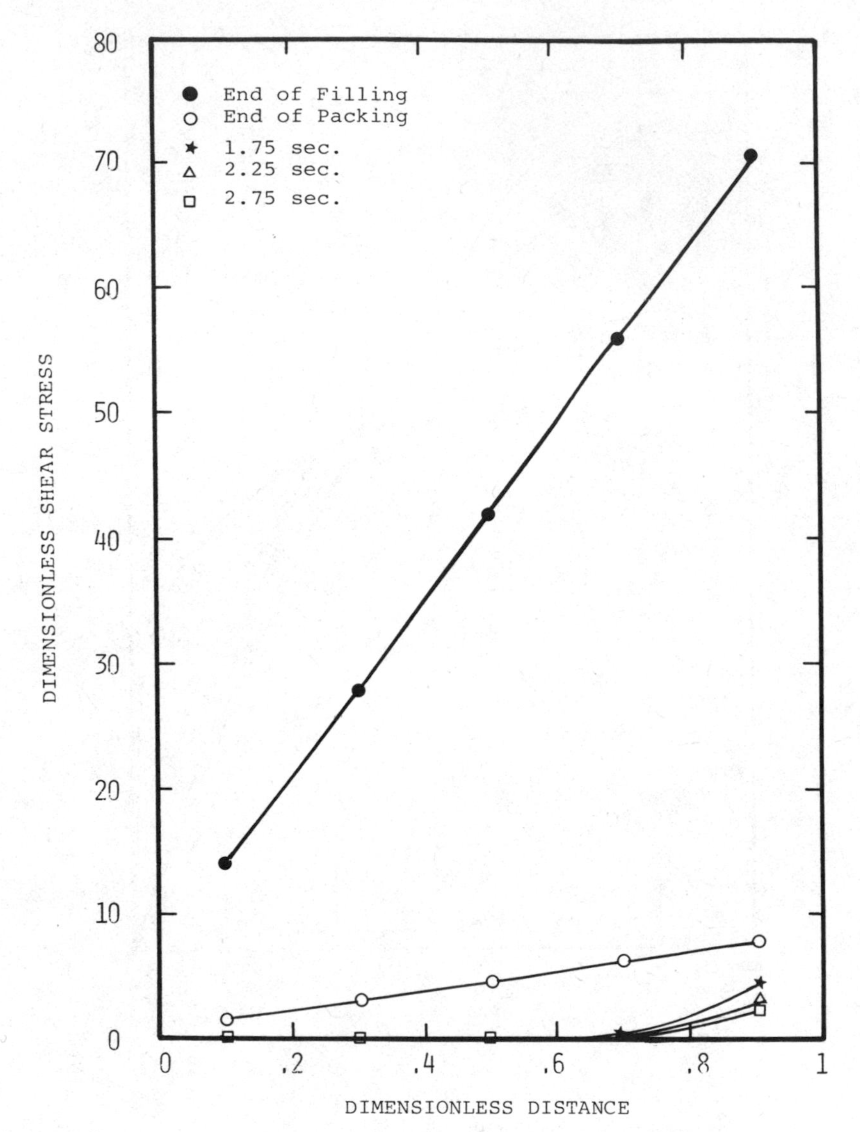

Figure 22: Variation of Shear Stress in the Cavity During Injection Molding of HDPE

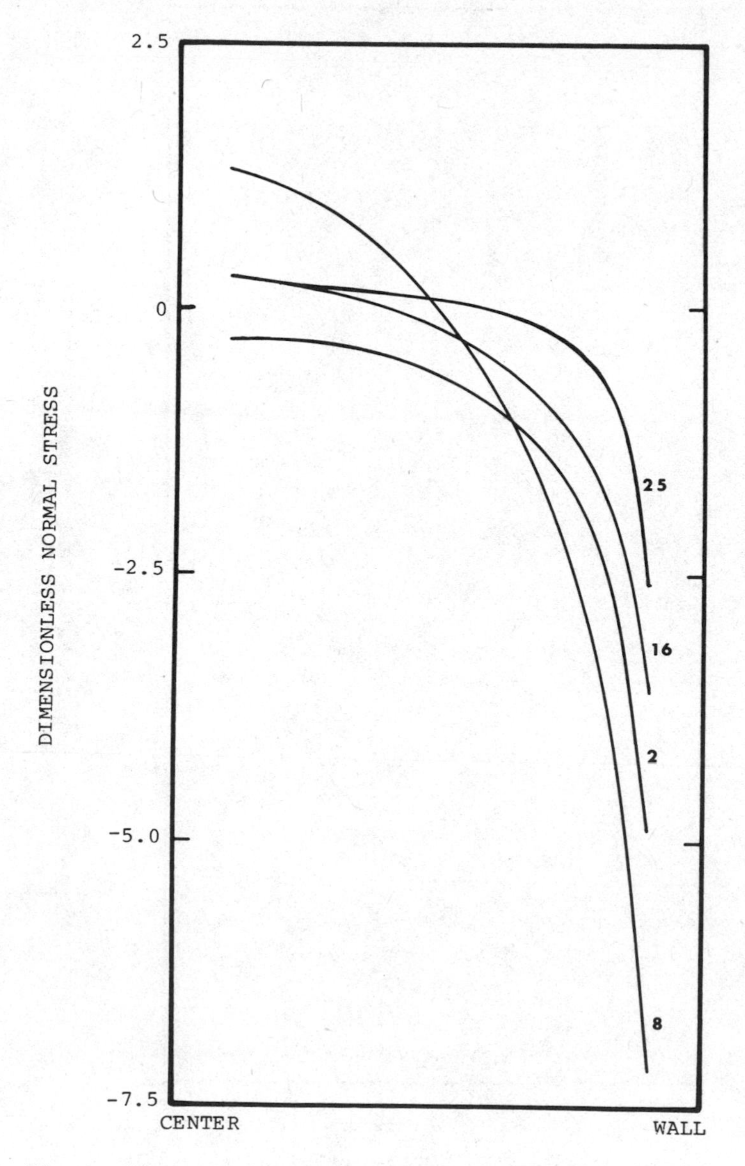

Figure 23: Normal Stress Distribution in the Cavity at the End of Packing (HDPE)

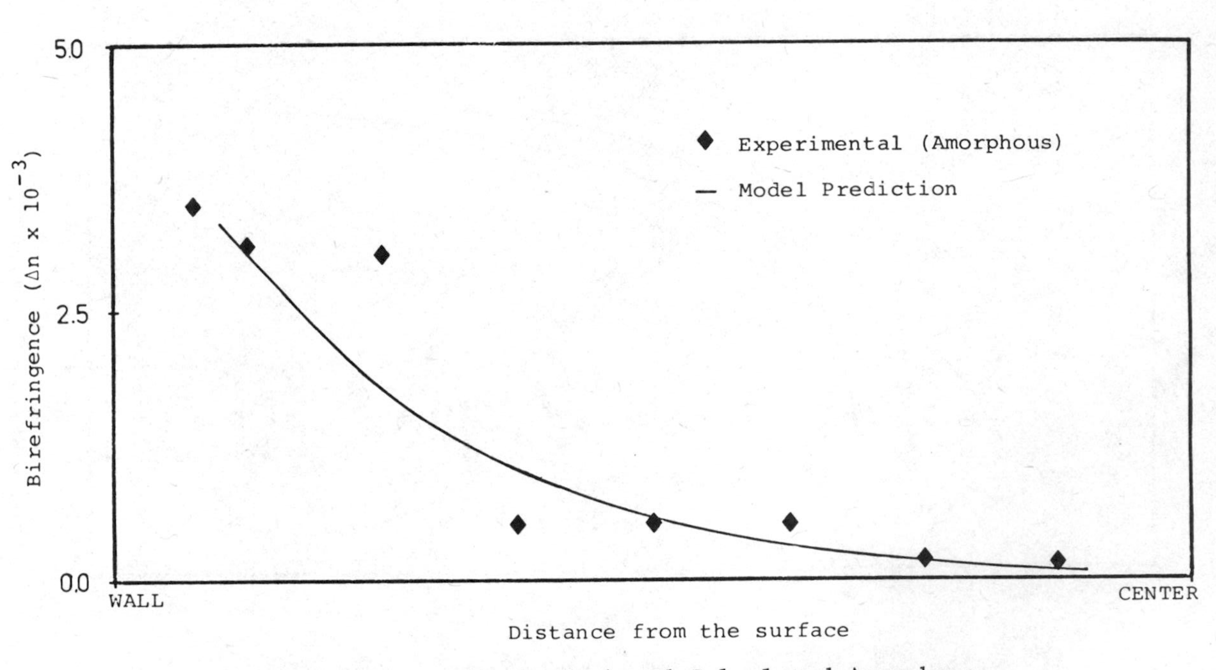

Figure 24: Experimental and Calculated Amorphous Birefringence in Injection Molded HDPE (85)

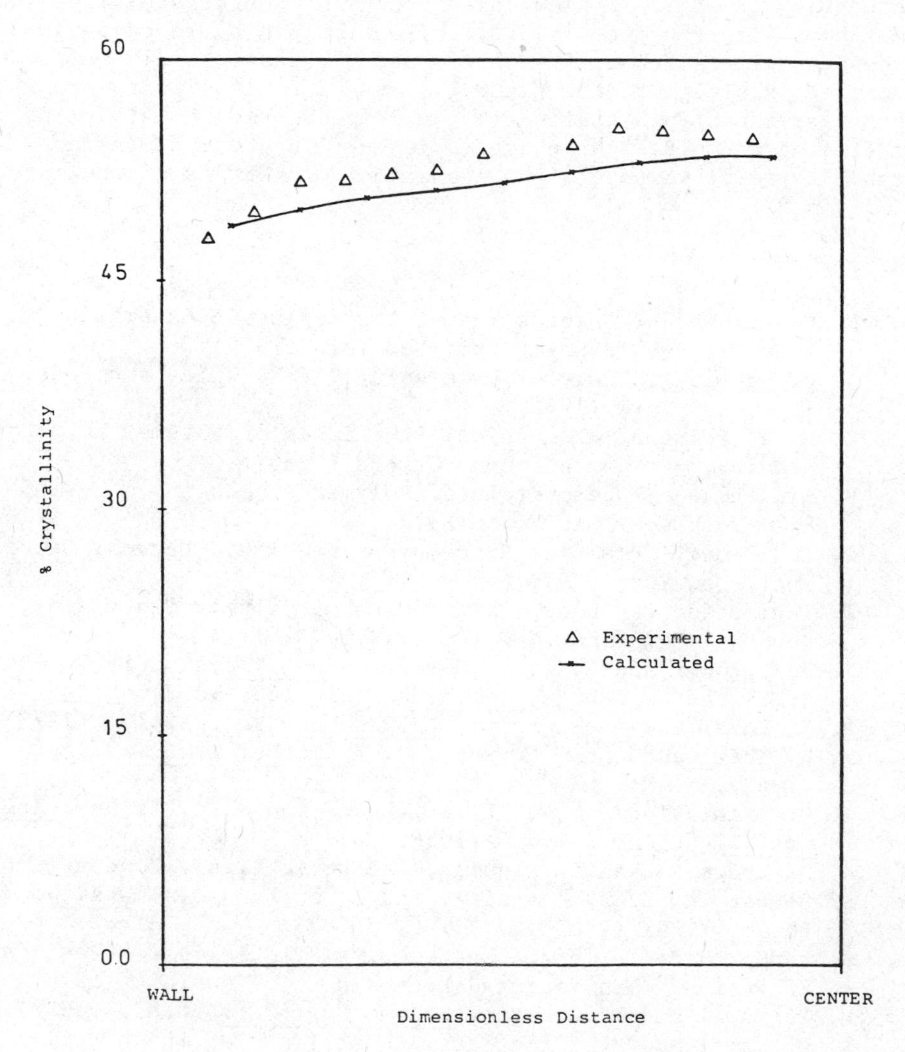

Figure 25: Gapwise Distribution of Crystallinity at the End of the Process (85)

corresponding distribution of morphology, crystallinity, density, and frozen stresses or orientation in the molded article. Recent models have attempted to predict the microstructure and resulting mechanical and optical properties of the moldings, with reasonable success (80,83-85). Typical predictions of the crystallinity and amorphous birefringence in polyethylene moldings are compared to experimental results in Figures 24 and 25, respectively (85). A detailed knowledge of the distributions of crystallinity and frozen stresses (or strains) in a molding should be helpful in predicting as-molded shrinkage (91), the long term dimensional stability (e.g. warpage), and mechanical properties of the material (85,92).

REFERENCES

1. E. C. Bernhardt, "Processing of Thermoplastic Materials," Reinhold Publishing Corp., New York (1959).
2. J. M. McKelvey, "Polymer Processing," John Wiley and Sons, Inc., New York (1962).
3. J. R. A. Pearson, "Mechanical Principles of Polymer Melt Processing," Pergamon Press, Oxford (1966).
4. S. Middleman, "Fundamentals of Polymer Processing," McGraw-Hill Book Co., New York (1977).
5. J. L. Throne, "Plastics Process Engineering," Marcell Decker, Inc., New York (1979).
6. Z. Tadmor and C. Gogos, "Principles of Polymer Processing," John Wiley & Sons, New York (1979).
7. R. S. Spencer and G. D. Gilmore, J. Appl. Phys. 20:502 (1949).
8. H. Breur and G. Rehage, Kolloid Z. 216:166 (1967).
9. M. R. Kamal and N. T. LeVan, Polym. Eng. Sci. 13:131 (1973).
10. J. L. White and A. B. Metzner, J. Appl. Polym. Sci. 7:1867 (1963).
11. W. O. Criminale, Jr., J. L. Ericksen, and G. L. Filbey, Arch. Rat. Mech. Anal. 1:410 (1958).
12. A. S. Lodge, "Elastic Liquids," Academic Press, London (1964).
13. B. Bernstein, E. A. Kearsley, and L. J. Zappas, Trans. Soc. Rheol. 7:391 (1963).
14. Z. Tadmor and I. Klein, "Engineering Principles of Plasticating Extrusion," Van Nostrand Reinhold Co., New York (1970).
15. M. R. Kamal and M. E. Ryan, Polym. Eng. Sci. 20:859 (1980).
16. M. R. Kamal and P. G. Lafleur, "Heat Transfer in Injection Molding of Crystallizable Polymers," Accepted for publication, Polym. Eng. Sci. 16:1077 (1972).
17. K. Nakamura, T. Watanabe, K. Katayama, and T. Amano, J. Appl. Polym. Sci. 16:1077 (1972).
18. W. Dietz, Colloid and Polym. Sci. 259:431 (1981).
19. C. F. Huang, Ph.D. Thesis, Stevens Inst. of Technology (1978).
20. K. R. Reddy and R. I. Tanner, J. Rheol. 22:661 (1978).
21. M. J. Crochet and M. Bezy, J. Non-Newt. Fluid Mech. 5:201 (1979).

22. M. J. Crochet and R. Keunings, J. Non-Newt. Fluid Mech. 7:199 (1980).
23. M. J. Crochet and R. Keunings, J. Non-Newt. Fluid Mech. 10:339 (1982).
24. P.-U. Chang, T. W. Patten and, B. Finlayson, Computers and Fluids &:285 (1979).
25. M. Viriyayuthakorn and, B. Caswell, J. Non-Newt. Fluid Mech. 6:245 (1980).
26. K. K. Wang et al., Injection Molding Project, Collect of Engineering, Cornell University, Progress Report No. 9 (Dec. 1982).
27. E. Mitsoulis, J. Vlachapoulos, and F. A. Mirza, "Finite Element Analysis of Isothermal and Non-Isothermal Polymer Melt Flows," Presented at IMRI/NRCC Mini-Symposium on "Mathematical Modelling of Plastics Processing and Control," Montreal (Feb. 1982).
28. R. E. Gaskell, J. Appl. Mech. 17:334 (1950).
29. C. Kiparissides and J. Vlachopoulos, Polym. Eng. Sci. 16:712 (1976).
30. W. W. Alston, Jr. and K. N. Astill, J. Appl. Polym. Sci. 17:3157 (1973).
31. T.-S. Chung, J. Appl. Polym. Sci. 25:967 (1980).
32. J. Vlachopoulos and A. N. Hrymack, Polym. Eng. Sci. 20:725 (1980).
33. I. Brazinsky
34. J. F. Carley, in "Processing of Thermoplastic Materials," E. C. Berhardt, editor, Reinhold Publishing Corp., New York (1959).
35. C. D. Han and D. A. Rao, Polym. Eng. Sci. 20:128 (1980).
36. J. F. Carley, T. Endo, and W. B. Krantz, Polym. Eng. Sci. 19:1178 (1979).
37. B. Vergnes, Materiaux et Techniques, p. 187 (juin-juillet 1981).
38. R. T. Fenner and J. G. Williams, Trans. J. Plast. Inst., p. 701 (Oct. 1967).
39. R. T. Fenner and F. Nadir, Polym. Eng. Sci. 19:203 (1979).
40. S. Basu, Polym. Eng. Sci. 21:1128 (1981).
41. B. Caswell and R. I. Tanner, Polym. Eng. Sci. 18:416 (1978).
42. M. R. Kamal and S. Petsalis, Unpublished results.
43. M. Viriyayuthakorn and R. V. DeBoo, SPE Tech. Papers 29:178 (1983).
44. N. Orbey, Ph.D. Thesis, McGill University (1983).
45. A. Garcia-Rejon, Ph.D. Thesis, McGill University (1980).
46. E. D. Henze and W. Wu, Polym. Eng. Sci. 13:153 (1973).
47. G. R. Cotten, Rubber Chem. Technol. 52:187 (1979).
48. D. H. Kalyon, Ph.D. Thesis, McGill University (1980).
49. C. E. Beyer and N. Sheptak, SPE Journal 21:190 (1965).
50. N. Wilson, M. Bentley, and B. Morgan, SPE Journal 26:34 (1970).

51. D. Kalyon, V. Tan, and M. R. Kamal, Polym. Eng. Sci. 20:773 (1980).
52. M. R. Kamal, D. Kalyon, and V. Tan, in "Rheology, vol. 3: Applications," edited by G. Astarita, G. Marrucci, and L. Nicolais; p. 149, Plenum Publishing Corp., New York (1980).
53. M. R. Kamal, V. Tan, and D. Kalyon, Polym. Eng. Sci. 21:331 (1981).
54. G. Ajroldi, Polym. Eng. Sci. 18:748 (1978).
55. H. Fukase, A. Iwaaki, and T. Kunio, SPE Tech. Papers 25:913 (1979).
56. M. E. Ryan and A. Dutta, Polym. Eng. Sci. 22:1075 (1982).
57. G. R. A. Pearson and S. Middleman, AIChE J. 23:714 (1977).
58. C. J. S. Petrie, Proc. Conf. on Polym. Rheol. and Plast. Processing, p. 307, The Plastics and Rubber Institute, London (1975).
59. M. R. Kamal and D. Kalyon, Polym. Eng. Sci. 23:503 (1983).
60. G. Menges, M. Kulik, and F. Rhiel, Plastverarbeiter, ___:685 (1973).
61. W. Dietz, Plastiques Modernes et Elastomeres 28:104 (1976).
62. M. F. Edwards, P. Suvanaphon, and W. Wilkinson, Polym. Eng. Sci. 19:911 (1979).
63. M. F. Edwards, S. Georghiades, and P. K. Suvanaphen, Plast. Rubber Processing Applications 1:162 (1981).
64. J. L. White and A. Agarwal, Polym. Eng. Rev. 1:267 (1981).
65. M. Cakmak, J. Spruiell, and J. L. White, SPE Tech. Papers 29: 394 (1983).
66. R. S. Spencer and G. D. Gilmore, J. Colloid Sci. 6:118 (1951).
67. R. L. Ballman, T. Shusman, and H. L. Toor, Ind. Eng. Chem. 51: 847 (1959).
68. R. B. Staub, SPE Journal 17:345 (1961).
69. D. H. Harry and R. G. Parrott, Polym. Eng. Sci. 10:209 (1970).
70. J. L. Berger and C. G. Gogos, Polym. Eng. Sci. 13:102 (1973).
71. P. C. Wu, C. F. Huang, and C. G. Gogos, Polym. Eng. Sci. 14: 223 (1974).
72. M. R. Kamal and S. Kenig, Polym. Eng. Sci. 12:294 (1972).
73. M. R. Kamal and S. Kenig, Polym. Eng. Sci. 12:302 (1972).
74. M. R. Kamal, Y. Kuo, and P. H. Doan, Polym. Eng. Sci. 15:863 (1975).
75. M. E. Ryan, Ph.D. Thesis, McGill University (1978).
76. E. Broyer, C. Gutfinger, and Z. Tadmor, Trans. Soc. Rheol. 19:423 (1975).
77. J. L. White, Polym. Eng. Sci. 15:44 (1975).
78. H. A. Lord and G. Williams, Polym. Eng. Sci. 15:569 (1975).
79. Y. Kuo and M. R. Kamal, AIChE J. 22:661 (1976).
80. K. K. Wang, et. al., Injection Molding Project, College of Engineering, Cornell University, Progress Report #6 (1979).
81. C. F. Huang, C. G. Gogos, and L. Schmidt, Paper 94e, 71st Annual AIChE Meeting, November 15, 1978.
82. M. E. Ryan and T.-S. Chung, Polym. Eng. Sci. 18:642 (1980).
83. A.I. Leonov, Rheol. Acta 15:85 (1976).

84. M. R. Kamal and P. G. Lafleur, Polym. Eng. Sci. 22:1066 (1982).
85. M. R. Kamal and P. G. Lafleur, SPE Tech. Papers, 29:386 (1983).
86. V. W. Wang, K. K. Wang, and C. A. Hieber, SPE Tech. Papers, 29: 663 (1983).
87. C. A. Hieber, R. K. Upadhyay, and A. I. Isayev, SPE Tech. Papers 29:698 (1983).
88. A. I. Isayev, SPE Tech. Papers 29:710 (1983).
89. R. K. Upadhyay and A. Isayev, SPE Tech. Papers 29:714 (1983).
90. Y. Kuo and M. R. Kamal, in "Science and Technology of Polymer Processing," edited by N. P. Suh and N.-H. Sung, p. 329, The M.I.T. Press, Cambridge, Mass. (1979).
91. W. B. Hoven-Nievelstein and G. Menges, SPE Tech. Papers 29: 737 (1983).
92. M. R. Kamal and F. H. Moy, Chem. Eng. Commun. 12:253 (1981).

APPENDIX

Parameters of Constitutive Equations for Two-Dimensional Flow

As indicated in the above discussion, there is a growing tendency to analyze polymer shaping operations in terms of two-dimensional viscoelastic flow. The convected Maxwell fluid, the White-Metzner model, and the CEF fluid are among those commonly employed in dealing with such problems. Also, they have been employed in some of the analyses discussed in this manuscript. Therefore, we list below the important parameters for two-dimensional flow for these models.

Convected Maxwell fluid (27):

$$\tau_{xx} + \lambda \left[V_x \frac{\partial \tau_{xx}}{\partial x} + V_y \frac{\partial \tau_{xx}}{\partial y} - 2 \frac{\partial V_x}{\partial x} \tau_{xx} - 2 \frac{\partial V_x}{\partial y} \right] = 2\eta \frac{\partial V_x}{\partial x}$$

$$\tau_{yy} + \lambda \left[V_x \frac{\partial \tau_{yy}}{\partial x} + V_y \frac{\partial \tau_{yy}}{\partial y} - 2 \frac{\partial V_y}{\partial y} \tau_{yy} - 2 \frac{\partial V_y}{\partial x} \tau_{xy} \right] = 2\eta \frac{\partial V_y}{\partial y}$$

$$\tau_{xy} + \lambda \left[V_x \frac{\partial \tau_{xy}}{\partial x} + V_y \frac{\partial \tau_{xy}}{\partial y} - \frac{\partial V_y}{\partial x} \tau_{xx} - \frac{\partial V_x}{\partial y} \tau_{yy} \right] = \eta \left(\frac{\partial V_y}{\partial x} + \frac{\partial V_x}{\partial y}\right)$$

CEF equation (Ψ_1, Ψ_2 are normal stress coefficients) (27):

$$\tau_{xx} = 2\eta \frac{\partial V_x}{\partial x} + (\tfrac{1}{2}\Psi_1 + \Psi_2)(\gamma_{xx}^2 + \gamma_{xy}^2) - \tfrac{1}{2}\Psi_1 \Big[(V_x \frac{\partial \gamma_{xx}}{\partial x} + \frac{\partial \gamma_{xx}}{\partial y}) - \gamma_{xy}(\frac{\partial V_x}{\partial y} - \frac{\partial V_y}{\partial x}) \Big]$$

$$\tau_{yy} = 2\eta \frac{\partial V_y}{\partial y} + (\tfrac{1}{2}\Psi_1 + \Psi_2)(\gamma_{yy}^2 + \gamma_{xy}^2) - \tfrac{1}{2}\Psi_1 \Big[(V_x \frac{\partial \gamma_{yy}}{\partial x} + V_y \frac{\partial \gamma_{yy}}{\partial y}) - \gamma_{xy}(\frac{\partial V_x}{\partial y} - \frac{\partial V_y}{\partial x}) \Big]$$

$$\tau_{xy} = \eta(\frac{\partial V_y}{\partial x} + \frac{\partial V_x}{\partial y}) + (\tfrac{1}{2}\Psi_1 + \Psi_2)\Big[\gamma_{xy}(\gamma_{xx} + \gamma_{yy})\Big] - \tfrac{1}{2}\Psi_1 \Big[(V_x \frac{\partial \gamma_{xy}}{\partial x} + V_y \frac{\partial \gamma_{xy}}{\partial y}) + \tfrac{1}{2}(\gamma_{xx} - \gamma_{yy})(\frac{\partial V_x}{\partial y} - \frac{\partial V_y}{\partial x}) \Big]$$

where

$$\gamma_{xx} = 2 \frac{\partial V_x}{\partial x}, \quad \gamma_{yy} = 2 \frac{\partial V_y}{\partial y} \text{ and } \gamma_{xy} = \frac{\partial V_y}{\partial x} + \frac{\partial V_x}{\partial y}$$

White-Metzner Model

$$\tau_{xx} + \frac{\eta}{G} (\frac{\partial \tau_{xx}}{\partial t} + V_x \frac{\partial \tau_{xx}}{\partial x} + V_y \frac{\partial \tau_{xx}}{\partial y} - 2 \frac{\partial V_x}{\partial x} \tau_{xy} - 2 \frac{\partial V_x}{\partial y} \tau_{xy})$$

$$= 2\eta \frac{\partial V_x}{\partial x}$$

$$\tau_{yy} = \frac{\eta}{G} (\frac{\partial \tau_{yy}}{\partial t} + V_x \frac{\partial \tau_{yy}}{\partial x} + V_y \frac{\partial \tau_{yy}}{\partial y} - 2 \frac{\partial V_y}{\partial x} \tau_{xy} - 2 \frac{\partial V_y}{\partial y} \tau_{xy})$$

$$= 2\eta \frac{\partial V_y}{\partial y}$$

$$\tau_{xz} + \frac{\eta}{G}\left(\frac{\partial \tau_{xy}}{\partial t} + u\frac{\partial \tau_{xy}}{\partial x} + v\frac{\partial \tau_{xy}}{\partial y} - \frac{\partial V_x}{\partial y}\tau_{yy} - \frac{\partial V_y}{\partial x}\tau_{xx}\right)$$

$$= \eta\left(\frac{\partial V_x}{\partial y} + \frac{\partial V_y}{\partial x}\right)$$

The second invariant of the rate of deformation tensor is

$$\vec{\vec{\Delta}} : \vec{\vec{\Delta}} = 4\left(\frac{\partial V_x}{\partial x}\right)^2 + 2\left(\frac{\partial V_x}{\partial y} + \frac{\partial V_y}{\partial x}\right)^2 + 4\left(\frac{\partial V_y}{\partial y}\right)^2$$

THE CONDUCTIVITY DISTRIBUTION IN INJECTION MOULDED LDPE AND HDPE CONTAINING CARBON BLACK

Carl Klason and Josef Kubát

Department of Polymeric Materials
Chalmers University of Technology
S-412 96 Gothenburg, Sweden

INTRODUCTION

This study reports on the influence of processing on the electrical resistivity of injection moulded samples of LDPE and HDPE containing a conducting type of carbon black (CB). The resistivity was found to increase with orientation, the resistivity levels in the skin and shear layers being much higher than in the core. No simple correlation, however, could be found between the resistivity and the orientation at varying distance from the surface of the samples. LDPE was more sensitive to processing than HDPE. The skin and shear zones (high orientation) showed a marked frequency dependence (resistivity decreasing with frequency). The high resistivity values at the surface were reduced upon annealing. The effects observed can be interpreted qualitatively in terms of a CB-network disrupted during the flow of the melt; upon annealing a recovery of the network takes place (1).

The effects observed are undesirable in several applications of conducting polymers, such as in PTC-materials for self-regulating heating elements and cables (2,3), conducting compounds for antistatic products (4), and materials for EMI/RFI shielding (5).

EXPERIMENTAL

The conducting carbon black (CB) type used was Vulcan XC-72 (BET 254 m^2/g, DBP 178 ml/100 g, porous structure) received from Cabot. The polymers used were low-density polyethylene (ICI type 017.040, density 917 kg/m^3, MFI 7 g/10 min) and high-density polyethylene (Unifos, DMDS 2215, density 953 kg/m^3, MFI 0.1 g/10 min). The CB-concentration was 20% giving a resistivity level of c. 10^3 ohm cm for both LDPE and HDPE.

Above a critical concentration of CB the conductivity of the samples fell by several orders of magnitude when the CB-content was increased by a few percent only; a further increase in the CB-content resulted in a minor conductivity change only. The CB-content used in this study (20%) produced a conductivity level just above the steep portion of the conductivity-concentration curve.

The polymer/CB-compounds were mixed in a Buss-Kneader (PR 46, L = 11 D, vented cylinder). The test bars (DIN 53455, effective length 75 mm, cross-section 10 x 3.5 mm) were injection moulded on a conventional injection moulding machine (Arburg 221E17R). Samples having normal and high degree of orientation were produced. The high degree of orientation was produced by changing the melt temperature from 200°C (LDPE) and 280°C (HDPE) to 160°C and 200°C, respectively. The bulk and surface resistivity (ρ_B and ρ_S) were measured at DC (Keithley electrometer 160 B) and at 50-10.000 Hz (General Radio Conductance Bridge 1630).

RESULTS

It is well known that injection moulded samples have a biaxially oriented surface, a shear zone mainly oriented in the flow direction, and a core with a low degree of orientation. The distribution of the orientation from the surface to the core is affected by the processing conditions. The following results show that this distribution is reflected in the conductivity values.

Bulk resistivity

The main result of this study is the observation that the bulk resistivity, ρ_B, depends on the orientation, an effect especially pronounced in the skin and shear zones of the injection moulded samples. Fig. 1 shows the bulk resistivity of a part (10 x 10 x 3.5 mm) cut from the middle of a test bar having normal orientation. The resistivity was measured in the three main directions, x, y, and z, as shown in Fig. 1. The highest ρ-values were always found in the directions perpendicular to the flow (y- and z-directions). The LDPE-samples had higher resistivity values than the HDPE-samples. For ρ-values exceeding 10^6 ohmcm the resistivity exhibited a pronounced frequency dependence, higher frequencies giving lower resistivity values. The effects were even more tangible in the samples having a higher orientation, Fig. 1.

Surface resistivity

The surface resistivity was measured at different distances from the surface (microtome cutting). The dependence of ρ_S (DC-values) in the flow direction on the distance from the surface is seen in Fig. 2. The highest ρ_S-values were found at the surface. The ρ_S-values had a frequency dependence similar to that of ρ_B, cf. Fig. 1. Again, it was found that LDPE was more sensitive to processing than HDPE. The high ρ_S-levels at the surface of the LDPE-samples decreased more gradually when approaching the core centre, especially for samples having a higher orientation this effect is clearly seen. The ρ_S-values at the surface exceeded the value in the core by several orders of magnitude.

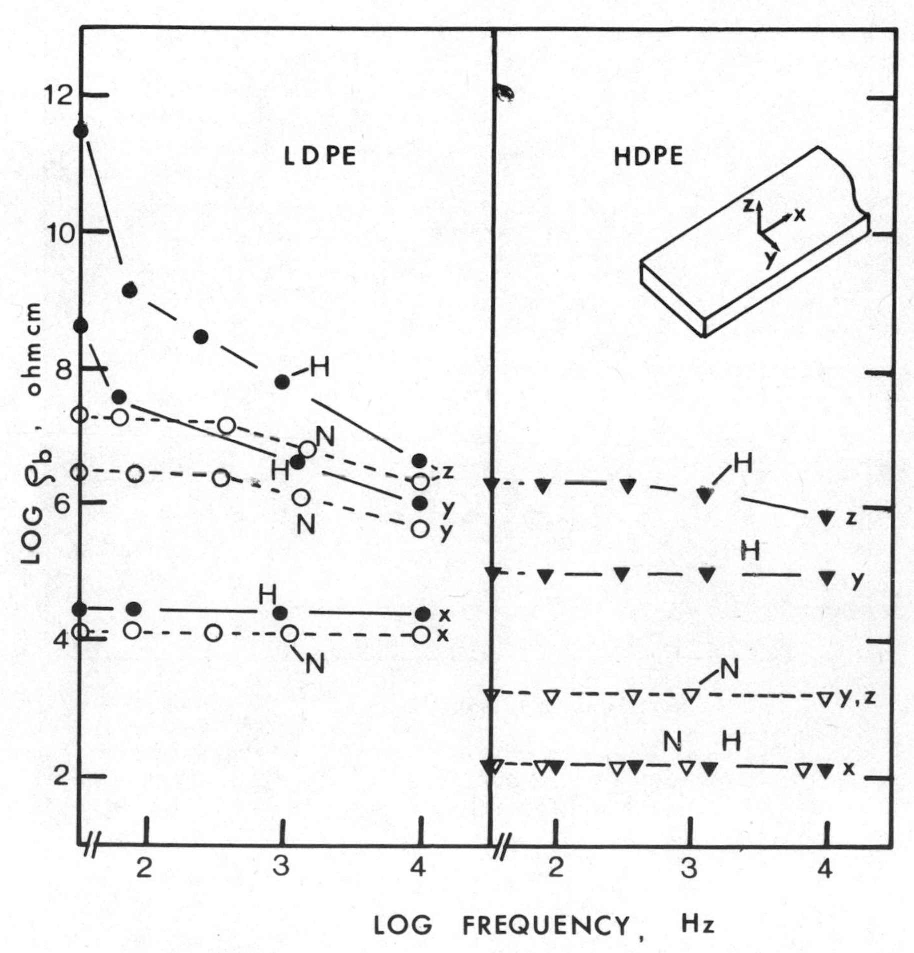

Fig. 1. Bulk resistivity of part cut from middle section of test bars. The resistivity was measured in the x, y, and z--directions. N - normal orientation, H - high orientation.

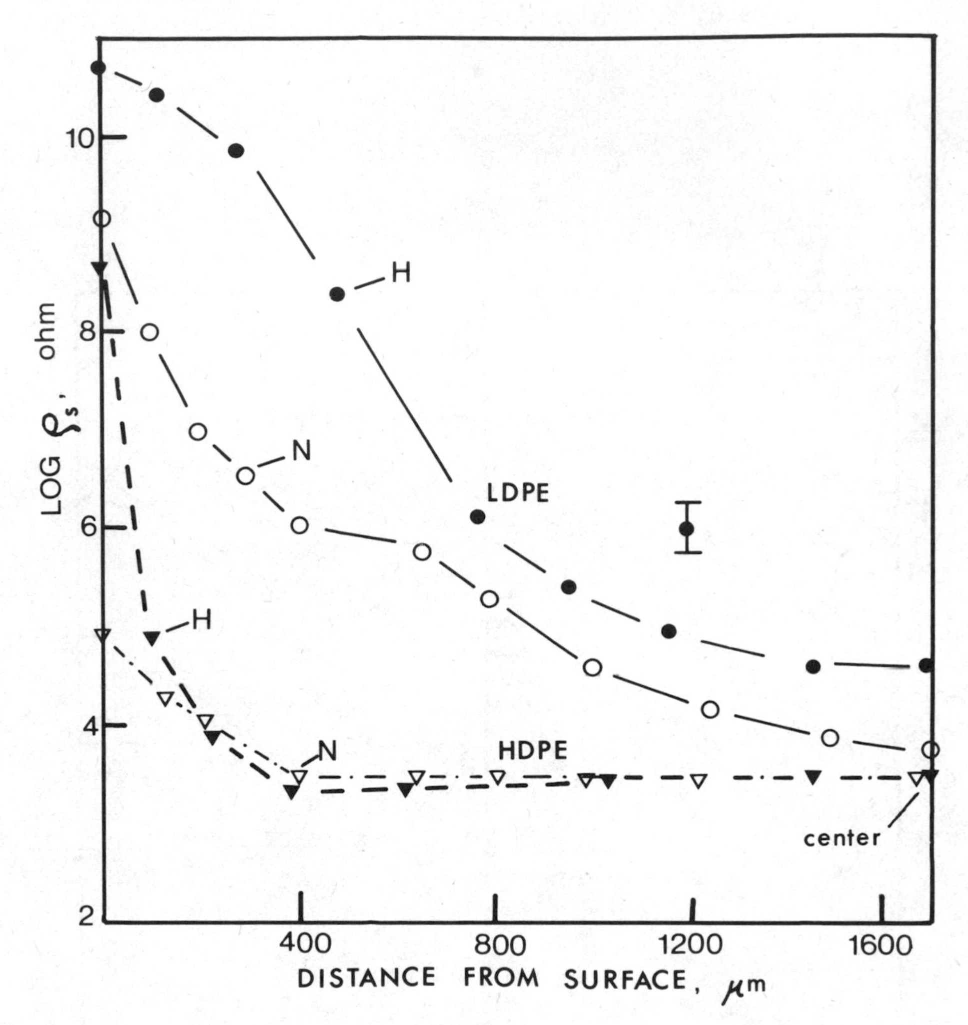

Fig. 2. The variation of the surface resistivity with the distance from the surface. N - normal orientation, H - high orientation.

FINAL REMARKS

The differences in orientation between the layers of injection moulded parts are well known (6). Not known, however, is the pronounced sensitivity of conductivity values to local changes in anisotropy in such parts where the polymer matrix has been rendered conductive by a particulate filler. This applies both to the bulk and surface values of the resistivity. Higher orientation always produces a higher resistivity; in the case of ρ_S the difference between the biaxially oriented skin and the practically unoriented core amounted to six decades for the LDPE-samples, cf. Fig. 2. There was, however, no one-to-one correspondence between the resistivity values and the orientation measured as thermal shrinkage of 50 μm thick microtome slices taken at varying distance from the surface. The high ρ-values at the surface apparently were not caused by the lower crystallinity normally found there and checked in the present case by DTA. This was concluded from experiments with samples quenched in liquid nitrogen where the crystallinity was found to be substantially lower, but where no significant change in ρ was recorded.

Qualitatively, the effects observed may be interpreted in terms of a conducting CB-network disrupted by the flow during the filling of the mould. Such disruption can be expected to result in an increased resistivity. When the network is allowed to reform, the resistivity recovers its normal value, characteristic of the unoriented polymer. This can, for instance, be achieved by annealing. A kind of "internal" annealing also takes place during the solidification of the core of an injection moulded part, as the time and temperature profiles for that region apparently allow the network to consolidate in its conducting state. This process takes place parallel with a desorientation of the matrix, the core becoming practically isotropic. Close to the cavity wall, and also in the shear zone, the CB-network is permanently frozen in its disrupted state, a recovery being possible only in connection with a subsequent annealing. The effects observed are reminiscent of the variation in resistivity when a CB-filled polymer is heated through the Tg- and Tm-regions (7). Also in that case the increase in ρ around Tg can be interpreted as resulting from a disruption process caused by the movement of the structure in the transition region.

Finally, it may be mentioned that the PTC-effect (increase in ρ around Tg or Tm of the matrix of a CB-filled material) is highly enhanced by increasing orientation. This can, however, not be utilized in practice, due to the irreversibility of such effect during repeated heating and cooling cycles.

The effects observed do not appear to be amenable to a quantitative interpretation in terms of recently proposed models for the conduction mechanism in CB-filled polymers (2,3,8). Despite of this, the variation of resistivity with the state of orientation appears to a highly sensitive tool to follow the changes in the state of processing conditions.

REFERENCES

1. C. Klason and J. Kubát, Advances in Chem. Series, No. 174:1 (1979).
2. A. Voet, Rubber Chem. Technol., 54:42 (1981).
3. J. Meyer, Polym. Eng. Sci., 13:462 (1973).
4. T. Flowers, A. Watkins, and D. Blevi, Plastics and Rubber Int., 7:5:174 (1982).
5. G. Smoluk, Modern Plastics Int., 12:9:46 (1982).
6. G. Wübken, Plastverarbeiter, 26:17 (1975).
7. C. Klason and J. Kubát, J. Appl. Polym. Sci., 19:831 (1975).
8. R. D. Sherman, L. M. Middleman, and S. M. Jacobs, Polym. Eng. Sci., 23:36 (1983).

HIGH PRESSURE INJECTION MOULDING OF POLYETHYLENE

J. Kubát and J.-A. Månson

Chalmers University of Technology
Department of Polymeric Materials
S-412 96 Gothenburg
Sweden

M. Rigdahl*

*Swedish Forest Products Research Laboratory
Paper Technology Department
Box 5604
S-114 86 Stockholm, Sweden

SUMMARY

By increasing the injection pressure up to 500 MPa the modulus and the strength of injection moulded high molecular weight, high-density polyethylene (HMWPE) can be increased several times. In this communication the effects of processing conditions and mould design on mechanical properties of this type of high pressure injection moulded material are reviewed. It is shown that the use of high injection pressures can increase the modulus of the HMWPE to more than 10 GPa. Some results on the effect of adding carbon black to HMWPE are also included.

INTRODUCTION

Injection moulding of thermoplastics is normally performed using injection pressures up to 150-200 MPa. The effect on the physical properties of an increase in injection pressure up to 500 MPa have been studied at our laboratory using a specially designed two-stage injection moulding machine. Among the effects observed when substantially raising the pressure level can be mentioned a decrease in mould shrinkage[1], an improvement in ductility of poly(ethylene terephtalate) and polyoxymethylene[2], a higher degree of crystallinity[2-4], etc. However, the most significant improvements are noted for high density polyethylene with a high molecular weight, for which significant increases in tensile modulus and strength are observed. The aim of the present communication is to review some of the results obtained in the mechanical testing of tensile test bars of high density polyethylene with a high molecular weight (HMWPE), injection moulded with an injection pressure of 500 MPa.

MATERIALS AND EXPERIMENTAL PROCEDURE

The high density polyethylene grade used in this study had a density of 0.953 g/cm^3 and a melt flow index of 0.1 g/10 min (MFI 190/2). The average molecular weights were $\bar{M}_w$ = 286.000 and $\bar{M}_n$ = 22.000.

In some cases the HMWPE-grades was filled with carbon black (Ketjenblack EC, AKZO Chemie).

The high pressure injection moulding, up to 500 MPa, was performed using a two-stage moulding machine with separate plasticating and high pressure injection units. The former is a conventional screw arrangement, which transports the melt into a chamber in the lower high pressure (injection) unit. The melt is then injected into the mould with a piston operating in the injection unit. A more detailed account of the high pressure injection moulding machine can be found in ref.[5].

The moulded tensile test bars had a gauge length of 25 mm and the thickness could be varied from 1 to 6 mm. The flow direction of the melt was along the gauge length and all reported mechanical properties refer to the flow direction.

RESULTS AND COMMENTS

Comparison with HMWPE-parts moulded at lower injection pressures

Fig. 1 shows tensile stress-strain curves for a tensile test bar (thickness 1.5 mm) of HMWPE moulded at 500 MPa and two others moulded at lower pressures. One of the two low pressure moulded bars was produced under conditions that resulted in a low degree of orientation (thickness 6 mm, injection pressure 100 MPa, mould temperature 30°C, barrel temperature 230°C) while the other was moulded under conditions chosen to produce a maximum degree of orientation in the flow direction (thickness 1.5 mm, injection pressure 160 MPa, mould temperature 30°C, barrel temperature 230°C. It is thus evident that the improvement in the stiffness and the strength of the high pressure injection moulded parts cannot only be accounted for by orientation effects of the normal type encountered in injection moulded parts. The high pressure level appears to be necessary for developing the high stiffness and strength.

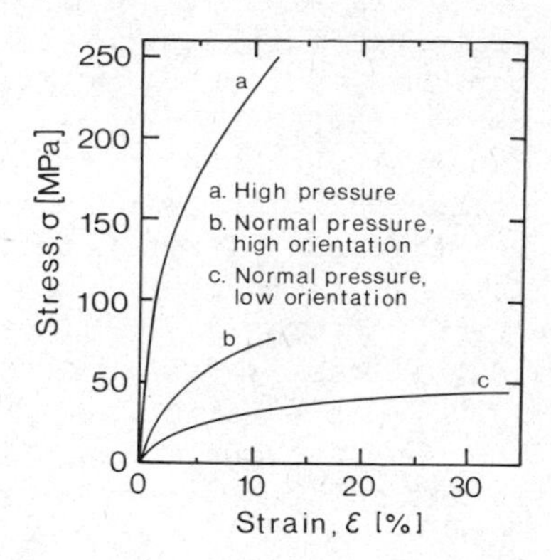

Fig. 1. Stress-strain curves for HMWPE moulded at high and normal pressure levels. One of the bars, moulded at lower pressure levels, was produced under conditions chosen to yield a high degree of orientation. From J. Kubát and J.-A. Månson, Plaste u. Kautschuk, 29(1982)91.

The improvement in mechanical properties of high pressure HMWPE is believed to be due to the formation of structural elements of another type than spherulites etc. The exact nature of these elements is, however, not known in detail at present. Shish-Kebab-structures or tie-molecules (or similar elements) connecting crystallites, which may be of the chain-extended type, may explain the improvement in mechanical properties[1,5]. It should be mentioned that the increase in stiffness and strength of the test bars is normally accompanied by the appearance of a second higher melting peak around 140°C. There is, however, not a one-to-one correspondance between the amount of the crystalline phase corresponding to this peak and the change in e.g. stiffness. This might indicate that elements which connect such crystallites are also of importance in developing the properties[5].

Recently Keller and Odell[6,7] devised a method for obtaining high modulus polyethylene from the melt, which appears to resemble high pressure injection moulding. In their experiments a polyethylene melt well over the melting point was made to flow in a glass capillary and if the flow rate was sufficiently high, thin birefringent fibrils, extending along the capillary length, appeared. If the flow rate was further increased the fibrils grew in thickness near the capillary entrance and eventually blocked the flow. The pressure then increased often with a consequent solidification of the material which was then slowly extruded through the capillary. Such thin fibrils are not capable themselves of increasing the stiffness of the material, but under high pressure they can act as nuclei for further crystallization resulting in high modulus structures. The material produced by Keller and Odell had a significantly higher modulus than the high pressure injection moulded HWMPE-bars studied here. This difference may be due to the non-stationary character of injection moulding, i.e. fewer fibrils may be formed during the rapid injection of the melt into the mould.

Influence of processing conditions on mechanical properties

Fig. 2 shows the influence of the barrel temperature (T_B) and the mould temperature (T_M) on the modulus and strength of HMWPE (thickness 1.5 mm) moulded with an injection pressure of 500 MPa. The shaded areas in the figure indicate the mechanical properties of high density polyethylene, injection moulded at conventional pressure levels, typically around 100 MPa. It is obvious that an increase in injection pressure up to 500 MPa improves both stiffness and strength several times. The highest values of modulus and strength (8 GPa and almost 200 MPa, respectively) are obtained for low mould and barrel temperatures, which indicates that the shear field during filling of the mould is an important factor.

If the thickness of the tensile test bars is increased from 1.5 mm to 3 mm this results in a substantial loss in stiffness and strength. For example, for a bar moulded at a mould temperature of 30°C and a barrel temperature of 170°C, the increase in thickness reduced the modulus from 8 GPa to ca 3 GPa and the tensile strength from 200 MPa to 125 MPa. This may also be interpreted in terms of a corresponding change in the shear field.

Properties of carbon-black filled HMWPE

In a series of experiments, HMWPE containing 1, 3 and 10 % by weight of carbon black (CB) was injection moulded at 500 MPa. Fig. 3 shows the tensile modulus of the test bars vs. the barrel temperature. The mould temperature was 30°C and the thickness of the bars 1.5 mm. Obviously, the CB-filling does not decrease the modulus substantially. At higher barrel temperatures the modulus of the CB-filled material is even higher than for unfilled HMWPE. The tensile strength decreases however when HMWPE is filled with CB (not shown in the figure).

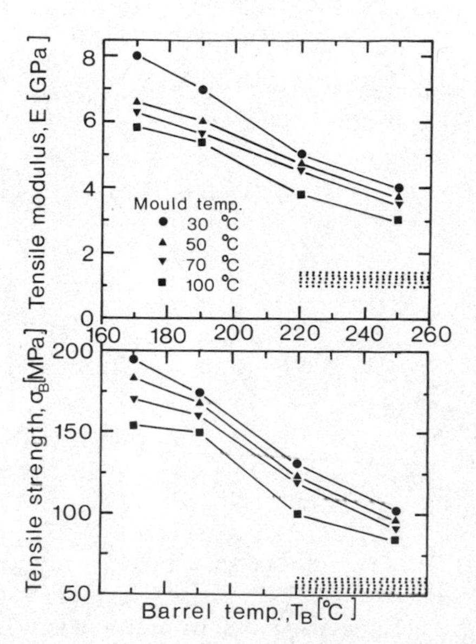

Fig. 2. Tensile modulus, E, and strength, σ_B, vs. barrel temperature, T_B, with different mould temperatures, T_M. From J. Kubát and J.-A. Månson, Polym. Eng. Sci., in print[5].

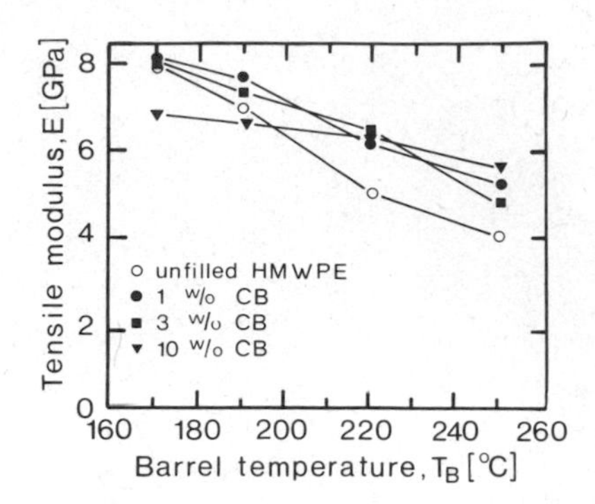

Fig. 3. Tensile modulus, E, for unfilled and CB-filled HMWPE vs. the barrel temperature, T_B.

One advantage of the addition of CB is that the reduction in stiffness and strength when increasing the thickness of the parts appears to be counteracted. For example, when the thickness is increased from 1.5 mm to 3 mm, the modulus decreases from ca 7.5 GPa down to only ca 5 GPa if the HMWPE-bar contains 1 % CB (T_B = 190°C). With no CB the modulus would have decreased to ca 3 GPa.

The effect of an exit cavity

It is reasonable to believe that if larger melt movements were allowed during the injection moulding cycle, the amount of high modulus structures, giving the improvement in mechanical properties, would increase[8]. To test this idea, the mould cavity was equipped with an exit cavity as shown in Fig. 4 (side view). In the following table 1 the effect of the exit cavity on the properties of HMWPE-bars is given. The barrel temperature was 190°C, the mould temperature 30°C, the injection pressure 500 MPa and two thicknesses of the bars, 1 and 4 mm, were used.

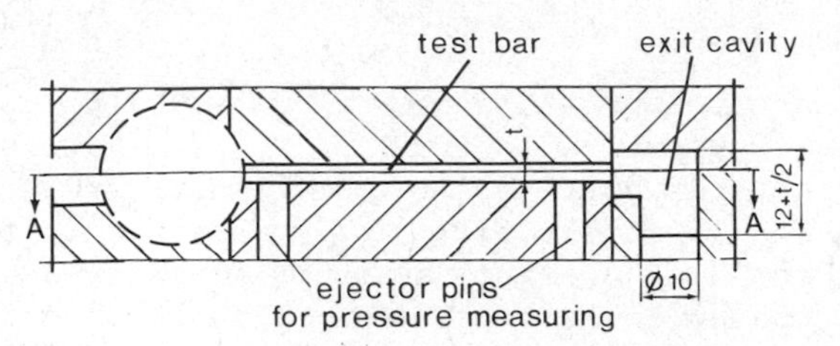

Fig. 4. Drawing (side view) of the mould cavity with the exit cavity. From J. Kubát et al., Polym. Eng. Sci., in print[8], and J. Kubát and J.-A. Månson, Plaste u. Kautschuk, 29(1982)91.

Table 1. Effect of the exit cavity on the tensile modulus (E) and the tensile strength (σ_B) of the HMWPE-bars.

		E (GPa)	σ_B (MPa)
No exit cavity	1 mm	9.5	210
	4 mm	2.8	70
With exit cavity	1 mm	11	265
	4 mm	4	110

Obviously the use of an exit cavity improves both the stiffness and the strength of the parts. Especially for thicker details the exit cavity must be useful in counteracting the decrease in mechanical properties.

The exit cavity also produces a more homogeneous structure along the gauge length of the test bars. This is to some extent illustrated by DTA-measurements. The second peak of the endotherms was then determined as a function of the distance from the gate. The peak temperatures have been for thin slices (0.5 mm) taken at various distances from the gate. For the bars moulded without the exit cavity the peak temperatures are always lower and decrease faster with increasing distance from the gate compared with parts moulded with the exit cavity.

FINAL REMARKS

It is obvious that high pressure injection moulding is a promising method for producing parts with relatively high values of the modulus and strength of polyethylene, which is a relatively inexpensive polymer. It should also be pointed out that high pressure injection moulding does not produce any increase in cycle time; the high production rate of injection moulding is still maintained with this technique.

REFERENCES

1. K. Djurner, Ph.D. Thesis, Chalmers University of Technology, Gothenburg, Sweden (1977).

2. K. Djurner and J. Kubát, Kunststoffe, 66:511 (1976).

3. K. Djurner, J. Kubát and M. Rigdahl, Polymer, 18:1068 (1977).

4. K. Djurner, J.-A. Månson and M. Rigdahl, J. Polym. Sci., Polym. Letters Ed., 16:419 (1978).

5. J. Kubát and J.-A. Månson, Polym. Eng. Sci., in print.

6. A. Keller, in: A. Ciferri and I.M. Ward (Eds.), "Ultra-high modulus polymers", Appl. Sci. Publ. Ltd., Barking, England, 1979.

7. A. Keller and J.A. Odell, J. Polym. Sci., Polym. Symp. No 63, 155 (1978).

8. J. Kubát, J.-A. Månson and M. Rigdahl, Polym. Eng. Sci., in print.

IRRADIATION OF HIGH DENSITY POLYETHYLENE FILLED WITH COLLOIDAL SILICA

G. Akay
School of Industrial Science
Cranfield Institute of Technology, Cranfield, England

F. Cimen and T. Tincer
Department of Chemistry
Middle East Technical University, Ankara, Turkey

INTRODUCTION

The addition of small particles can greatly enhance the mechanical and transport properties of polymers and elastomers. The irradiation of polymers by high energy radiation can also result in similar reinforcements. In addition, the irradiation of reinforced rubber is also used to understand the nature of reinforcement in these materials[1]. Omelchenko, et al[2] investigated the bonding of resin binders to glass fibre fillers by radiation curing and concluded that stronger adhesion could be obtained by comparison with heat curing. Fornes, et al[3] have shown that the irradiation of graphite fibre reinforced composites (used in space applications) at very high doses hardly effects the mechanical properties. These studies are typical of the purpose of radiation processing in filled polymers.

The reinforcement of the mechanical properties in filled polymers requires polymer-filler compatibility which can be achieved by coating the filler surface with a suitable silane coupling agent,or by radiation induced graft polymerisation onto the filler surface[4,5]. Melt mixing of polymers with ultra-fine particles may also result in binding macromolecules with the filler particles. The thickness of this so called "bound polymer" layer is 2-4 nm and therefore a substantial amount of polymer will be bound on the filler surface if the particle diameter is comparable with the thickness of bound polymer.

Investigations of the mechanical properties of polymers containing substantial amounts of bound polymer indicate that fracture probably occurs not at the filler/polymer interface but at the bound polymer/bulk polymer interface[6]. The reason for the weakness of this second interface is the distinction between surface and bulk molecules; a non-crystallising molecule near the filler surface is in a different state from a molecule in the bulk. The filler surface effectively segregates the molecules in its vicinity and reduces entanglements. Disentanglement must therefore be easier near the surface unless the surface chains are chemically crosslinked to bulk chains, or unless the surface chains are sufficiently long to extend some distance from the filler, as in the case of silane treated fillers.

In this study, we initially investigate the nature of the bound polymer in two types of high density polyethylene (HDPE) with similar molecular weights using three different types of silicas. These filled polymers are then subjected to γ-radiation in vacuum and then their tensile properties are evaluated. It is shown that the relative amount of bound polymer decreases with increasing filler concentration and the presence of fillers causes large reductions in ultimate tensile properties while high radiation doses eliminate the effect of the filler. Degradation of the ultimate tensile properties in HDPE is reduced in the presence of short chain branching. At low strains the mechanical properties of filled HDPE are improved with increasing radiation dose.

EXPERIMENTAL

High density polyethylenes were supplied by BP Chemicals. They are Rigidex H060-45P and Rigidex H020-54P. The former is an ethylene hexene-1 copolymer and contains one butyl short branch per 1000 carbon atoms, while the latter is a homopolymer.

Pyrogenic colloidal silica fillers were supplied by Degussa:
AE-1: Aerosil 380, with an average primary particle size of 7 nm and a BET surface area of 380 m^2/gr.
AE-2: Aerosil 130, with particle size of 16 nm and surface area of 130 m^2/gr.
AE-3: Aerosil TT600, with particle size of 40 nm and surface area of 200 m^2/gr.

Two types of mixing are used:
A) A Show twin-roll intermix is used at 180°C for 15 minutes at 80rpm.
B) A Brabender Roller Mixer (type W50H) is used to prepare small quantities of filled material. Mixing is conducted at 180°C for 15 minutes at 80 rpm.

At the end of mixing, samples are compression moulded at 190°C for 5 minutes and cooled at a rate of 20°C per minute. Dumbbell-shaped samples are cut using a small die. Some of these samples are exposed to γ-radiation under a high vacuum in a Co-60 source with a dose rate of 0.14 Mrad/hr. After completion of irradiation, samples are annealed at 115°C for five hours in the sealed glass tubes.

The determination of bound polymer is conducted by solvent extraction using xylene in a Soxhlet apparatus. Extraction is continued for as long as four to five days. The undissolved part of the samples are dried in a vacuum oven at 80°C and then pyrolysed at 700°C to determine the amount of bound polymer.

The number average and weight average molecular weights (M_n and M_w) of the polymers are determined by Gel Permeation Chromatography. For the filled polymers, we only determine the molecular weight of the free polymer, although that of the bound polymer can be determined after extracting silica from the bound polymer.

An Instron tensile testing machine is used to measure the mechanical properties of the dumbbell-shaped samples. Two cross-head speeds are used; 0.05 cm/min when measuring small-strain behaviour and 5 cm/min when determining ultimate tensile properties.

RESULTS AND DISCUSSION

Evaluation of Bound Polymer

In Table-I, we summarize the properties of the compositions used in the present study. As seen in this table, the molecular weights of the extracted polymer are smaller than those of the base polymer. This is partly due to degradation of the base polymer during mixing. The reduction in the molecular weight increases with increasing filler concentration and decreasing filler size. With the exception of the highly filled mixtures, heterogeneity index M_w/M_n is larger in the extracted polymer than in the base polymer. The mixing of highly filled material yields a large reduction in molecular weights and heterogeneity index, probably due to degradation of the base polymer. Binding of the polymer on the filler surface appears to be selective and long chain molecules are bound preferentially. These results are in agreement with the findings of Maurer, et al.[7]

Variation of the weight ratio of bound polymer and filler, W (W=weight of bound polymer/weight of filler) with the weight fraction of the filler in the original mixture, C (C=weight of the filler/total weight of the polymer and filler) is shown in Fig. 1. It indicates that as the concentration of the filler approaches zero the maximum amount of bound polymer per unit area of filler surface is obtained. This is probably due to a high degree of filler dispersion at low filler concentrations and low levels of macromolecular degradation, thus making a larger number of long chains available for binding on the filler surface.

The bound polymer and filler weight ratio,W for sample CB-2 is 0.24. For this sample the particle size is 16 nm and surface area

Table-I. Description of the samples. Here, C is the weight fraction of the filler in the original mixture, W is the weight ratio of the bound polymer and filler. Unbound polymer is used for molecular weight determination.

Sample	Base Polymer	Mixer Type	Filler		M_n $\times 10^4$	M_w $\times 10^5$	$\frac{M_w}{M_n}$	W
			Type	C				
CP	CP: Copolymer Rigidex H060-45P	—	—	—	2.3	2.8	12	—
CA-1		A: Roller Mixer	AE-1	0.20	1.7	2.4	14	1.0
CA-2			AE-2	0.23	1.8	2.7	15	0.24
CA-3			AE-3	0.24	2.0	2.7	14	0.10
CB-1		B: Brabender Mixer	AE-1	0.02	1.5	2.7	18	3.7
CB-2			AE-1	0.044	1.4	2.6	18	3.3
CB-3			AE-1	0.085	—	—	—	2.1
CB-4			AE-1	0.19	—	—	—	1.4
CB-5			AE-1	0.26	1.1	0.66	5.9	0.95
HP	HP: Homopolym. H020-54P	—	—	—	2.7	3.7	14	—
HA-1		A	AE-1	0.26	1.5	2.7	18	0.65
HB-1		B	AE-1	0.088	—	—	—	2.6
HB-2		B	AE-1	0.26	1.2	0.90	7.5	0.72

is 130 m^2/gr. In Fig.1, we have normalised W for CA-2 by multiplying 0.24 by 380/130 and the result is in agreement with the other data. However, for sample CA-3, the filler size is 40 nm while the surface area is 200 m^2/gr. Similar normalisation (to compensate for the difference in the surface area) for this case does not yield results consistent with the rest of the data in Fig.1. This probably indicates that the secondary surface structure in the filler Aerosil TT600 is not available for large macromolecules.

Yield Behaviour

In Fig.2, the initial part of the stress-strain curves of unfilled and filled copolymer H060-45P are reproduced. Similar results are also obtained for the unfilled and filled homopolymer. The draw rate is very small (0.05 cm/min) and therefore the yield stress is small and the drop in the stress after reaching the yield stress is very small for the unfilled polymer. Irradiation initially reduces the yield stress but with increasing radiation dose, yield is restored as shown in Fig.2(a). Unlike the unfilled polymer, all of the

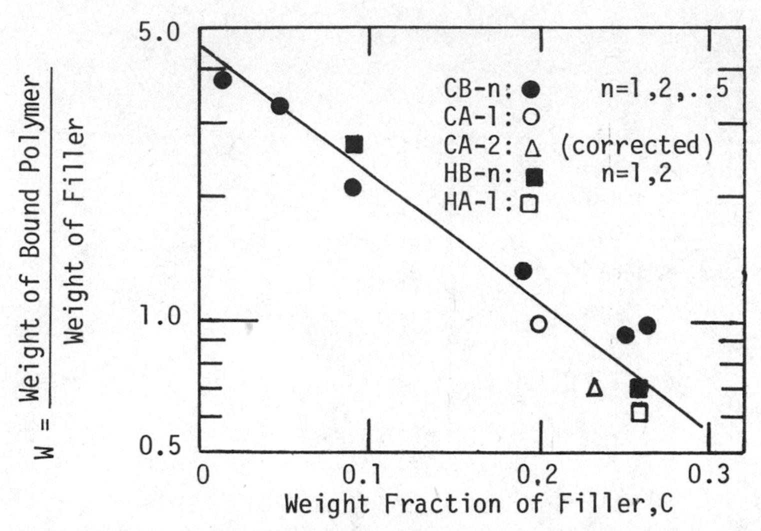

Fig.1. Variation of the weight ratio of bound polymer and filler(W) with concentration of the filler in the original mixture (C). Data for CA-2 is corrected for the surface area as explained in the text.

filled polymers yield a large drop in stress after reaching a maximum as shown in Figs.2 (b),(c) and (d). However, unlike the case in unfilled polymers, radiation appears to increase the yield stress and decrease the post-yield stress drop.

This relatively large post-yield stress drop in filled polymers is probably due to phase separation at the bound polymer/bulk polymer interface,forming microvoids. Only a small amount of straining is required to cause phase separation and after this initial phase separation, the filler-bound polymer phase no longer has a significant strengthening effect on the matrix; in fact these microvoids act as stress risers, causing premature fracture, at strains and stresses well below that of the unfilled material. This appears to be particularly true for the filled homopolymer, HO20-54P.

In polyethylene, doses below that required to induce gelation (usually 1-5 Mrad) the effect of radiation is to increase the average molecular weight and degree of long chain branching while some degradation is also possible[8]. The effect of low levels of radiation is to increase the entanglements at the bound polymer/bulk polymer interface since radiation induced changes are confined to the amorphous phase. As a result,elastic modulus increases with increasing radiation dose as seen in Fig.2. At high doses, cross-linking dominates and this results in further enhancement of the strength of the amorphous phase. At this stage we have not investigated possible changes at the filler-bound polymer interface.

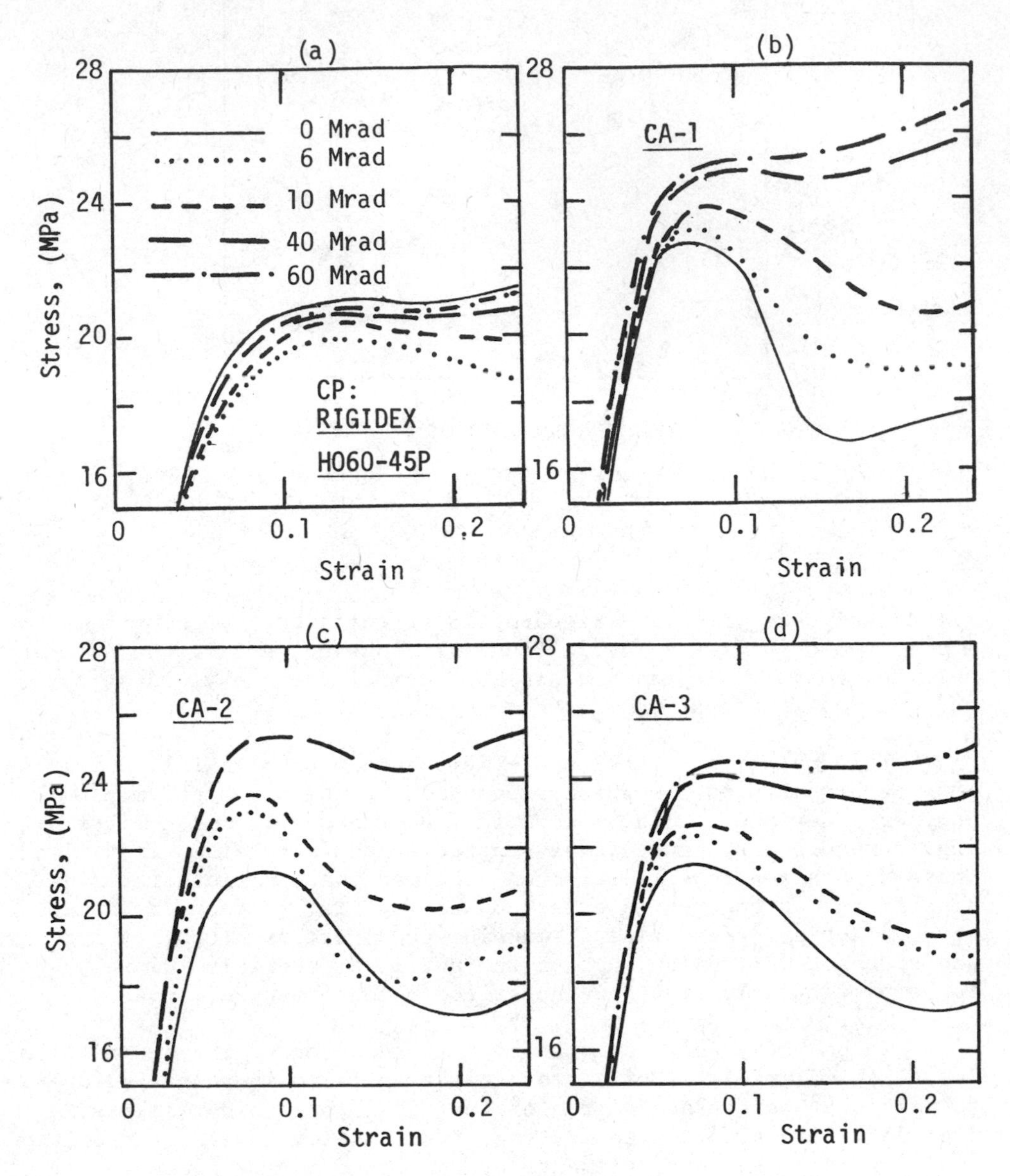

Fig.2. Initial parts of the stress-strain curves of various compositions based on copolymer Rigidex H060-45P, drawn at 25°C and at a rate of 0.05 cm/min. Radiation doses: 0 Mrad (———); 6 Mrad (.......); 10 Mrad (- - - -); 40 Mrad (— — —); 60 Mrad (—·—·)

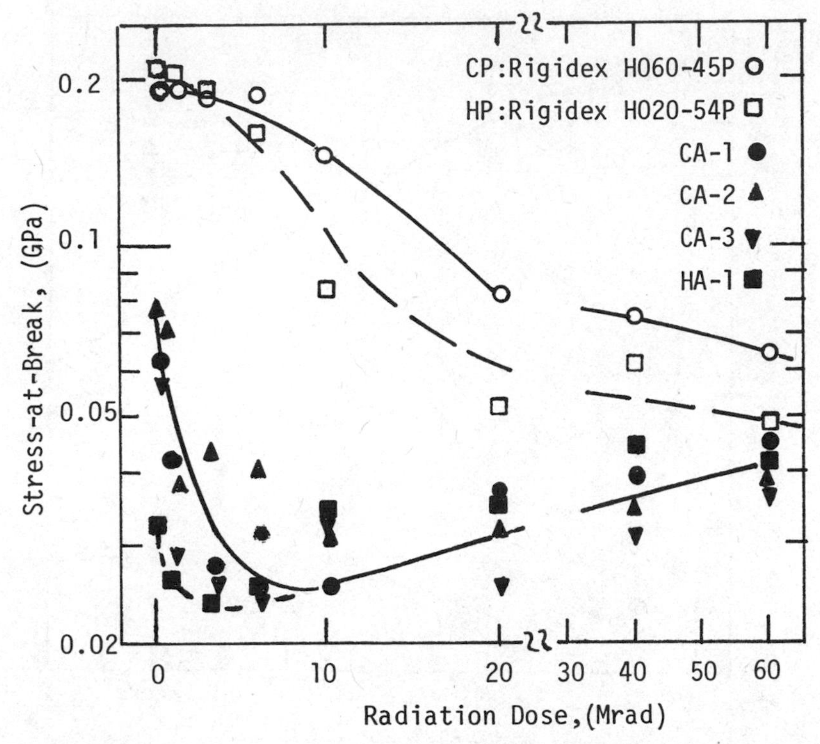

Fig.3. Variation of stress-at-break with radiation dose. Draw rate is 5 cm/min and draw temperature is 25°C.

Ultimate Tensile Properties

The variation of stress- and strain-at-break with radiation dose for the unfilled and filled polymers are shown in Figs.3 and 4. In this case, the draw rate is 5 cm/min. As seen in these figures, until a radiation dose of 6 Mrad, there is hardly any change in strain-at-break and the reduction in stress-at-break is small. But, corresponding reductions in the filled copolymer are very fast. Within a dose of 3 Mrad, strain-at-break may be reduced by a factor of 8 while the stress-at-break may be reduced by a factor of 3. In the case of filled homopolymer (i.e., sample HA-1) the ultimate properties are totally degraded even before irradiation. Nevertheless, radiation further decreases strain- and strss-at-break initially.

Ultimate tensile properties of the unfilled polymers continue to decrease with increasing radiation dose while those of the filled polymers appear to increase, especially stress-at-break. This is particularly true for samples CA-1 and HA-1, polymers filled with the smallest size particles. Both stress- and strain-at-break in the filled polymers appear to approach those of the unfilled polymers as the radiation dose is increased.

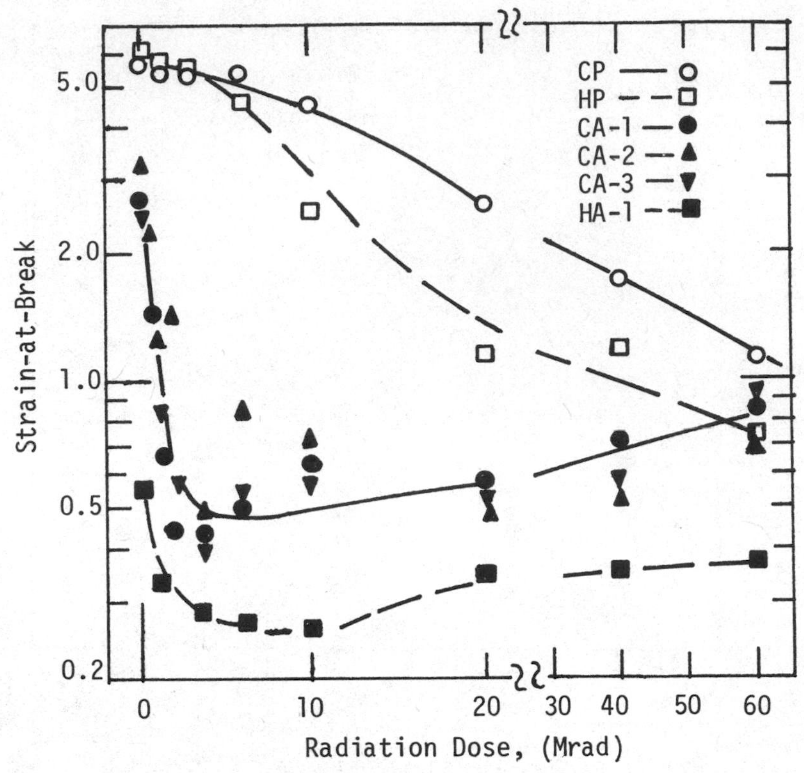

Fig. 4. Variation of strain-at-break with radiation dose.

At low radiation doses, degradation and long chain branching at the bound polymer/bulk polymer interface may increase the entanglement density, thus resulting in a higher yield stress and smaller post-yield stress drop and higher elastic modulus. However, these changes do not effect the ultimate tensile behaviour of the filled polymers since they effectively increase secondary bonding at the interface. At higher radiation doses, primary bonding in the form of cross-linking, results in a stronger bound polymer/bulk polymer interface.

CONCLUSIONS

The presence of bound polymer in ultra-fine particle filled polymers may be desirable since it eliminates the necessity to coat the filler surface. However, the mechanical failure in these systems may now be initiated at the bound polymer/bulk polymer interface as a result of phase separation at relatively low strains. Low levels of gamma-radiation enhance the decay of ultimate tensile properties while the presence of butyl branches appear to be useful in reducing this decay.

ACKNOWLEDGEMENTS

We would like to thank the National Physical Laboratory (U.K.) for supporting this research and BP Chemicals for the donation of HDPE's used in the experiments.

REFERENCES

1. A. Charlesby, Radiation effects in polymers, in: "Polymer Science" A. D. Jenkins, ed., North-Holland Publishing Company, Amsterdam (1972).
2. S. I. Omel'chenko, V. V. Shlapatskaya, G. A. Bakalo, O. G. Ponomarenko, V. N. Miryanin, N. I. Shchepetkina, G.N.P'yankov, and S. P. Polovnikov, Int. Polym. Sci. Tech., 8:65 (1981).
3. R. E. Fornes, J. D. Memory, and N. Naranong, J. Appl. Polym. Sci., 26:2061 (1981).
4. K. Hashimoto, T. Fujisawa, M. Kobayashi, and R. Yosomiya, J. Appl. Polym. Sci., 27:4529 (1982).
5. T. Fujisawa, M. Kobayashi, K. Hashimoto, and R. Yosomiya, J. Appl. Polym. Sci., 27:4849 (1982).
6. K. Kendall and F. R. Sherliker, Brit. Polym. J., 12:85 (1980) and 12:111 (1980).
7. F. H. J. Maurer, H. M. Schoffeleers, R. Kosfeld, and Th. Uhlenbroich, Analysis of polymer-filler interaction in filled polyethylene, presented at the Fourth International Conference on Composite Materials, Tokyo, 25-28 October, 1982.
8. M. Dole, Radiation chemistry of polyethylene, in: "Advances in Radiation Chemistry", Vol.4, M. Burton and J. L. Magee, eds., John Wiley and Sons, Inc., New York (1974).

THE ROLE OF MELT RHEOLOGY IN PARISON FORMATION

J.M. Dealy and N. Orbey

Department of Chemical Engineering, McGill University

Montréal, Canada H3A 2A7

INTRODUCTION

In extrusion blow molding, the shape of the parison and its thickness distribution at the moment the mold closes are of central importance.

Factors governing the geometry of the parison include the die design, the resin properties and the extrusion conditions. With regard to the resin, it is the rheological properties of the melt that govern the parison shape, and it is highly desirable to be able to relate parison behavior to laboratory rheological measurements.

It has been customary to simplify this problem by considering parison formation to be the result of two phemonena, extrudate swell and sag or drawdown. For an annular die, two swell ratios are required to describe the relationship between the parison cross section and that of the die at its exit. While there are several ways of defining swell ratios, we have chosen to use a diameter swell (B_1) and a thickness swell (B_2). These quantities are defined as follows, using the dimensions illustrated in Figure 1.

$$B_1 \equiv D_p/D_o$$

$$B_2 \equiv h_p/h_o$$

Many attempts have been made to relate extrudate swell to rheological properties, but none of these approaches has led to a reliable predictive method valid at high flow rates.

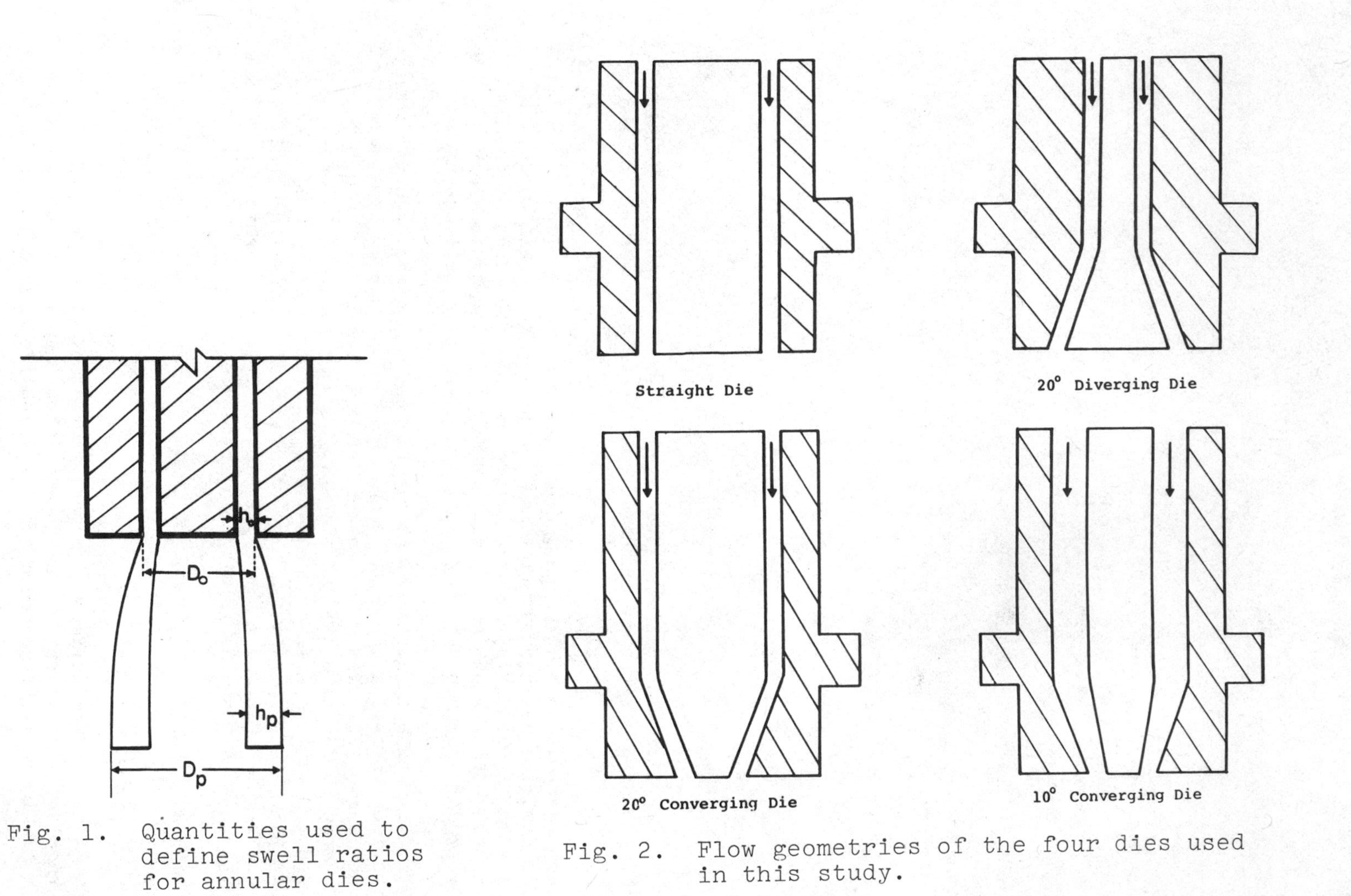

Fig. 1. Quantities used to define swell ratios for annular dies.

Fig. 2. Flow geometries of the four dies used in this study.

It would appear, then, that for the present, we must manage without a method for predicting swell. Taking this point of view, the goal of the work reported here was to develop a method for calculating parison geometry, starting from measured swell values together with a rheological analysis of the sag process. The results reported here are taken from a larger work (1) in which a number of blow molding resins were examined. We report here only data for a single resin, Union Carbide Canada DMDJ-5140 high density polyethylene, having a melt index of 0.96, with $M_w = 1.25 \times 10^5$ and $M_N = 1.76 \times 10^4$.

The major features of the flow geometry of the four dies used in this study are shown in Figure 2. In all cases the annular gap at the die lips was 1.5875 mm and the total die length was 69.85 mm.

SWELL MEASUREMENTS

A number of methods have been proposed to measure parison swell values (2). However, the only method that permits the accurate determination of swell as a function of time, isothermally and in the absence of sag, is the one reported by Garcia-Rejon and Dealy (3). Their procedure was to extrude into a thermostatted oil bath, with the oil having the same temperature and density as the melt. Using this method, it was found that about 70% of the swell occurs in the first 5 seconds, with an equilibrium value being reached only after several minutes. Figures 3 and 4 show equilibrium values of the diameter and thickness swells, respectively, as functions of flow rate and die design.

PARISON MODEL

In developing a model to predict parison behavior, we established several basic criteria. First, all the resin properties required by the model must be obtainable using readily available or easily constructed laboratory equipment. Second, extensive computation should not be required. Indeed, it should be possible to run through one set of conditions in a matter of minutes using an advanced microcomputer.

In the modelling of nonuniform flows of viscoelastic liquids, it is necessary to track the motion of each fluid element so that the effect of strain history can be accounted for by use of a constitutive equation. This generally leads to the introduction of elaborate computational algorithms requiring a large storage and high speed. To avoid this, we developed a lumped-parameter model in which the parison is considered to consist of a number of cylindrical elements of finite length, L_i, each characterized by a single time-dependent value of each of the two swell ratios, B_1 and B_2. Element No. 1 is

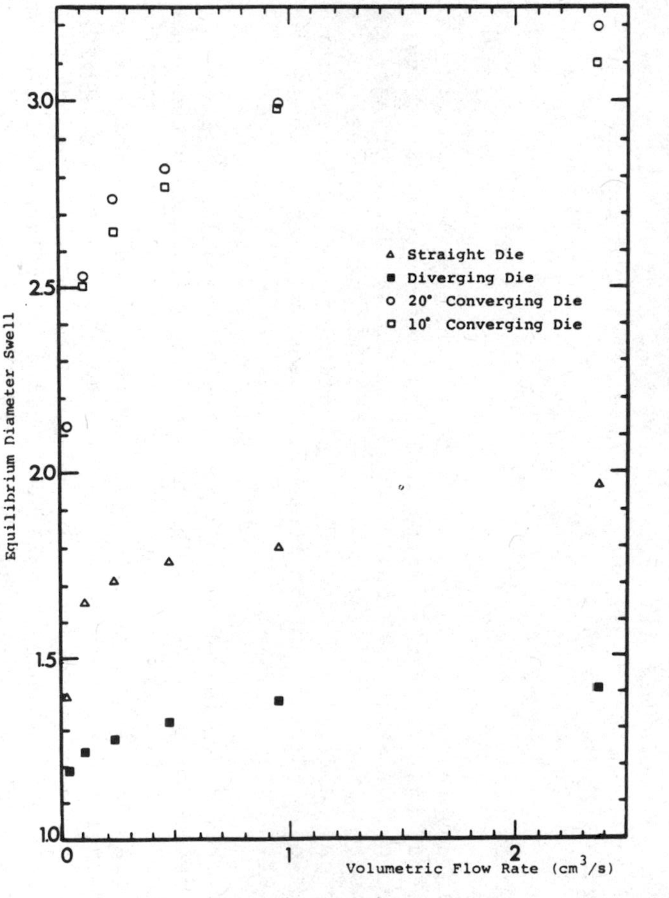

Fig. 3. Equilibrium diameter swell as a function of flow rate for each of the four dies.

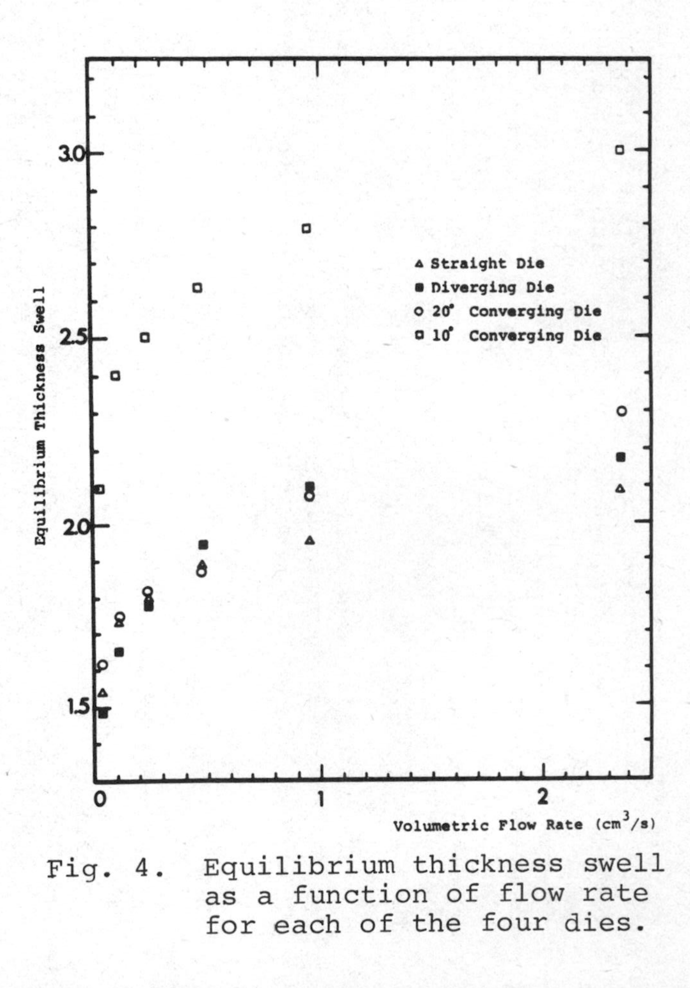

Fig. 4. Equilibrium thickness swell as a function of flow rate for each of the four dies.

the first one extruded, while at any instant of time, N is the index of the most recent, complete element to be extruded. Thus, the element currently being extruded corresponds to $i = N + 1$.

We consider first the period of time during which the parison is being extruded i.e. for which $t > t_e$ when t_e is the parison extrusion time. The length of the element currently being extruded increases by an amount dL during a time interval dt. Taking the density to be constant, dL is related to the average melt velocity at the die exit v_e, the area of the die exit, Ae, and the area of the parison element, $A_p(t)$, as follows:

$$dL_{N+1} = \frac{A_e v_e dt}{A_p(t)} \tag{1}$$

If the parison thickness is small compared to its radius, the ratio Ap/Ae can be approximated as the product of the two swell ratios. Using this approximation and integrating:

$$L_{N+1}(t) = \int_{N\Delta t}^{t} \frac{v_e dt'}{B_1(t') \, B_2(t')} \tag{2}$$

The time interval, Δt, is the time required to form one element of the parison. It is chosen to be an integral fraction of the total extrusion time, te, so that at the end of the parison extrusion stage of the process, the number of elements will be an exact integer.

Noting that about 70% of the swell occurs very quickly, with the remainder taking place much more gradually, we simplify the integral in equation 2 by using approximate average values of the swell ratios.

$$L_{N+1}(t) = \frac{v_e(t-N\Delta t)}{B_1(\Delta t/2) B_2(\Delta t/2)} \tag{3}$$

In order to model the time dependency of the swell, we used an empirical equation that was found to fit the data quite closely.

$$B(t) = B_\infty - (B_\infty - B_0)e^{t/\lambda} \tag{4}$$

Different values of the three constants, B_∞, B_0 and λ, were used for diameter and thickness swell.

The element formation time, Δt, is sufficiently small that we can neglect the sag that occurs during this period. Once an element is fully formed, however, we must allow for its sag due to the weight of the elements below it. If t_i is the time that has elapsed since element i was fully extruded, and t is the total time since parison extrusion commenced:

$$t_i = t - i\Delta t \tag{5}$$

Now, we assume that the sag of an element occurs independently of swell and that it is equivalent to uniaxial extension. We let Z_i be the hypothetical length that element i would have if there had been no swell, and identify the Hencky strain rate, $\dot{\varepsilon}_i$, of the element as follows:

$$\dot{\varepsilon}_i(t_i) = \frac{1}{Z_i}\frac{dZ_i}{dt_i} \tag{6}$$

Since sag is neglected during the formation of the element, at time $t_i = 0$,

$$Z_i(0) = v_e \Delta t \tag{7}$$

The Hencky strain rate will be related to the net stretching stress by the rheological constitutive equation of the melt. The net stretching stess is the weight of the (i - 1) elements below, divided by the hypothetical area A_i' of the unswollen element.

$$\tau_{zz} - \tau_{rr} \equiv \tau_i = \frac{(\rho v_e A_e \Delta t)(i-1)g}{A_i'} \tag{8}$$

From continuity, with the density assumed constant:

$$Z_i A_i = v_e A_e \Delta t \tag{9}$$

Thus;

$$\tau_i = \rho g(i-1)Z_i \tag{10}$$

To complete the system of equations for calculating $Z_i(t_i)$, a constitutive assumption must be made to relate τ_i and some functional of ε_i, evaluated over past times. We found the calculated parisor behavior to be very sensitive to the constitutive model used. In evaluating constitutive assumptions, we looked for the simplest possible model that would give reasonable agreement with experimental observations. We found that the Lodge rubber-like liquid model with a spectrum of relaxation times was sufficient to give good agreement in many cases. Where the agreement was not satisfactory, we believe it was due to some of the assumptions used to simplify the basic model rather than to the choice of constitutive model.

To find the actual length of an element at any time after it has been formed, we must take account of the swell that has occurred.

$$L_i(t_i) = \frac{Z_i(t_i)}{B_1(t_i + \Delta t/2)B_2(t_i + \Delta t/2)} \tag{11}$$

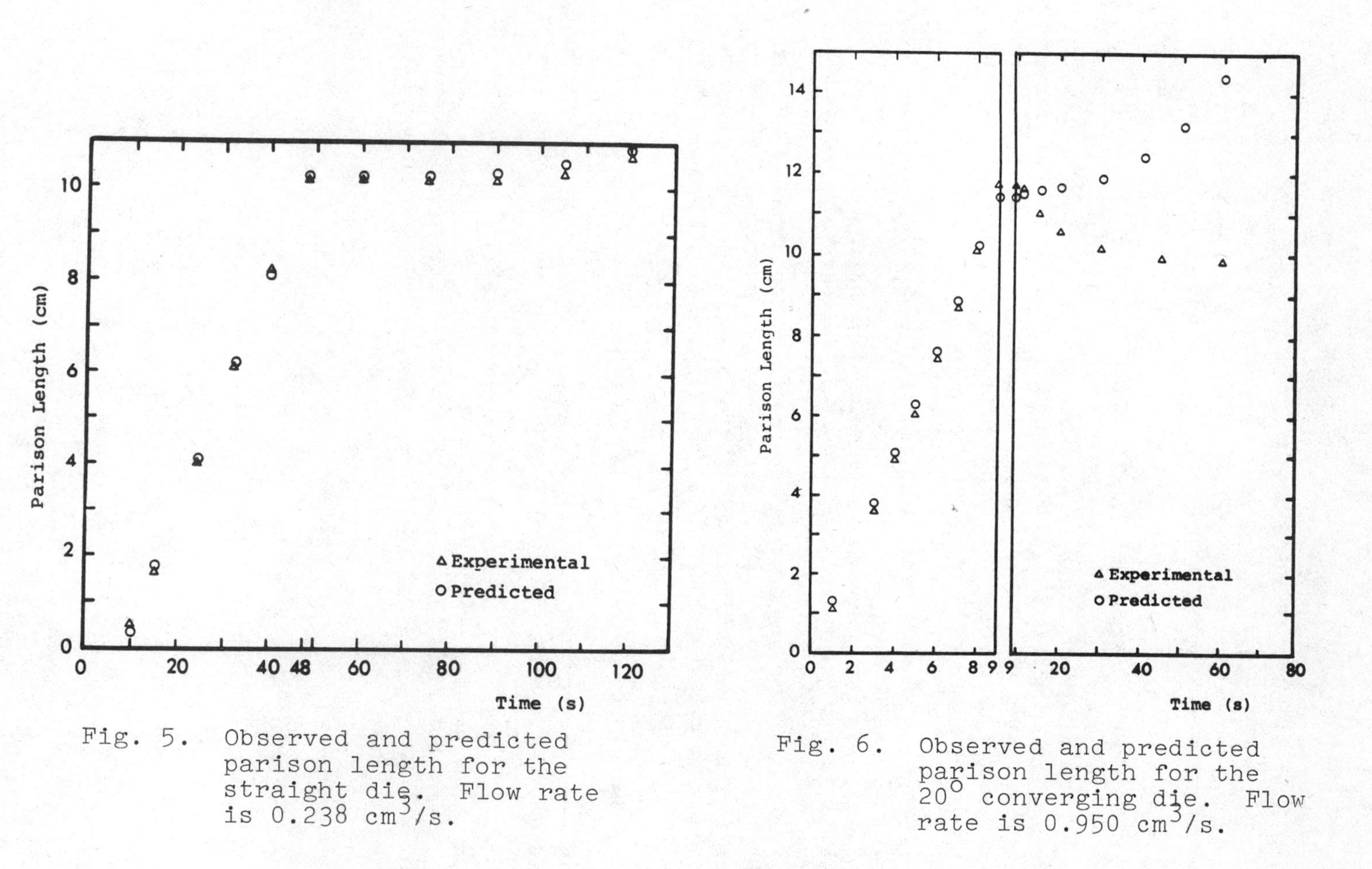

Fig. 5. Observed and predicted parison length for the straight die. Flow rate is 0.238 cm^3/s.

Fig. 6. Observed and predicted parison length for the 20° converging die. Flow rate is 0.950 cm^3/s.

$$L(t) = \frac{v_e(t - N\Delta t)}{B_1(\Delta t/2)B_2(\Delta t/2)} + \sum_{i=1}^{N} L_i(t_i) \qquad (12)$$

Note that once the parison is fully formed ($t \geq t_e$) the first term on the right hand side disappears.

COMPARISON WITH PARISON LENGTH MEASUREMENTS

To provide a preliminary evaluation of the model, we made measurements of parison length versus time for isothermal extrusion into air, using the dies already described. A temperature-regulated air oven was used, which had windows to permit lighting and photography.

Figure 5 shows experimental results and model predictions for the straight die at a relatively low extrusion rate. The agreement is very good, both during parison extrusion and in the post-extrusion period. For higher rates of extrusion, the observed length at large times was somewhat larger than that predicted by the model. We believe that this reflects the substantial shear thinning that occurs in the die gap at large shear rates. It is now recognized (4) that the "disentanglement" that accompanies shearing at high rates continues to affect the rheological behavior of a melt for a long period of time.

The predicted length was also in good agreement with experimental results for the diverging die at low flow rates. As is shown in Figure 6, however, in the case of the 20° converging die, the agreement is very poor in the post extrusion period. We believe that this results from the very large diameter swell that occurs with this die. This has the effect of rendering invalid the assumption that swell and sag are independent phenomena. If this is the case, no simple model of the type described here will be able to describe the behavior of parisons with very large diameter swell.

REFERENCES

1. N. Orbey, Doctoral Dissertation, McGill University, 1983
2. D. Kalyon, V. Tan and M.R. Kamal, Polym. Eng. Sci., 20, 773(1980)
3. A. Garcia-Rejon and J.M. Dealy, Polym. Eng. Sci., 22, 158 (1982)
4. J.M. Dealy and W. K-W Tsang, J. Appl. Polym. Sci., 26, 1150(1981)

MELT RHEOLOGY AND PROCESSABILITY OF POLYETHYLENES

L.A. Utracki*, A.M. Catani*, J. Lara* and M.R. Kamal**

*NRCC-IRMI - **Dept. Chem. Eng., McGill University

INTRODUCTION

The modern methods of the elongational viscosity, η_e, measurement have been developed during the last 30 years (1). In spite of a fair number of publications on the subject (2-6), the importance of the elongational flow in plastics processing so far has not been adequately realized by the industry. In this paper the correlation between the dynamic, shear and the uniaxial extensional viscosities of commercial polyethylenes, PE, is examined. In addition, the extrudability of these resins is analyzed.

EXPERIMENTAL

MATERIALS. Four commercial polyethylene, PE, samples: (1) linear low density, (2) high density and (3,4) low density resins, supplied through the courtesy of Union Carbide Canada Ltd., were used. Their properties are listed in Table 1.

TABLE 1: PROPERTIES OF POLYETHYLENE RESINS

N°	CODE	Density (g/cm^3)	Melt index[a] (g/10') Nominal	Measured	10^{-3}x Molecular Weights[b] M_n	M_w	M_z
1	LLDPE	0.918	1.0	0.71	97.5	261	624
2	HDPE-1	0.953	0.2	0.20	43.5	306	1021
3	LDPE-1	0.921	0.15	0.14	55.6	264	625
4	LDPE-2	0.922	5.5	4.06	39.1	210	668

Notes: (a) at 190°C with 2160g; (b) by GPC at 135°C in TCB

EQUIPMENT. The shear properties of the samples at 150, 170 and 190°C were determined in the dynamic and rotational modes using the Rheometrics Mechanical Spectrometer, RMS, Model 605. The samples were premolded at 170°C to fit the RMS tooling. For each sample at each temperature, first the strain, γ, and time, t, sweep tests, at two frequencies ω = 0.1 and 10 (rads/s), were done. In all cases the linear viscoelastic region of γ, measured over a period of stable response was found. The flow at a high range of $\dot{\gamma}$ was studied using the Instron Capillary Rheometer, Model 3211, ICR. The Rabinowitch and Bagley corrections were used. It was found that in some cases, the die slip occurred (e.g. low temperature extrusion of LLDPE) and the Mooney correction had to be applied. For details see (8-9).

The elongational properties were determined in the Rheometrics Extensional Rheometer, RER. The instrument, described in the literature (10-12), was used in the constant elongation rate, $\dot{\varepsilon}$, and constant extensional stress, σ_{11}, modes. The samples were transfer molded under vacuum, annealed and then affixed to the aluminum ties. The initial length and diameter of the samples were respectively: L_o = 21.0 ± 0.5 and D_o = 5.54 ± 0.05 (mm). The specimens did not show any shape change on immersion in hot silicon oil. For the test, the samples were mounted in the instrument and immersed in poly(dimethylsiloxane) oil. The densities of PE's at 150 to 190(°C) were calculated (13) as ρ = 0.781 to 0.751 (g/cm^3). The oil-polymer density difference, $\Delta \lesssim 0.08$, was too small to cause any detectable deformation of specimens. The testing started when the maximum temperature difference did not exceed 0.2 (°C) (7 to 10 min). In order to determine the effect of the residence time, t_R, the samples were evaluated in RER for up to 30 min, and tested for weight gain/loss after exposure to silicon oil for up to 23 hrs (14-15).

The molecular parameters of the samples were determined by gel permeation chromatography using a Waters Sci. GPC-150C instrument at 135°C with 1,2,4-trichlorobenzene, TCB, as a solvent. The four columns (10^3, 10^4, 10^5 and 10^6 Å) were calibrated using 24 polystyrene, PS, standard samples. The results are listed in Table 1. To prevent oxidative degradation of PE, 0.1 (wt %) of antioxidant (Topanol CA, Nonox DLTDP; 50:50) was added. There was no change in GPC traces for samples molded and sheared on RMS for 45 min. at 190°C (17).

The PE resins (as received) were extruded through a 19 mm-diameter laboratory extruder attachment to the Brabender Plastograph Model EPL-V5501. Screw N° 043-3 (diameter D = 18.3 mm, length L = 25D, helix angle 20° and a compression ratio of 2.85) and commercial slit and rod dies were used. Their dimensions were respectively: (width W = 76.04, length L = 25.14, thickness T = 0.52 mm), and (diameter d = 5.80, L = 43.07 mm). The pelletized HDPE and LDPE resins were led by gravity while the granules of LLDPE by means of a force-feeder.

RESULTS

In the text the following nomenclature will be used:

- dynamic viscosity: $\eta' = G''/\omega$, (where G'' is the loss modulus and ω-the frequency);
- elongational viscosity: $\eta_e = \sigma_{11}/\dot{\varepsilon}$, (where σ_{11} is the steady state extensional stress and $\dot{\varepsilon} = d\varepsilon/dt$);
- non-steady state elongational viscosity: η_e^t
- Hencky strain: $\varepsilon = \dot{\varepsilon}t = \ln(L/L_o)$, (where L_o and L are sample lengths initially and at time-t).

I. **Effect of silicon oils on PE.** The molded PE specimens were placed in the oils at 150°C and successively taken out, dried and weighed. For LDPE and LLDPE the weight remained constant within ± 0.05% (or ± 0.2 mg) and FTIR analysis detected neither a Si-O band in PE, nor CH_2 bands in the oil. On the other hand, a systematic weight loss, Δ, of HDPE samples was observed. For $0.25 \leqslant t(hrs) \leqslant 10$:

$$\Delta = a_o + a_1 \ln t \qquad (1)$$

where the constants (a_o = 0.2801 and a_1 = 0.1204), were found to be slightly decreasing functions of M_w (14). For HDPE the FTIR detected CH_2 bands in oil but not Si-O in the polymer. For the HDPE sample of Table 1 the M_w was not affected by the immersion. Addition of 0.1 wt % of Nonox DLTDP to the oil reduced the weight loss by about 50%. In agreement with the GPC results, the η_e^t vs ε of HDPE was independent of the residence time t_R < 30 (min). The maximum spread of η_e for LLDPE, HDPE and LDPE-2 was ± 6%. For LDPE-1 at 190°C, the behaviour near breaking strain, ε_c, was not reproducible.

II. **Results at $\dot{\varepsilon}$ = const.** For HDPE, log η_e^t vs log t plot depends slightly on temperature, but strongly on molecular weight (14). The behaviour of LLDPE is similar to that of the HDPE (Fig. 1), and quite different from the LDPE (Fig. 2), where a strain-hardening occurs. The temperature dependence of η_e^t for both low and high density PE allows computation of activation energy of flow as E = 24.4 and 24.8 (kJ/mole) respectively. In Figs. 1 and 2, the broken lines indicate stress growth in a transient shear deformation at $\gamma = 10^{-2}$; agreement with η_e^t is observed. For LDPE, agreement is obtained only before the onset of stress hardening.

III. **Results at σ_{11} = const.** It is known (10-12) that the steady state can be reached sooner under this condition than at $\dot{\varepsilon}$ = const. This mode is particularly useful for HDPE, for which at σ_{11} = const. a wider range of constant rate of deformation can be observed than for LDPE (Fig. 3).

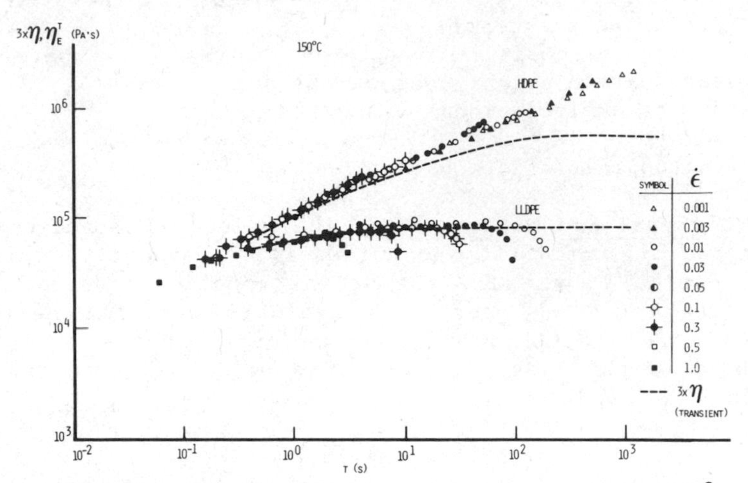

Fig. 1. Stress growth in elongation and shear ($\dot{\gamma} = 10^{-2}$) for HDPE and LLDPE at 150°C.

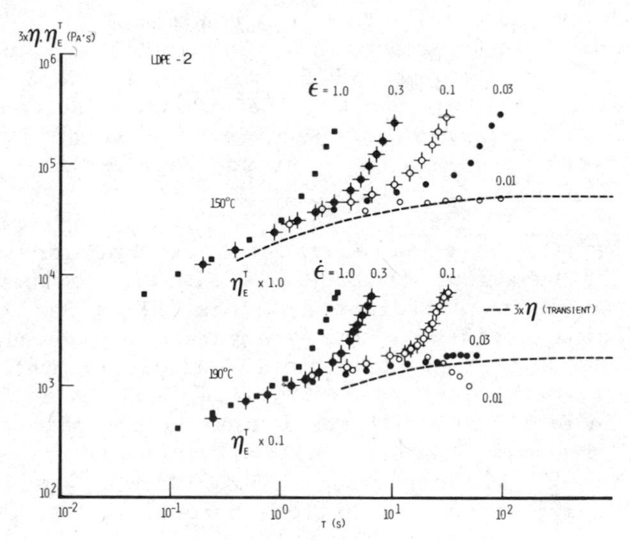

Fig. 2. Stress growth in elongation and shear ($\dot{\gamma} = 10^{-2}$) for LDPE-2 at 150 and 190°C (curve lowered by one decade).

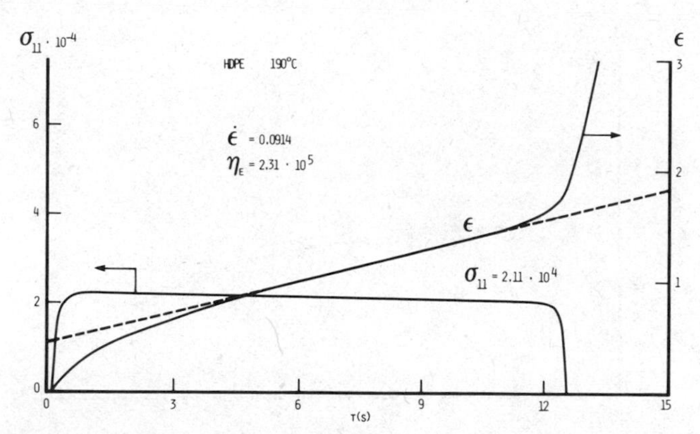

Fig. 3. Stress, σ_{11}, and strain, ε, vs time; HDPE-1 at 190°C. Constant stress experiment.

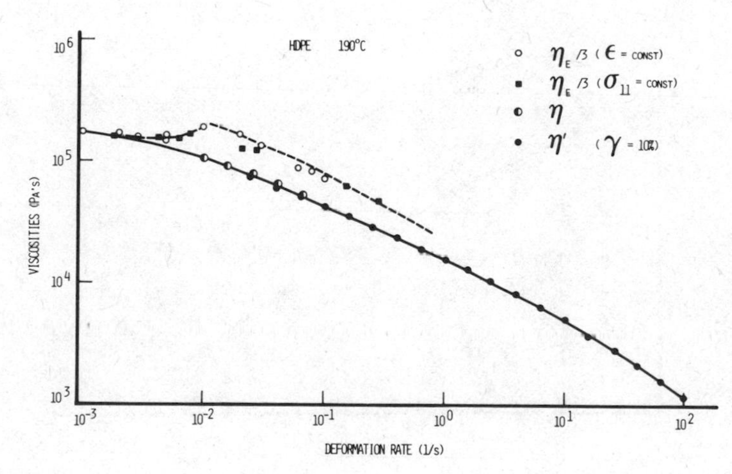

Fig 4. Deformation rate dependent viscosities of HDPE at 190°C.

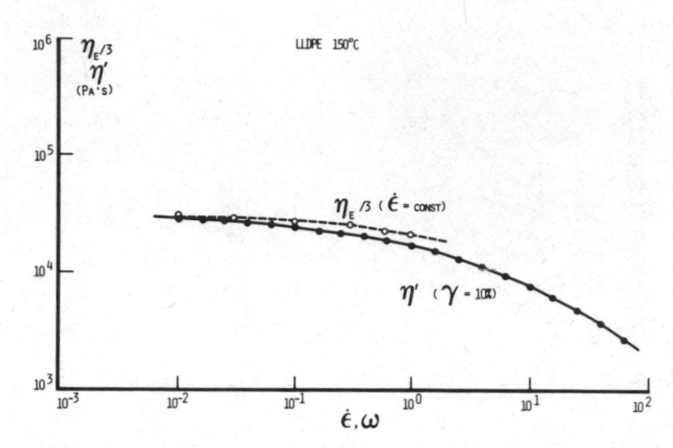

Fig. 5. Deformation rate dependent viscosities of LLDPE at 150°C.

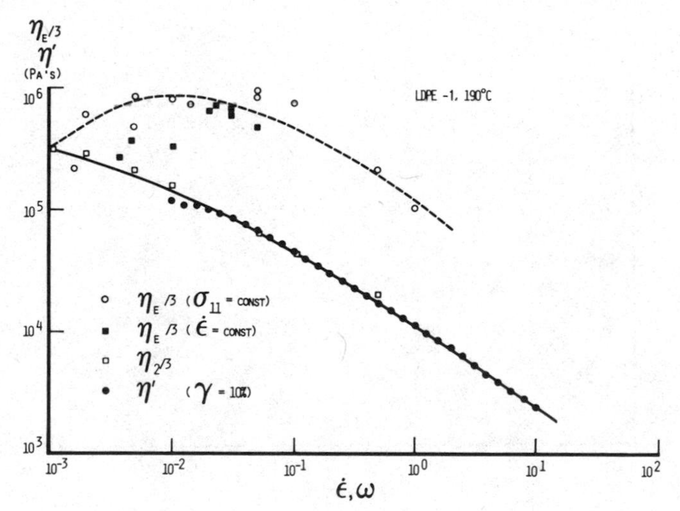

Fig. 6. Deformation rate dependent viscosities of LDPE-I at 190°C.

IV. RMS The results were accepted only if the difference between two runs was $\lesssim$ 5%. Plotting η' vs ω and $\eta_e/3$ vs $\dot{\varepsilon}$ at T = const, the following results were obtained: (i) for HDPE and LLDPE the two functions were similar (Figs. 4,5); the largest difference for HDPE was observed at the lowest temperature - in this case $R_T = \eta_e/3\eta'$ was about 2.5, for LLDPE $R_T \lesssim 1.2$. (ii) for LDPE the two functions were quite, different; while η' is a monotonically decreasing function of ω, the η_e passes through a maximum at a high rates of deformation and at lower temperatures; the strain-hardened LDPE had R_T = 10 to 15 (Fig. 6).

V. Definition and properties of η_2. For LDPE melts, the stress-strain, or η_e^t ($\dot{\varepsilon}$ = const) vs t, dependence seems to consist of two separate parts (Fig. 7), each in turn resembling the HDPE correlation. If the derivatives:

$$(\partial\sigma_{11}/\partial\varepsilon)_{T,\varepsilon} = (\partial\eta_e^t/\partial t)_{T,\varepsilon} = E \tag{2}$$

are a measure of modulus, than its stability indicates a stable structure. There are two regions where E $\simeq$ const. Within these, η_e^t changes relatively little. The η^t at the onset of the first region is labeled η_2. It was found that plot of $\eta_2/3$ vs $\dot{\varepsilon}$ superimposes on the η' vs ω.

VI. Extrusion Output, Q, torque, C, die pressure, P, and the extrudate surface roughness are reported in Fig. 8. Extrusions of the four resins were made at 190°C over a range of screw speeds 0 to 90RPM, with and without die (Table 2). These data show differences in the extrusion characteristics between the four resins, associated with the rheological properties. Work is in progress to explain these differences.

DISCUSSION AND CONCLUSIONS

The steady and dynamic shear, and elongational flow properties of four PE resins were determined. A number of observations are worth noting:

1. There are specific effects of silicon oil on PE.
2. The equilibrium elongational viscosity, η_e, defined as a maximum and/or plateau value of η_e^t for constant $\dot{\varepsilon}$ experiments agrees well with its analog calculated from the strain slope in a constant σ_{11} experiment.
3. At vanishingly small rates of deformation, $\eta = \eta_0$.
4. The maximum value of R_T was the lowest for LLDPE (1.2) and the highest for LDPE (15) with HDPE (2.5) in between.
5. At low rates of deformation: $\eta = \eta'$ and $N_1 = 2G'$ (14). At high: $\eta \simeq \eta^*$ and $N_1 > 2G'$ (9) were found.

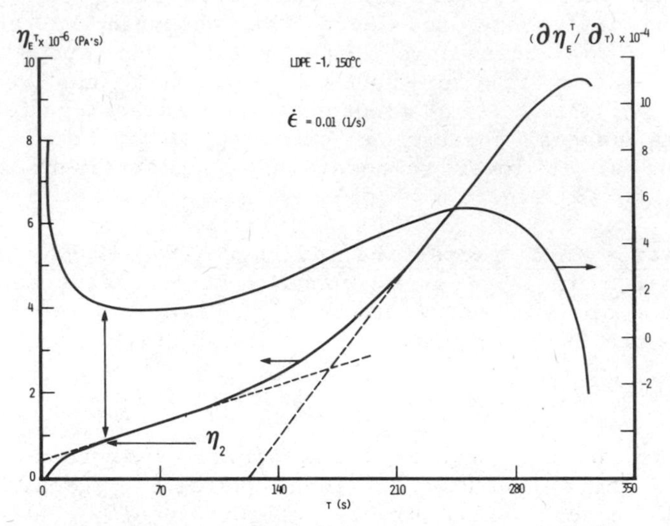

Fig. 7. Determination of η_2 using either the $\eta_e^t(\dot{\varepsilon}$ = const) vs t or its derivative.

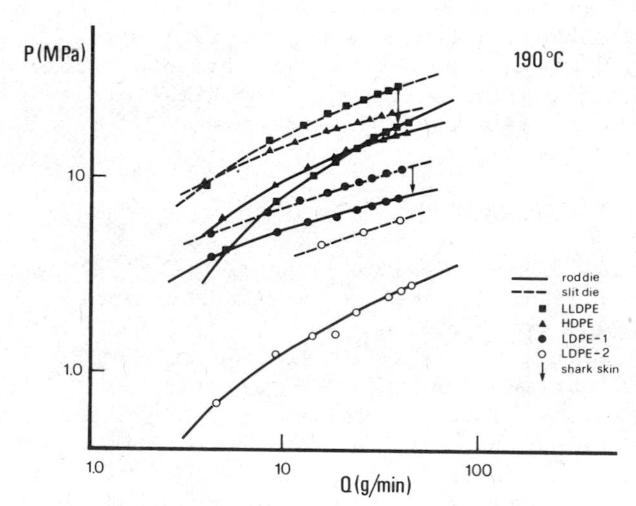

Fig. 8. Extruder output vs pressure; rod and slit dies, 190°C, four PE's. See text.

6. The maximum strain at break: ε_c = 1.5 for HDPE and ε_c > 3 for LLDPE and LDPE.
7. During extrusion the torque for LLDPE and HDPE was found to be nearly independent of the presence or absence of a die. The output (excepting the force-fed LLDPE) followed the sequence of the η. From Q, P and die geometry, the apparent viscosity η_a vs $\dot{\gamma}$ was calculated. Next, the Bagley correction was computed as $(\eta_a - \eta_{ICR})\dot{\gamma} = \sigma_e$ where η_{ICR} was determined in ICR. It was found that $\sigma_e \approx \Delta P_e$ where ΔP_e is the Bagley correction for ICR flow.

TABLE 2. MEASURED PROPERTIES OF PE RESINS AT 190°C

N°	Property	LLDPE N° 1	HDPE N° 2	LDPE-1 N° 3	LDPE-2 N° 4
1	C; rod/slit die	42/41	34/33	23/22	13/11
2	Q; rod/slit die	30/25	28/23	29/24	30/25
3	P; rod/slit die	25/17	21/16	10/7.6	6.0/2.6
4	Bulk density (g/L)	452	623	568	565
5	σ_e (MPa)	0.12	0.028	0.27	0.22
6	$\eta_o \times 10^{-3}$	8.9	127	197	4.97
7	$\eta_{e,o} \times 10^{-3}$	(37.7)	490	660	16
8	η	2660	2340	744	176
9	$\eta_e(\text{at}\,\sigma_{11}=10^4)$	(36.0)	360	2000	4.5
10	$G' \times 10^{-3}$ (Pa)	45.8	75.8	88.6	0.846
11	$G_x \times 10^{-3}$ (Pa)	1597	37.7	6.3	15.4
12	ω_x (rad/sec)	2981	6.3	0.25	39.6
13	ε_c	3.0	1.5	>3.5	3.5
14	% Die swell (L/D = 1)	50	70	124	5

Notes: C at 70 RPM (N-m); Q (g/hr.RPM); P at 70 RPM (MPa); σ_e (MPa) see text; $\eta_o, \eta_{e,o}$ - zero deformation rate, shear, η, and elongational viscosity, η_e (Pa.s); η_e and $\eta_{e,o}$ in parentheses calculated from 150°C results using the activation energy of flow E_η = 34.95 (kJ/Kmole); G', η and η_e taken at constant stress $\sigma^* = \sigma_{12} = 10^5$ and $\sigma_{11} = 10^4$ Pa respectively. G_x, ω_x define G' = G" condition.

ACKNOWLEDGEMENT

The help of Mr. R. Gendron in evaluation of the extruder flow data is deeply appreciated.

REFERENCES

1. C.J.S. Petrie and J.M. Dealy, Proceed. 8th Intl. Congress Rheology, G. Astarita, G. Marrucci and L. Nicolais (Eds.), Vol. 1 pg. 177, Plennum Press, New York 1980.
2. C.F. Denson, Polymer Eng. Sci., 13, 125 (1973).
3. J.A. Brydson, "Flow Properties of Polymer Melts", G. Godwin Ltd., London 1981, Chapter 6.
4. E. Deprez and W.J. Bontinck, in "Polymer Rheology and Plastics Processing", Eds P.L. Clegg et al, Chameleon Press, London, 1975.
5. K. Walters, "Rheometry", Chapman and Hall, London, 1975.
6. Z. Tadmor and C.G. Gogos, "Principles of Polymer Processing", J. Wiley & Sons, New york 1979.
7. L.A. Utracki and M.R. Kamal, Proceed. AIChE meeting, Los Angeles, CA. November 14-19, 1982 Paper No. 3b.
8. L.A. Utracki and M.M. Dumoulin, The 40th ANTEC'82, San Francisco, California, May 10-13, 1982; SPE Techn. Papers, 28, 51-4 (1982).
9. L.A. Utracki and R. Gendron, Proceedings Annual Meeting CSChE, Toronto, 1983.
10. H. Munstedt, J.Rheology, 23,421 (1979).
11. H.M. Laun and H. Munstedt, Rheologica Acta, 15, 517 (1976).
12. V.S. Au-Yeung and C.W. Macosko, Mod. Plastics, 58, 84 (1981); J. Rheology, 25, 455 (1981).
13. O. Olabisi and R. Simha, Macromolecules, 8, 211 (1975).
14. L.A. Utracki and J. Lara, 2nd Intl. Workshop on Extensional Flows, 24-28 January 1983, Mulhouse-La Bresse (France).
15. L.A. Utracki and J. Lara, NRCC-IMRI, mini-symposium "Composites-82" Nov. 30, 1982; Polymer Composites, 4, (1983).
16. L.A. Utracki and M.M. Dumoulin, NRCC-IMRI, mini symposium "Polyblends-83", April 12, 1983; Polymer Eng. Sci., 23, (1983).
17. L.A. Utracki and M.M. Dumoulin, ACS Div. Org. Coat. Plastics Chem., 48, 859 (1983); Polymer Plastics Techn. Eng., in press.

VOID FORMATION IN EXTRUDED BARS

G. Titomanlio(°), S. Piccarolo(°) and G. Marrucci(*)

(°)Istituto di Ingegneria Chimica, Palermo, Italy
(*)Istituto di Principi di Ingegneria Chimica
Napoli, Italy

INTRODUCTION

In the extrusion of thick polymeric manufacts, voids may form during the cooling process as a result of volume contraction of the hot "core" which is restricted by the outer "shell", already solidified. The effect is well known in the manufacture practice (see e.g. Ref. 1) and patents have been issued for systems which attempt to overcome the difficulty[2,3]. Void or microvoid formation is particularly relevant in cable covering, where it adversely affects both the plastics-metal adhesion and the electrical properties of the polymer, such as dielectric strength and permittivity[4-6].

With the aim of better understanding the influence of processing conditions on void formation, we have undergone an experimental program which makes use of a simple laboratory technique to simulate the extrusion of bars under controlled conditions.

The experimental procedure and the results so far obtained are briefly illustrated in the following. A simplified analysis of the phenomena which take place is also presented.

EXPERIMENTAL TECHNIQUE

The apparatus used in this work is shown schematically in fig.1. A glass cylinder, filled with the polymer, is placed vertically in an oven where the polymer is molten and compacted, with a piston,

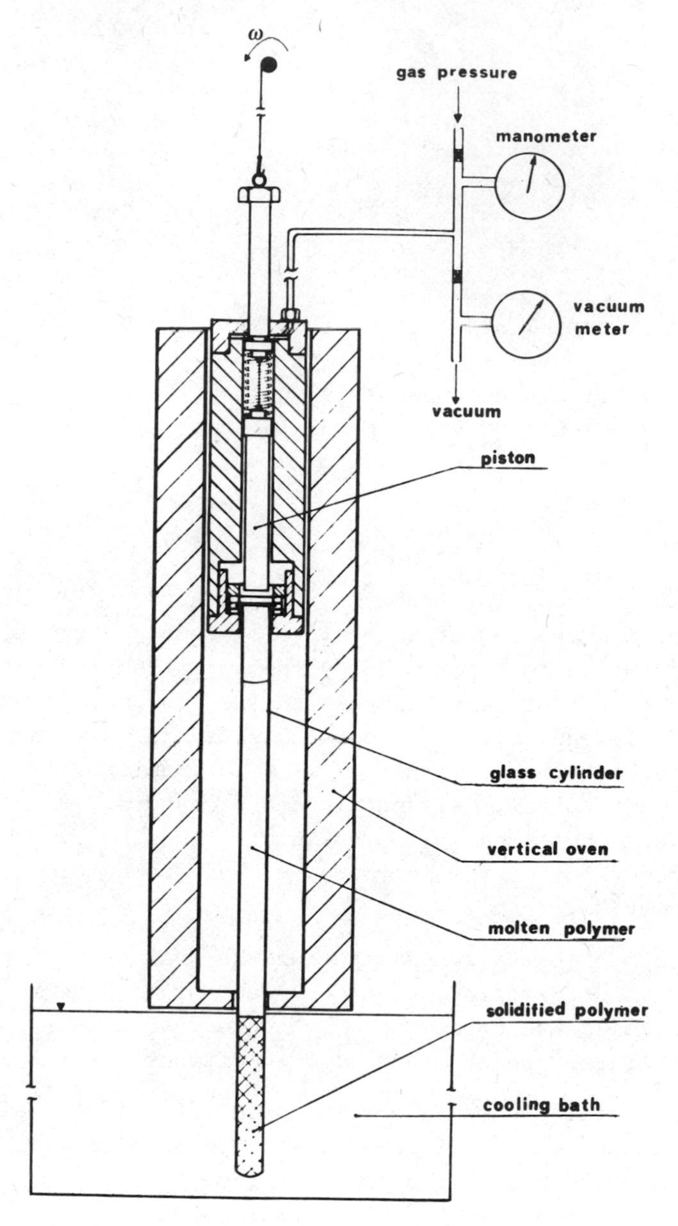

Fig. 1 Schematic drawing of the "extrusion" apparatus

under vacuum in order to avoid gas bubbles in the melt. When this operation is completed, i.e. the melt is degased and the temperature is uniform, the compacting piston is removed and a gas pressure is applied. Positive pressures up to 9 atm have been used in this work, but negative ones may be employed as well. Then the cylinder starts moving downwards with a fixed speed. At the exit from the oven the cylinder enters a cooling bath where a constant temperature is maintained.

Since the polymers adopted for the tests (various grades of HDPE) were not transparent in the solid state, the presence of voids was detected by X-rays photographs of the bars obtained. When present, voids were always located along the bar axis. The resolution of the X-rays images was not very high however. It allowed certain detection only of the voids larger than about 0.03 cm.

It is apparent that this "extrusion" technique allows to change all "processing" conditions in a very simple way. A possible limitation is to be found in the maximum allowable length of the sample.

RESULTS

The polymers used in this work were three grades of HDPE having a 190 °C zero-shear viscosity of $2x10^4$, $2x10^5$ and $6x10^6$ poises. The complete viscosity curves of these polymers are reported in Ref. 7 where they are indicated as samples A, B and F respectively.

In all the tests performed the melt and the cooling bath temperatures were fixed at 200 °C and 0 °C respectively. Tests were run at absolute pressures of 1, 4 and 9 atm and two cylinder internal diameters were used, .74 and 1.24 cm. The glass wall thickness for the two cylinders was 0.06 and 0.12 cm respectively, i.e. approximately proportional to cylinder diameter.

For each choice of polymer, pressure and diameter, the "extrusion" velocity was varied in a suitable range. In all cases no voids were found at sufficiently low velocities whereas they always formed at high velocities. A critical velocity, V_c, could be determined which corresponds to the onset of void formation as revealed by the X-rays photographs.

Figs. 2 and 3 show values of this critical velocity. The vertical segments indicate the range of uncertainty of the data points. Fig. 2 gives V_c for the three polymers and a fixed diameter of .74 cm. Fig. 3 shows the dependence of V_c on the zero-shear viscosity of the polymer for the two diameters and a fixed pressure, P =1 atm.

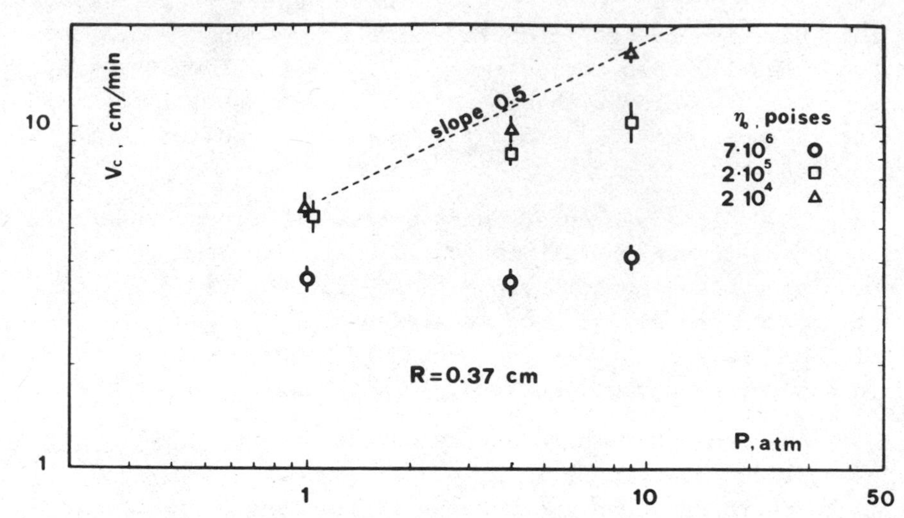

Fig. 2 Critical velocity, V_c, vs. operating pressure, P. The vertical segments indicate the range of uncertainty of the data points

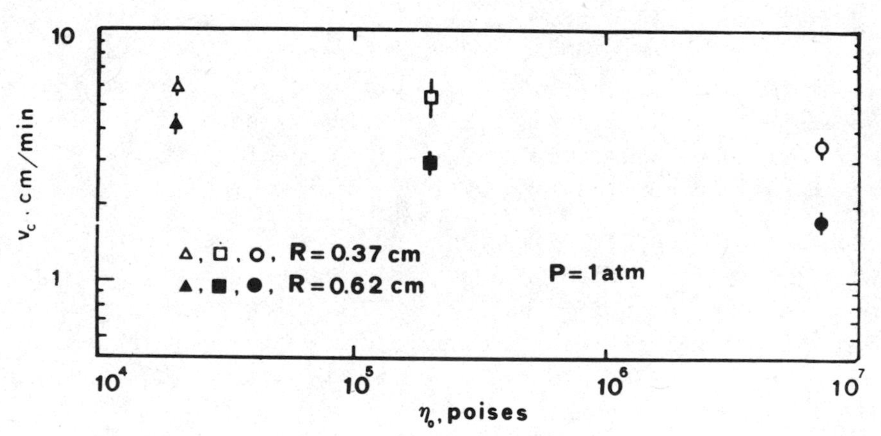

Fig. 3 Critical velocity, V_c, vs. viscosity η. The vertical segments indicate the range of uncertainty of the data points

ANALYSIS AND DISCUSSION

We have attempted an interpretation of the experimental results by assuming that the following phenomena are the relevant ones.

Fig. 4 shows schematically the solidification of the sample. Though the sample is moving along z with the velocity V, the solidification front, which bounds a sort of conical "cavity", is of course fixed in space.

Due to restricted volume contraction in the solidification process, it is required that a compensative flow of the melt relative to the solid takes place within the cavity. If the pressure drop associated with this flow exceeds the applied pressure, then a region of negative pressures is created towards the tip of the cavity. In this region voids may form and grow up to detectable dimensions.

On the basis of the above description the following analysis is developed. The flow rate through a section of the cavity required to compensate for the downstream contraction is calculated as

$$Q = \pi r^2 V \beta \qquad (1$$

where β is the relative volume change upon solidification ($\beta \simeq 0.2$ in the present case). By assuming Newtonian behavior, the pressure gradient is related to Q through the Hagen-Poiseuille equation, thus obtaining

$$-\frac{dP}{dz} = \frac{8\,\eta V \beta}{r^2} \qquad (2$$

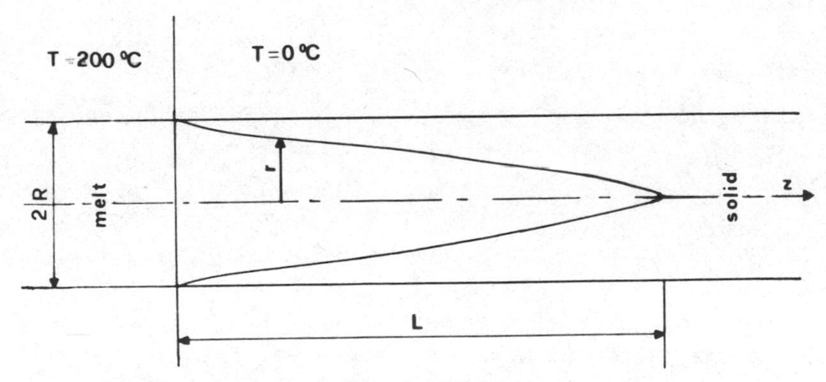

Fig. 4 Scheme of the sample solidification process. R is the sample radius; r(z) and L are radius and length of the solidification front.

In order to integrate eq. 2, the shape of the cavity is required. This was obtained by solving numerically the heat transfer problem, where the flow in the cavity, which depends on β, was also accounted for. It was found that the profile of the cavity can be described to a good approximation as

$$\frac{r}{R} = \left(1 - \frac{z}{L}\right)^m \qquad (3$$

where L, the length of the cavity, is given by

$$L = \lambda \frac{V R^2}{\alpha} \qquad (4$$

In eq. 4, α is the thermal diffusivity and λ is the dimensionless cavity length. Since in our case the latent heat, the heat capacity and all relevant temperatures are fixed,as well as β and the respective glass wall thickness, λ and m are constant. The numerical results are $\lambda \simeq 0.6$, $m \simeq 0.8$.

Eq. 4 allows L to be estimated. For the values of V_c reported in figs. 2 and 3, L ranges from 5 to 30 cm. In all cases the sample length was longer than L and the analysis of the X-rays images was limited to the sample region which corresponded to a fully developed cavity.

By using eq. 3, eq. 2 is readily integrated to obtain the pressure profile along the cavity, P(z). In particular, the value of z_0 at which the pressure becomes zero, is obtained as

$$\frac{z_0}{L} = 1 - \left(1 + \frac{2m-1}{8\beta\lambda}\,\frac{P\alpha}{\eta V^2}\right)^{-\frac{1}{2m-1}} \qquad (5$$

Finally, the growth process of the voids was described through the equation

$$\frac{4\eta V}{D}\,\frac{dD}{dz} = -P(z) \qquad (6$$

where D is the diameter of a spherical void. Of course, eq. 6 only applies in the region of negative pressures, from z_0 onwards (we are neglecting surface tension). Eq. 6 requires an initial value for D, D_1, which is unknown. However, D_1 only appears in the result through a logarithmic term.

In the previous equations the velocity V has the meaning of critical velocity V_c when the upper value of D, D_2, in the integration of eq. 6 is the minimum detectable void dimension.

Further details of the analysis will be given elsewhere. Here, we indicate the following results:

i. - For small values of the dimensionless group $\eta\alpha/PR^2$, i.e. at low viscosities, the group which appears in eq.5, $\eta V_c^2/P\alpha$, is approximately proportional to $(D_2/R)^{(2m-1)/m}$. Thus, for a fixed viscosity and cylinder diameter, the critical velocity is predicted to increase as the square root of P.

ii.- For large values of $\eta\alpha/PR^2$, the group V_cR/α becomes constant. Thus V_c is predicted to be independent of pressure and inversely proportional to cylinder diameter.

As shown by figs. 2 and 3, these predictions favorably compare with the experimental results. For $\eta = 2x10^4$ poises the critical velocity appears to increase as $P^{1/2}$, whereas for $\eta = 7x10^6$ poises it is virtually independent of pressure. In the latter case, V_cR has the same value for both cylinder diameters whereas, by decreasing the viscosity, V_c becomes less dependent on R.

The analysis briefly described here does not account, among other things, of the non-Newtonian character of the polymer. However, two limiting behaviors, similar to those identified above, are also obtained for a power-law fluid. In particular, a critical velocity independent of pressure is still predicted for large viscosities. Other details which are shown by the experimental results also might play some role. For example the shape of the voids is not always spherical. For the polymer with the smallest viscosity used in this work the voids appeared more as "filaments" along the axis of the bar. However, this observation is consistent with the indication of the analysis whereby nucleation of voids in low viscosity liquids occurs very close to the tip of the cavity. Conversely when the viscosity is large, the point of zero pressure is very close to the cavity entrance.

REFERENCES

1) Fisher E.G., "The extrusion of tube, rod and profile", chap 20 in "Polythene", Renfrew A. and Morgan P. Eds, Interscience, N.Y.1960
2) British Patent: 689252 (6,7,1951)
3) Goldman C., U.S. Patent: 4154893 (15,5,1979)

4) Fairfield R.M. and Cox F.J., "The extrusion of polythene cable covering", chap. 19 in "Polythene", Renfrew A. and Morgan P. Eds, Interscience, N.Y. 1960
5) Lemainque H. and Terramorsi G., Conference Internationale des Grandes Réseaux Electriques, 1972, paper 21-07
6) Heckman W. and Schiag J., Kunststoffe 72 (1982) 2, p.96
7) La Mantia F.P., Valenza A. and Acierno D., Rheol. Acta, in press (1983)

UTILIZATION OF RHEOLOGICAL REAL TIME AND SAMPLING MEASUREMENTS FOR QUALITY CONTROL OF POLYMER MATERIALS AND PROCESSES (Abstract)

J. M. Starita

Rheometrics, Inc.

Elizabeth, NJ, USA

Zero shear viscosity relates to molecular weight while elastic measurements such as recoverable compliance and dynamic storage moduli show strong dependence on molecular weight distribution and branching. This paper concerns itself with new and convenient methods for measuring dynamic mechanical properties in real time extrusion and in batched sampling.

Alternatively, a new batch type constant stress or creep device which automatically measures zero shear viscosity and the time dependent recoverable compliance is discussed.

PART II

PROCESSING: OPERATIONS OTHER THAN SHAPING

POLYMER PROCESSING OPERATIONS OTHER THAN SHAPING

Costel D. Denson

Department of Chemical Engineering
University of Delaware
Newark, Delaware 19711

INTRODUCTION

Polymer processing operations which do not involve shaping are defined here as those which are concerned with the synthesis of high molecular weight polymers, and those which are concerned with affecting physico-chemical changes in the nature of polymeric materials. These operations, which are most often conducted upstream from the shaping operations (i.e., extrusion, injection molding, blow molding, tubular film blowing, fiber spinning), include by way of example: vapor-liquid stripping operations when the liquid phase is a molten polymer or polymer solution (devolatilization), liquid-liquid stripping operations where one liquid is a molten polymer or polymeric solution, gas absorption in molten polymers, polymerization and grafting reactions, mixing, pumping and pressurization, and filtration.

The scientific and engineering principles which govern behavior in these 'non-shaping' polymer processing operations have not been well articulated, and certainly not as well as those which govern behavior in the shaping operations. Consequently, it has not always been completely clear as to how to proceed in the analysis and design of these operations. What is particularly noteworthy about this point is the fact that these 'non-shaping' operations closely parallel many of the classical unit operations in the chemical process industries where analysis and design procedures are based on well-established engineering concepts, which leads one to believe that these same concepts might be applied to the analysis and design of the 'non-shaping' polymer processing operations. The extent to which this may be true is a question which is largely unanswered at the present time.

In this paper we wish to draw attention to the fact that relatively little research has been conducted toward developing a scientific and engineering understanding of the 'non-shaping' polymer processing operations, and that there is a need for further research in this area. Polymeric materials have two properties which, more than other set of properties, serve to distinguish them from the materials of usual interest in the chemical process industries: inordinately high viscosities, and the property of elasticity. As a result, these materials must be processed in ways that are different from those traditionally used in the chemical process industries. New and "unusual" flow geometries are required and this means that the process equipment will be different. This, in turn, suggests that even though the scientific principles which govern behavior may be somewhat the same, the specific application of these principles may lead to quite different results. To illustrate this point, we have selected three 'non-shaping' operations for discussion. From this perspective we will show where uncertainties exist in our understanding of the basic concepts that should be used in analysis and design and how these uncertainties can be traced to the unique properties of polymeric materials.

STRIPPING OPERATIONS WITH POLYMERIC SOLUTIONS

Vapor-liquid stripping of polymeric solutions, or devolatilization as it is known in the polymer processing industries, is conducted in a wide variety of process equipment depending on the viscosity of the polymeric solution being processed. Solutions having low viscosities are often processed in wetted wall towers or flash chambers depending on the solvent concentration. Solutions having viscosities of, say, 500 poise or greater, must be processed in falling strand devolatilizers or in geometries which mechanically generate a wiped film (screw extruders, for example), and it is in the processing of these solutions where much uncertainty exists in analysis and design procedures.

When devolatilization processes are conducted in screw extruders, the screw channels are only partially filled with the polymeric solution to be stripped of the volatile component (Figure 1) while the unoccupied portion of the screw channel serves to carry away the evaporated liquid. Because of the finite helix angle on the screw, a component of motion exists in the down channel direction (Tadmor and Klein, 1970), and the solution is caused to flow from the extruder inlet to the outlet which is out of the plane of the paper. The cross channel component of motion has two effects. First, it causes a circulation of the fluid in the nip, and because of the continual exposure of fresh material at the surface of the fluid in this nip, mass is transferred to the gas phase when the fugacity of the volatile component in the gas phase is less than that in the solution.

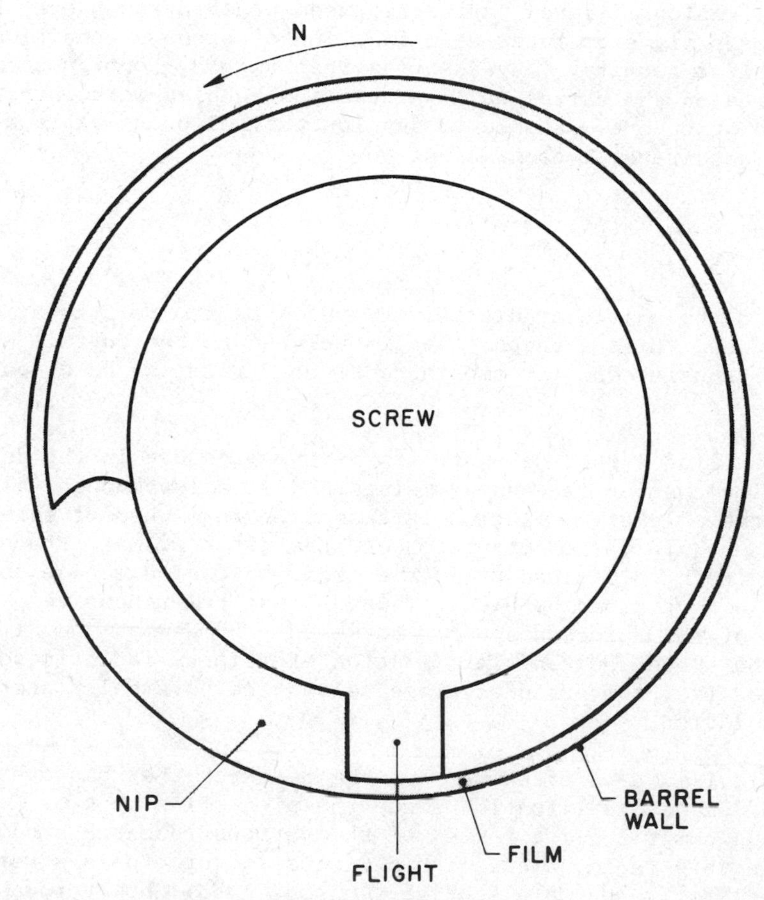

Figure 1. Cross section of a single screw extruder that is partly filled with liquid.

The second effect which results from the cross channel component of motion is to generate a wiped film of the polymeric solution as the solution is dragged from the nip in an adjacent screw channel through the clearance between the flight tip and the barrel wall. Since this film is continually regenerated, mass is also transferred to the gas phase from this surface.

Latinen (1962), who was the first to analyze the problem of devolatilization in screw extruders, used penetration theory to predict mass transfer rates as a function of process conditions and equipment geometry. By assuming that smooth, continuous films were formed on the barrel wall and that no bubbles were entrained in the solution, he was able to develop the following expression for the mass transfer coefficient

$$(k_L a)_{av} = \beta (DN)^{1/2} \tag{1}$$

where D is the molecular diffusivity and N is the rotational speed of the screw. β is a factor that is related to the surface areas for mass transfer and for smooth continuous film can be computed exactly.

In falling strand devolatilizers, the pressure in the devolatilization chamber is usually maintained at a low enough value so that the polymeric solution is caused to foam when it enters the chamber. In these circumstances, the rates of mass transfer can be quite high because of increases in surface area due to the presence of entrained bubbles. The earliest comprehensive analysis of this process appears to be that of Newman and Simon (1980) who showed how devolatilization efficiency is influenced by bubble size, temperature, pressure and the physical properties of the solution.

Table 1 is a brief summary of the research that has been conducted on devolatilization since the pioneering work of Latinen (1962). A comprehensive review of the various research studies listed in this table, along with a discussion of problems yet to be solved, is presented elsewhere (Denson, 1983); however, a few points are worth noting here.

First, in two experimental studies in which the same polymer/monomer system was used and identical operating conditions were employed (Latinen, 1962; Beisenberger and Kessidis, 1982), contradictory results were obtained. Second, in experimental studies where the pressure in the extraction zone in an extruder was low enough so that bubble nucleation was a distinct possibility, no dependence on screw speed was observed in some cases, and in others, a square root dependence on screw speed was observed in accordance with the Latinen theory. Third, the study conducted

Table 1. Summary of Research on the Stripping of Polymeric Solutions

Investigator	Polymer System	Operating Conditions T(°C)	P(mm)	w_1	Equipment Type	Main Conclusions
Latinen (1962)	polystyrene/styrene	200-250	5-20	0.008	single screw extruder	Developed devolatilization model for screw extruders using penetration theory. Experimental data varied as $N^{1/2}$ in accordance with model. Computed D from experimental data.
Coughlin and Canevari (1969)	polypropylene/methanol polypropylene/xylene	260 260	0.0(?) 0.0(?)	0.003-0.010 0.003-0.010	single screw extruder	Measured mass transfer coefficient experimentally. Mass transfer coefficient independent of screw speed.
Todd (1974)	polystyrene/ethylbenzene polyethylene/cyclohexane	- 210	- 0.5-100	0.01-0.15 0.05-0.10	twin screw extruder	Good agreement between experimentally measured and predicted exit concentrations using Latinen theory. Diffusion coefficient and surface area were unknown and were determined experimentally.
Werner (1980)	polymethylmethacrylate/ methylmethacrylate ethylene/polyethylene (low density)	250 240	15-760 23-760	0.013 0.0014	twin screw extruder	Determined values for the mass transfer coefficient from experimental measurements. Independent of screw speed for at least one system (MMA/PMMA).
Newman (1980)	polystyrene/styrene	200-250	5-10	0.0015-0.0125	falling strand	Developed bubble model for devolatilization and applied results to falling strand devolatilization.
Beisenberger and Kessidis (1982)	polystyrene/styrene	200 200	1 760	0.0054 0.0054	single screw extruder	Exit concentration varied with $N^{1/2}$ at atmospheric pressure. Computed D from experimental data. Exit concentration independent of screw speed at reduced pressure.
Collins, Denson and Astarita (1983)	polybutene/freon	20	760	0.05	twin screw extruder	Mass transfer coefficient determined from experimental data. Varied as $N^{1/2}$. Difference between theory and experimental data attributed to uncertainty in surface area of films. Diffusion coefficient was known.

by Collins et al. (1983) is the only one in which the diffusion coefficient was known from independent measurements, and thus uncertainty about the value of the diffusion coefficient was eliminated as a possible cause for the discrepancy between theory and experiment.

On the basis of the research conducted thus far it is quite obvious that our understanding of the stripping of polymeric solutions is in a primitive stage and that much research remains to be done. One area which certainly needs attention is understanding how bubbles are formed in concentrated, viscoelastic polymeric solutions.

CONTINUOUS STRIPPING OF POLYMERIZING SOLUTIONS

A problem that is closely related to the stripping of concentrated polymeric solutions is the one which deals with the stripping of a viscous, polymerizing solution when a volatile by-product is formed during the course of the reaction. Typically, this situation arises in condensation polymerizations where it is important that the by-product be removed in order to obtain a polymer of high molecular weight.

To illustrate this point, consider the simple condensation polymerization reaction that is represented by Equation (2).

$$HO[RCOOR]_x COOH + HO[RCOOR]_y COOH \rightleftarrows HO[RCOOR]_{x+y} COOH + H_2O \quad (2)$$

To further fix ideas, let us assume that the polymerization reaction is being conducted in a stagnant film that is exposed to a vapor phase that has a by-product partial pressure P_v

Mass balances on the polymer and by-product result in the following set of equations

$$\frac{\partial C_p}{\partial t} = k\, C_p^2 - \frac{k}{K} C_{P_o} C_v \quad (3)$$

$$\frac{\partial C_v}{\partial t} = k\, C_p^2 - \frac{k}{K} C_{P_o} C_v + D \frac{\partial^2 C_v}{\partial y^2} \quad (4)$$

where the major assumptions which have been made are:

-- the polymer molecules are immobile
-- the rate constant is independent of chain length
-- the degree of polymerization $\bar{x}_n$ is much greater than unity

Equations (3) and (4) can be written in dimensionless form as

$$\left(\frac{D}{kC_{P_o}\delta^2}\right)\frac{\partial}{\partial\tilde{t}}\left(\frac{1}{\bar{x}_n}\right) = \left(\frac{1}{\bar{x}_n}\right)^2 - \frac{\tilde{C}_v}{K} \tag{5}$$

$$\frac{\partial\tilde{C}_v}{\partial\tilde{t}} = \left(\frac{kC_{P_o}\delta^2}{D}\right)\left[\left(\frac{1}{\bar{x}_n}\right)^2 - \frac{\tilde{C}_v}{K}\right] + \frac{\partial^2\tilde{C}_v}{\partial\tilde{y}^2} \tag{6}$$

where

$$\bar{x}_n = \frac{C_{P_o}}{C_P} \qquad \tilde{t} = \frac{Dt}{\delta^2}$$

$$\tilde{C}_v = \frac{C_v}{C_{P_o}} \qquad \tilde{y} = \frac{y}{\delta}$$

The appropriate initial and boundary conditions are:

$$\bar{x}_n = \bar{x}_{n1}\ ;\ \tilde{C}_v = \tilde{C}_{v1} \qquad \tilde{t} = 0 \tag{7}$$

$$\tilde{C}_v = \tilde{C}_{v_i} = \frac{P_v}{HC_{P_o}} \qquad \tilde{y} = 1 \tag{8}$$

$$\frac{\partial\tilde{C}_v}{\partial\tilde{y}} = 0 \qquad \tilde{y} = 0 \tag{9}$$

The important point to note about these equations is not so much the exact solution to them, but rather, the very important role that mass transfer plays in obtaining a polymer with a high degree of polymerization. In the particular example chosen here, the importance of mass transfer is reflected in the dimensionless group

$$Da = \left(\frac{D}{kC_{P_o}\delta^2}\right)$$

which is the ratio of the characteristic time for the polymerization reaction to the characteristic time for mass transfer. In the extreme case when the value of this group is considerably less than unity, the polymerization reaction will be in chemical equilibrium at all times. In these circumstances, Equation (5) reduces to

$$\bar{x}_n = \left(\frac{K}{\tilde{C}_v}\right)^{1/2} \tag{10}$$

and Equation (6) reduces to

$$\frac{\partial \tilde{C}_v}{\partial \tilde{t}} = \frac{\partial^2 \tilde{C}_v}{\partial \tilde{y}^2} \tag{11}$$

which means that the synthesis of a high molecular weight polymer depends directly on the ability to strip the by-product from the reacting mixture.

As judged by the published literature, there has been little research of a systematic and fundamental nature on stripping operations where a polymerization reaction is involved. And consequently, our ability to conduct rational engineering analyses and to develop sound design procedures has suffered. Pell and Davis (1973) published an important paper in which they demonstrated, both theoretically and experimentally, the significance of mass transfer in the polymerization of polyethylene terephthalate for the case of a stagnant film. Amon and Denson (1980) addressed the problem of continuous polymerization and stripping in screw extruders and showed quantitatively how the screw speed, residence time, surface area for mass transfer from the wiped film (Figure 2) and the vapor phase partial pressure (Figure 3) influence the molecular weight of the polymer which discharges from the extruder. From an analysis and design point of view, it is important to note that the degree of polymerization increases dramatically with increases in the surface area for mass transfer (Figure 3), and, that a point of diminishing returns is reached in reducing the partial pressure of the by-product in the vapor phase (Figure 3).

In commercial practice it is generally advantageous to introduce gas or vapor bubbles into a condensation polymerization reaction so as to obtain the desired degree of polymerization. And thus, it seems that we are faced with finding answers to the same kinds of problems that arose in the analysis and design of the stripping operations. One additional complication exists here, though, and that is the presence of the chemical reaction. It thus seems that research is needed not only on how bubbles are formed, but also on how mass transfer is affected by changing rheological

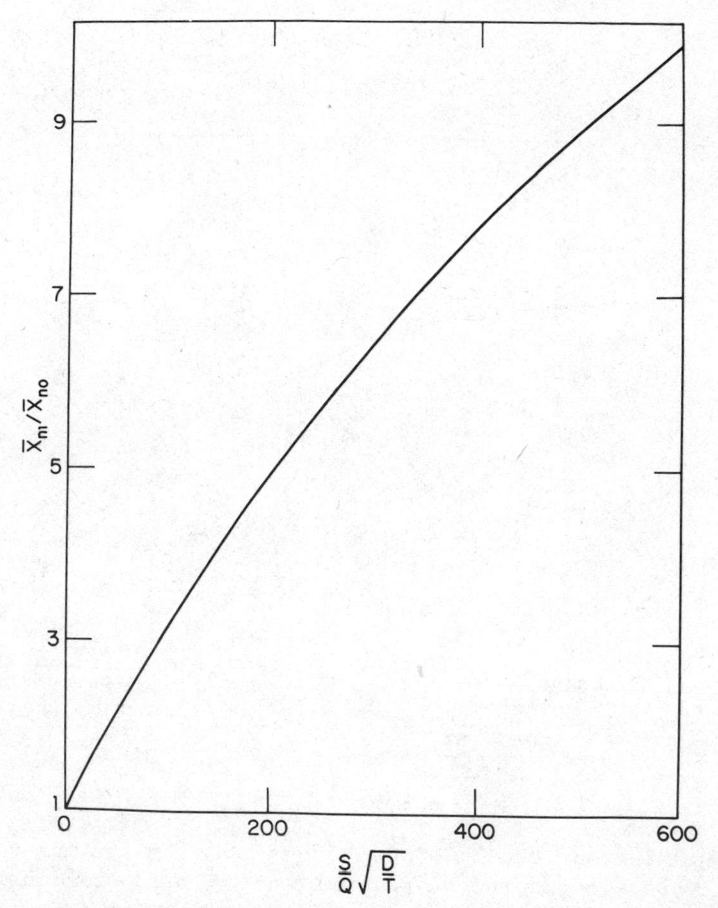

Figure 2. Effect of the interfacial surface area per unit length (S) on the exit degree of polymerization for a condensation polymerization. Q is the volumetric flow rate, D is the molecular diffusivity and T is the reciprocal of the screw speed.

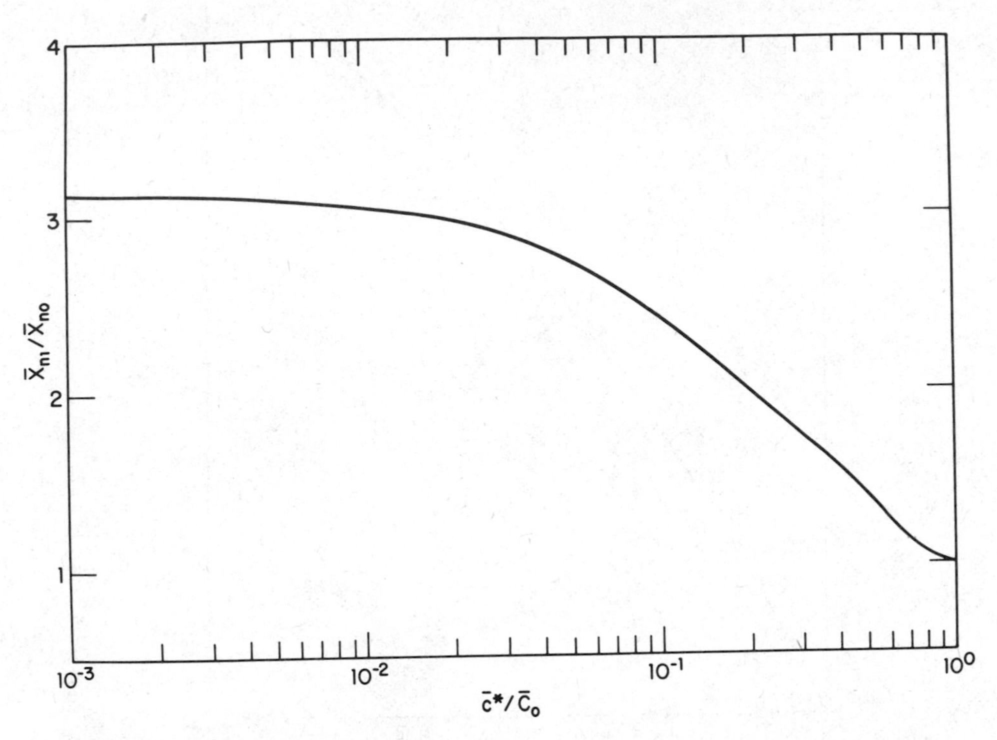

Figure 3. Effect of the by-product concentration in the vapor phase on degree of polymerization for a condensation polymerization. C* is the concentration in the liquid that is in equilibrium with the vapor phase composition. C_o is a scaling factor.

properties and how mass transfer is affected by diffusion coefficients which vary as the reaction proceeds.

GAS ABSORPTION IN POLYMER MELTS

In commercial practice gas absorption in polymer melts is usually encountered in the fabrication of foamed thermoplastic structures when a "physical blowing agent"--that is, a gas not generated by a chemical reaction--is used to create the foam. In processes such as the injection molding of thermoplastic structural foam or the extrusion foaming of coaxial cables (Randa, 1981), gas is injected under pressure directly through the wall of the barrel of the extruder (or injection molding machine) into a section of the screw that is only partly filled with the polymer melt. When the gas-polymer melt solution is subsequently injected into a mold cavity or extruded into an environment, say, at atmospheric pressure, gas bubbles form in the molten polymer and a foamed structure results. The void fraction of this foam depends directly on the concentration of dissolved gas in the polymer; and, since the performance properties of the foamed structure are critically dependent on the void fraction, control of the dissolved gas concentration is important to the successful operation of the process. From an engineering analysis and design point of view, the problem reduces to one of understanding the interrelationships between the geometry of the screw, the operating conditions and the physical properties of the gas-polymer melt solution.

The published literature contains no evidence as to how this problem should be treated. On an intuitive basis, though, it would seem that the Latinen analysis should be appropriate, in which case computing the dissolved gas concentration would be a simple matter. The picture is really not quite this clear, however, for the problem can also be viewed as one of a moving contact line at a gas-solid-liquid interface.

A closely related problem is the one which concerns behavior of a gas-liquid interface at the surface of a rotating cylinder that is partly immersed in a liquid bath of infinite extent. Wilkinson (1975), among others, has conducted research on this problem, and above a certain rotational speed gas entrainment occurs at the moving contact line. He found that his results could be correlated in accordance with the following equation

$$Ca = \frac{U_o \mu}{\sigma} = 1.2 \qquad (12)$$

The implications concerning gas absorption in polymer melts in the vicinity of a rotating surface and the use of the Latinen analysis are obvious.

CONCLUSIONS

In this paper we have drawn attention to the importance of polymer processing operations that do not involve shaping, and to the fact that our understanding in relation to analysis and design is in a primitive state. We have also suggested a few areas where further research is needed with the idea of stimulating interest in this area.

At the University of Delaware we have directed much of our own research so as to address problems related to polymer processing operations other than shaping. We have also brought the ideas and concepts related to the analysis and design of these operations into the classroom and have restructured the content of our polymer processing courses so to reflect the importance of the 'non-shaping' operations.

REFERENCES

Amon, M., and Denson, C. D., 1980, Simplified Analysis of the Performance of Wiped-Film Polycondensation Reactors, Ind. Eng. Chem. Fundam., 19:415.

Biesenberger, J. A., and Kessidis, G., 1982, Devolatilization of Polymer Melts in Single Screw Extruders, Poly. Eng. Sci., 22:832.

Collins, G. P., Denson, C. D., and Astarita, G., Determination of Mass Transfer Coefficients for Bubble-Free Devolatilization of Polymeric Solutions in Twin Screw Extruders, AIChE J., in press (1983).

Coughlin, R. W., and Canevari, G. P., 1969, Drying Polymers During Screw Extrusion, AIChE J., 15:560.

Denson, C. D., 1983, Stripping Operations in Polymer Processing, in: "Advances in Chemical Engineering, Vol. 12," J. Wei, ed., Academic Press, New York.

Latinen, G. A., 1962, Devolatilization of Viscous Polymer Systems, ACS Adv. Chem. Series, 34:235.

Newman, R. E., and Simon, R. H. M., A Mathematical Model of Devolatilization Promoted by Bubble Formation, 73rd Annual A.I.Ch.E. Meeting, Chicago, IL (November 1980).

Pell, T. M., Jr., and Davis, T. Q., 1973, Diffusion and Reaction in Polymer Melts, J. Poly. Sci., Poly. Phy. Ed., 11:1671.

Randa, S. K., Extrusion Foaming of Coaxial Cables of Melt-Fabricable Fluorocarbon Resins, 35th International Wire and Cable Symposium, Cherry Hill, NJ (November 1981).

Tadmor, Z., and Klein, I., 1970, "Engineering Principles of Plasticating Extrusion," Van Nostrand Reinhold, New York.

Todd, D. B., 1974, Polymer Devolatilization, Soc. Plast. Eng., Technical Papers.

Werner, H., 1980, Devolatilization of Polymers in Multi-Screw Devolatilizers, Devolatilization of Plastics, Verrin Deutscher Ingenieure Dusseldorf, UDI-Gessellschaft Kunststofftechnik.

Wilkinson, W. L., 1975, Entrainment of Air by a Solid Surface Entering a Liquid/Air Interface, Chem. Eng. Sci., 30:1227.

MECHANOCHEMICAL DEGRADATION OF GLASS FIBRE REINFORCED POLYPROPYLENE MELTS DURING FLOW

G. Akay and D. W. Saunders
School of Industrial Science
Cranfield Institute of Technology, Cranfield, England

T. Tincer and F. Cimen
Middle East Technical University, Ankara, Turkey

INTRODUCTION

In this work, a study of the mechanochemical degradation of short glass fibre reinforced polypropylene is undertaken. The motivation for this study is two fold. Firstly, it is necessary to assess the extent and the origin of degradation during processing as it results in the deterioration of the physical properties of the finished product and if these materials are to be recycled after processing. Secondly, reinforced polymers contain two additional ingredients, namely, glass fibres and silane coupling agents which are used to coat the fibres. The presence of glass fibres is likely to modify the stress field experienced by the macromolecules and therefore mechanically induced chain scission will be effected. It is well known that the distribution of stress on a macromolecule is not uniform and stress concentration either directly causes chain scission, or thermally activated chain scission may be accelerated by stress. Both thermal and mechanical degradation occur during the processing of polymers. At low processing temperatures, mechanical degradation may be significant, but there are difficulties in distinguishing thermal and stress-induced degradation. However, there are some polymers where the dominant degradation process is either stress-induced (i.e., polyisobutene) or thermally activated as in polystyrene. Random chain scission occurs during thermal degradation while selective chain scission usually takes place during stress induced degradation, in which macromolecular chains rupture mainly near the middle of the chain where the stress is maximum. Stress-induced degradation proceeds until a critical limiting molecular weight is reached. A review of mechanochemical degradation of polymers in general and that of polypropylene in particular may be found in books by Casale and Porter[1].

The function of the silane coupling agents in polymer processing and rheology is well known. However, depending upon the type of polymer and coupling agent, a number of thermally or mechanically activated chemical reactions may take place, affecting the rate of degradation and causing cross-linking or chain branching. Some of these possibilities have been studied recently by Lunt[2].

A number of techniques exist to study the stress-induced degradation of macromolecules. One such method which is more relevant to polymer processing, is repeated extrusion through a capillary. In such a flow, there is a distribution of shear stress and hence, there exists the possibility of stress-induced diffusion which will result in the migration of higher molecular weight fractions and also that of the fibres into the low-shear-rate regions. The migration of macromolecules will reduce the rate of macromolecular rupture as discussed recently by Akay[3]. The effect of modified stress field on stress-induced degradation due to the presence of fibres is not clear.

In this study, we investigate the mechanochemical degradation of four different polypropylene (PP) melts using a capillary viscometer. Two of the melts are unfilled and the others are filled with short glass fibers. The variation of viscosity, molecular weight and its distribution and the development of the carbonyl groups are determined as a function of number of passes through the capillary. This initial study indicates that the degradation of glass fibre filled PP is slower compared with that of the unfilled material and infact at the early stages of the extrusion, molecular weight increases.

EXPERIMENTAL

Materials

All the polymers are supplied by ICI. These are:
HF22: Unfilled PP powder which does not contain any stabilizer.
GXM43: Same as HF22 but contains stabilizer. Supplied as granules.
HW60 GR/20: A commercially available short glass fibre reinforced PP supplied in a granular form and contains 20 percent by weight of well dispesed glass fibres having diameter of 10μm and length of ½mm.
HW60 GR/30: Same as above but contains 30 percent glass fibres with modal length of 350μm.

Silane coupling agents are supplied by Pilkington Brothers Limited and they are:
PTMO: Propyltrimethoxysilane, $CH_3CH_2CH_2Si(OCH_3)_3$.
A-1100: γ-Aminopropyltriethoxysilane, $NH_2CH_2CH_2CH_2Si(OC_2H_5)_3$, produced by Union Carbide.

Repetitive Extrusion

A Davenport plunger-type viscometer is used for repetitive extrusion experiments. A two-stage capillary assembly is fitted to the reservoir of the viscometer, as described previously by Akay.[4] The first capillary is 70mm long while the second capillary is 10mm long and both capillaries are 1mm in diameter. Pressures in the reservoirs are recorded. The pressure in the second reservoir is denoted by P and P^o denotes the reservoir pressure when only a 10mm long capillary is present. The extrudate from this experiment (single capillary) is not recycled. The ratio P/P^o yields the reduction in viscosity due to degradation. The temperature of the reservoirs and capillaries is kept at 200°C. The flow rate is kept constant so that the apparent shear rate in the capillaries is 1500 sec^{-1}. The number of extrusions is denoted by N.

Gel Permeation Chromatography (GPC)

After each extrusion, some of the extrudate is removed for determining number average and weight average molecular weights M_n and M_w using Gel Permeation Chromatography GPC. Initial molecular weights are denoted by M_n^o and M_w^o.

IR-Spectroscopy

A Perkin-Elmer 177 grating IR-spectrophotometer is used to determine changes in the absorbance of the carbonyl groups at 1725 cm^{-1}. Films of 65 µm are compression moulded at 180°C using the extrudate and undegraded polymers which are used as reference and therefore difference spectra is obtained. This technique is especially necessary for the filled polymer.

Brabender Mixing

Silane coupling agents are allowed to polimerize thus forming siloxane homopolymer and they are blended with the unfilled PP (GXM43) using a Brabender Roller Mixer, W50H. The total weight of the mixture is 30 grams and the weight percent of the siloxane polymer is three. The initial temperature of the mixer is 180°C. Mixing is continued for 25 minutes at 60 rpm while circulating oil at 180°C through the mixer. During mixing, a steady temperature of 188°C is obtained after the first five minutes. As the polymer melts, torque, T, increases rapidly and reaches a maximum (denoted by T_m) then starts decreasing.

RESULTS AND DISCUSSION

Viscosity Reduction

In Fig.1, we illustrate the reduction of melt viscosity with the number of extrusions. As expected, degradation of the unstabilised polymer, HF22 is the fastest and after seven extrusions, viscosity is reduced by a factor of two. The viscosity reduction of unfilled but stabilised PP (GXM43) is slower but that of the 30

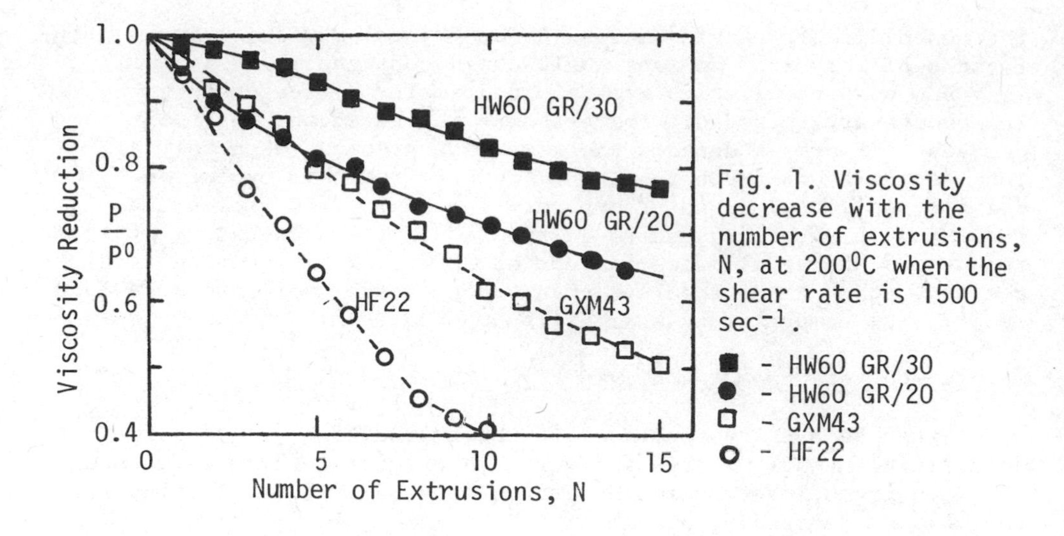

Fig. 1. Viscosity decrease with the number of extrusions, N, at 200°C when the shear rate is 1500 sec^{-1}.

percent glass fibre reinforced PP (HW60 GR/30) is slowest. During repeated extrusion, glass fibres also undergo mechanical breakage and this should result in further reduction in the viscosity with N. Since the fibres in HW60 GR/20 are long, they are more prone to breakage and as a result, initial decrease in the viscosity is faster than that in GXM43.

Both of the unfilled polymers show melt instability initially and the resulting extrudate is wavy and has die swell. However, with the increasing number of extrusions, extrudate distortion

Table I. Variations of molecular weights an heterogeneity index with the number of extrusions, N.

N	HF22			GXM43			HW60 GR/20			HW60 GR/30		
	x10^4 M_n	x10^5 M_w	$\frac{M_w}{M_n}$	x10^4 M_n	x10^5 M_w	$\frac{M_w}{M_n}$	x10^4 M_n	x10^5 M_w	$\frac{M_w}{M_n}$	x10^4 M_n	x10^5 M_w	$\frac{M_w}{M_n}$
0	4.47	5.43	12.2	4.43	3.93	15.6	5.10	3.64	7.14	5.37	4.35	8.10
3	3.93	3.15	8.02	4.27	3.40	10.5	6.12	3.07	5.02	5.46	4.71	8.63
7	3.48	2.40	6.90	3.62	2.83	8.74	5.03	2.45	4.87	4.77	3.66	7.67
10	2.25	1.57	6.98	3.37	2.36	8.52	5.21	2.58	4.95	6.08	3.66	6.02
11	—	—	—	3.23	2.07	5.91	—	—	—	—	—	—
13	—	—	—	—	—	—	4.17	2.02	4.84	3.85	3.11	8.08
15	—	—	—	2.12	1.26	5.03	—	—	—	3.55	2.73	7.67

disappears and die swell becomes small, indicating large reduction in melt elasticity. In fibre reinforced polymers the extrudate is smooth with very small die swell at all times and the reduction in die swell is negligible.

Changes in Molecular Weights

In Table I, changes in the molecular weights during the repetitive extrusion of various polymers are shown. Amongst the undegraded polymers, HF22 has a considerably higher M_w than GXM43. This difference is probably due to the degradation of GXM43 during compounding. The base polymer in fibre filled compounds also has greater molecular weights than GXM43. As may be expected, molecular weights M_n and M_w and heterogeneity index M_w/M_n decrease with increasing N in HF22 and GXM43. These trends are in confirmation of the previous results of Schott and Kaghan[5]. Molecular weight distributions indicate that there is no increase in the high molecular weight fraction as a result of the degradation of the unfilled polymers. In the case of filled polymers, both M_n and M_w appear to increase initially. High molecular weight end of the distribution is also increased. This is more marked in HW60 GR/30 for which the heterogeneity index decreases initially and then increases after the tenth extrusion. Fluctuations in the molecular weight distribution and in M_n have also been reported by Baranwal and Jacobs[6] during the millong of styrene rubber which is attributed to the recombination reaction.

The term (M_n^0/M_n-1) has been defined as equal to the average number of polymer bonds scissioned per original polymer molecule[7] and has been widely used in degradation studies[7,8]. The variation of the number of scissions per molecule with N is shown in Fig.2. Unfilled polymers appear to yield exponential increase in chain scission while the filled grades yield chain addition initially, although there is considerable scatter in the data.

Development of the Carbonyl Groups

Determination of the carbonyl groups gives a quantitative extent of the oxidative degradation in the samples. The amount and type of stabilizer in the polymers used in the present experiments are different and therefore the extent of oxidation will not be the same. Nevertheless, development of the carbonyl groups with N follows the same trend as in the viscosity reduction. In Fig.3, we plot P/P^0 against the carbonyl absorption. A similar correlation has been used previously relating carbonyl absorption to changes in mechanical properties[9]. As seen in this figure, carbonyl absorption appears to follow the reduction in viscosity. Therefore, there are two important structural changes in glass fibre filled polymers as a result of repeated extrusion; initial molecular weight increase and relatively slow oxidation and slower molecular weight decrease when compared with the base polymer.

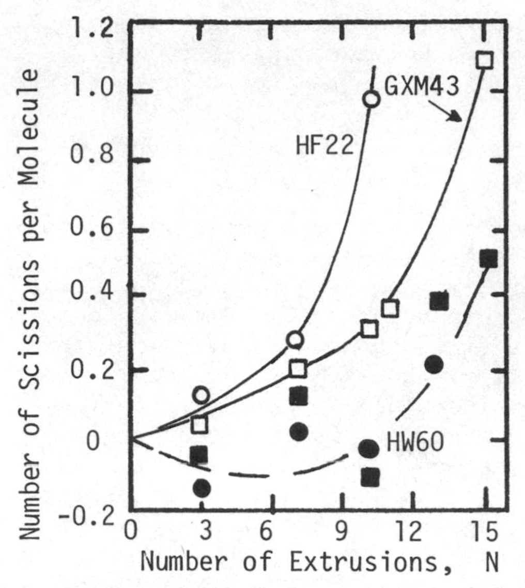

Fig.2. The number of scissions per molecule as a function of the number of extrusions. Identification of symbols are as in Fig.1.

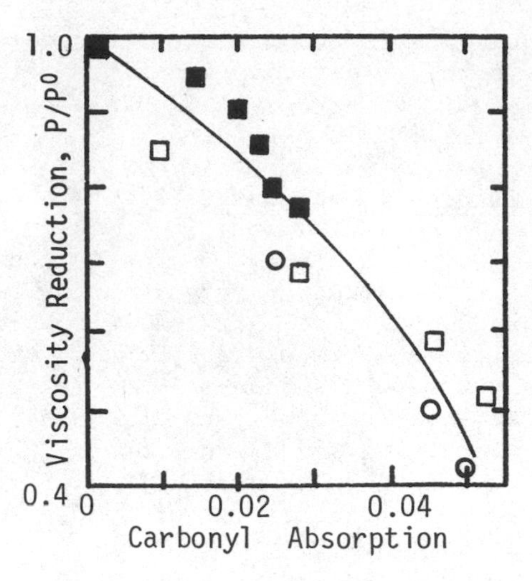

Fig.3. Variation of viscosity reduction with carbonyl absorption at 1725 cm^{-1}. Identification of symbols are as in Fig.1.

Degradation by Mixing

One of the most important functions of the silane coupling agents is to reduce the melt viscosity. There are even cases when the use of the coupling agent reduces melt viscosity of filled polymer to below that of the base polymer[10]. The presence of small amounts of oligomeric polyethylsiloxane in PP appears to reduce the melt viscosity by a factor of ten or more at low melt temperatures[11]. However, the required concentration of the siloxane homopolymer is low, some 0.1 percent and the mixing should be done by co-dissolving both phases. There are also other reports where viscosity decreases with the presence of an incompatible phase.

In Fig.4, the mixing curves for the unfilled PP are reproduced. The initial maximum torque, T_m is the same for all the samples. In this figure, we plot dimensionless torque, T/T_m against the mixing time. Both mixtures containing siloxane homopolymer yield lower torque as mixing is continued, suggesting lower melt viscosity. When the mixing is completed, samples are compression moulded for IR-spectroscopy. Very small amounts of carbonyl absorption is detected in all cases. The GPC analysis of these samples (as shown in Fig.4) indicates that reduction in M_w in siloxane homopolymer containing samples is in fact smaller than that in pure GXM43.

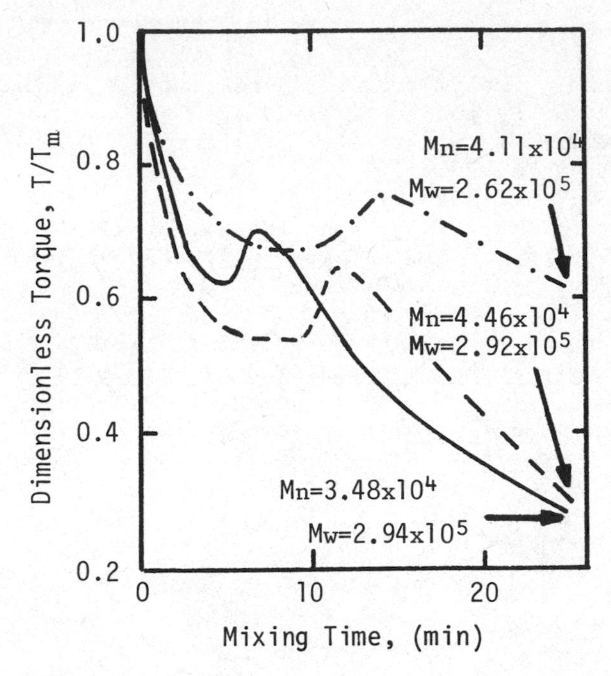

Fig. 4. The effect of siloxane homopolymer on the mixing characteristics of unfilled polypropylene.

-·-·- GXM43,

----- GXM43+3% siloxane homopolymer from PTMO,

——— GXM43+3% siloxane homopolymer from A-1100.

Therefore, reduction in the torque can not be atrributed to decrease in the molecular weight. It is probable that reduction in the torque is due to structure formation in the melt in the presence of siloxane homopolymer.

CONCLUSIONS

Reduction in the melt viscosity and improvements in the mechanical properties of fibre reinforced polymers as a result of using coupling agents are attributed to the creation of a new phase on the fibre surface which provides compatibility with the base polymer. As pointed out by Piggott[12] claims in favour of significant amounts of covalent bonding on the glass surface should not necessarily be taken at face value. Present experiments indicate that free siloxane homopolymer or the silane coupling agents can modify the base polymer imparting a reduction in melt viscosity and improving mechanical properties by increasing molecular weight.

ACKNOWLEDGEMENTS

We would like to thank the National Physical Laboratory (U.K.) for supporting this research and ICI and Pilkington Brothers Limited for the donation of materials used in the experiments.

REFERENCES

1. A. Casale and R.S. Porter, "Polymer Stress Reactions", Academic Press, New York, Vol.1 (1978) and Vol.2 (1979).
2. J.Lunt, Doctoral Dissertation, University of Liverpool, (1980).
3. G. Akay, Polym. Eng. Sci., 22:798 (1982).
4. G. Akay, Polym. Eng. Sci., 22:1027 (1982).
5. H. Schott and W.S. Kaghan, Soc.Plast. Eng. Trans., 3:145 (1963).
6. K. Baranwal and H.L. Jacobs, J. Appl. Polym. Sci., 13:797 (1969).
7. J.H. Adams, J. Polym. Sci., A-1, 8:1077 (1970).
8. K.B. Abbas, Polymer, 22:836 (1981).
9. G. Akay, T. Tincer and H.E. Ergoz, Europ. Polym. J., 16:601 (1980).
10. S.H. Morrell, Plastics and Rubber Processing and Applications, 1:179 (1981).
11. G.P. Andrianova, J. Polym. Sci., Polym. Phys. Ed., 13:95 (1975).
12. M.R. Piggott, Polymer Composites, 3:179 (1982).

APPLICATIONS OF A RIM PROCESS MODEL

S. R. Estevez and J. M. Castro

PLAPIQUI
UNS - CONICET
Bahía Blanca, Argentina

INTRODUCTION

Reaction injection molding (RIM) is a relatively new processing technique that is rapidly taking its place alongside the more established plastic processes. During a RIM molding cycle, two highly reactive monomers or prepolymers (in the case of urethanes a diisocyanate and a polyol) are brought into intimate molecular contact by impingement mixing. From the mixing head they flow into the mold and react rapidly to form a solid part. RIM thus combines the polymer synthesis and final shaping into a single operation. Its appeal is clear, most of the polymerization occurs after the reactive mixture is in the desired shape. The low viscosity reactants can be fairly easily mixed and require low pressures to fill the mold. The only way pressures can become large is if the material gels before filling is completed in which case, a short shot occurs.

For convenience of analysis the RIM process can be divided into mixing, filling and curing stages as outlined by Broyer and Macosko[1]. The filling and curing stages form most of the process and, if mixing is complete, are the stages where changes can be made to improve the final product. As RIM molds become larger, and as more reactive systems are used, understanding the filling step takes on greater importance. Premature gelling has been reported to give problems in complex parts[2]. Knit lines are also a potential problem. If the extent of reaction of the fluid streams meeting each other, is higher than a critical value, the material will not interpenetrate and the knit line will be visible and weak. Criteria are needed to predict incomplete filling and knit line problems[3].

In this paper, after a brief preliminary discussion of some pertinent results from the model of the RIM process developed by Castro and Macosko[4], we present a way to correlate guidelines to avoid premature gelling. They are correlated using the relevant dimensionless groups of the process. The numerical calculations are done using the model just mentioned which has been shown to agree with experimental results.

BASIC EQUATIONS

During mold filling a well mixed low viscosity material is fed into the mold. As the material progresses, the viscosity increases due to chemical reaction. Neglecting entrance effects the material particles during the filling of a narrow gap mold can be considered to have straight path lines, except on the vicinity of the flow front where, the flow is of a spreading or smearing nature (fountain flow). This region constitutes a small portion of the flow field, except at the beginning of the filling process. The pressure drop is therefore negligible compared to the pressure drop in the main flow. Thus, when analyzing the filling step for nonreacting materials with a uniform temperature field, the flow front can be neglected.

For reactive materials or when nonisothermal effects are important, the effect of the flow front must be considered, because it governs final location of each material element. The relatively fast moving central region has a shorter than average residence time. It is this material which arrives at the front which fountains out to become wall material. The transverse fountain flow carries with it the temperature and extent of reaction of that fluid element. For mathematical modelling of the filling step, the flow field is then divided into two subdomains. Main flow, where the material particles have straight path lines and the flow front.

When the mold is full, flow stops. There is no packing as is done with thermoplastic injection molding. In the commercial process, the entrance nozzle is plugged off, usually with a ram. There is no attempt to maintain a specific pressure, but rather the ram is used to clean out the mixing chamber. Shrinkage is accounted in the case of urethanes by foaming. The material is left in the mold reacting until it is dimensionally stable so that it can be removed without loosing its shape. This stage of the RIM process is called curing. The field equations for the filling and curing steps and their numerical solution are discussed in detail by Castro and Macosko[4]. Here, we need to consider only the filling stage and will discuss only what is relevant for our goals.

Filling

For mathematical modelling the general approach consists of treating the reactive mixture as a material continuum, where the only effect of the chemical reaction is to change the viscosity and to generate heat. For our purposes, we need only to discuss the balance equations for the main flow. They are written for a rectangular mold with an end gate.

The assumptions of the model are

1) Constant density (ρ) and thermal properties ($\hat{k}$, C_p)
2) Constant flow rate (Q)
3) Negligible molecular diffusion
4) Neglect side wall (W/H >> 1)
5) Second order kinetics with Arrhenius temperature dependence

$$R_a = k_o \, e^{-E/RT} \, C_a^2 \qquad (1)$$

6) Neglect entrance length
7) Laminar flow
8) Newtonian fluid, that is, the viscosity is only an explicit function of the extent of reaction and temperature

$$\eta = \eta \, (c^*, T) \qquad (2)$$

Castro and Macosko[5] and Perry et al[6], have found that the viscosity rise for RIM materials can be represented with the following equation

$$\eta = A_n \, e^{E_\eta/RT} \left(\frac{c_g^*}{c_g^* - c^*}\right)^{A + Bc^*} \qquad (3)$$

9) The flow is considered to be one dimensional thus, the only non zero velocity is in the x direction ($v_y = 0$; $v_z = 0$)
10) Negligible heat conduction in x and y direction (W/H >> 1 and L/H >> 1)

With these assumptions, the balanced equations in terms of dimensionless variables are shown below. The balance of linear momentum

$$\left(\frac{\partial p}{\partial x}\right)^* = \frac{\partial}{\partial z^*} \left(\eta^* \frac{\partial v_x^*}{\partial z^*}\right) \qquad (4)$$

The component balance

$$\frac{\partial c^*}{\partial t^*} + v_x^* \frac{\partial c^*}{\partial x^*} = G\, k^* \,(1 - c^*)^2 \qquad (5)$$

The dimensionless parameter G, the gelling number, is the most important parameter for the filling step. It is defined as

$$G = \frac{\text{filling time } (t_f)}{\text{isothermal gel time at } T_o (tg_o)} \qquad (6)$$

where the filling time is just the mold volume divided by the volumetric flow rate

$$t_f = \frac{HWL}{Q}$$

and tg_o is the time to reach the gel point conversion, c_g^*, at the initial temperature (T_o).

$$tg_o = (k_{To}\, C_o)^{-1} \frac{c_g^*}{1-c_g^*}$$

The balance of energy reduces to

$$\frac{\partial T^*}{\partial t^*} + v_x^* \frac{\partial T^*}{\partial x^*} = G_z^{-1} \frac{\partial^2 T}{\partial z^{*2}} + G\, k^* \,(1 - c^*)^2 \qquad (7)$$

where the Graetz number is defined as

$$G_z = \frac{\text{heat transport by convection}}{\text{heat transport by conduction}} = \frac{\frac{\bar{v}_x}{L}\, \rho C_p}{\frac{\hat{k}}{(H/2)^2}} \qquad (8)$$

Note that the balance equations are coupled since, the viscosity is a function of the extent of reaction and temperature. To obtain the temperature field, the conversion field and the pressure rise during filling, the above set of equations must be solved together with the model for the flow front. The details can be found in reference four.

PREMATURE GELLING

As discussed before, during filling, the viscosity of the flowing mixture increases due to chemical reaction. The filling stage must be over, before this rise becomes too large so as to avoid premature gelling. If the filling stage were isothermal, the upper bound for the filling time would be the isothermal gel time of the material. Since during filling temperature is a function of position and furthermore, the paths followed by the different material particles are not straight due to the flow front, stablishing an upper bound for the filling time is not that simple. We will consider that premature gelling occurs when ever any transversal section of the mold has conversions above the gelling point in 50 % of it. This criteria was choosen because it seems to coincide with the experimental observations of Castro[7].

It is not practical, to try to obtain guidelines valid for any chemical system in any molding situation because, there are to many parameters to consider. It is more usefull to produce, a set of graphs for each chemical system. The ones developed here are valid for the experimental RIM 12-56 system described by Castro and Macosko[5]. It is not difficult to generate graphs for any other chemical system once we know its kinetics and rheology. This aspect is discussed in the last section. Fixed the chemical system, we have only four relevant dimensionless groups (G, G_z, T_0^* and T_w^*).

The most usefull way to correlate the guidelines for avoiding premature gelling, we think, is as shown in Figure 1, where we show a plot of the Gelling number for premature gelling, which we call the critical gelling number (G_c), as a function of the inverse of the Graetz number, for several values of the wall dimensionless temperature for a fixed initial temperature. Thus, a point on any line coincides with a condition where premature gelling occurs. For example, suppose we want to fill a mold with H = 0.345 cm., L = 20 cm. and W = 10,1 cm. with the experimental RIM type chemical system using a Q = 13.88 cm^3/s flow rate and an

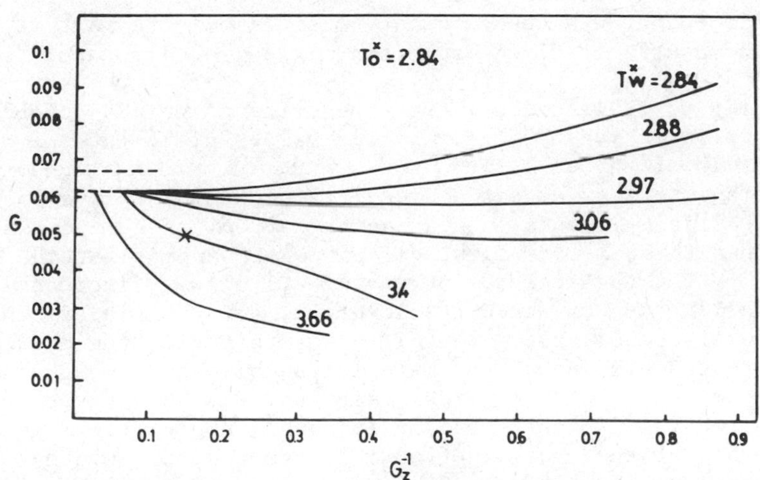

Fig. 1. Gelling number for premature gellig vrs. inverse of Graetz number for several values of the wall dimensionless temperature.

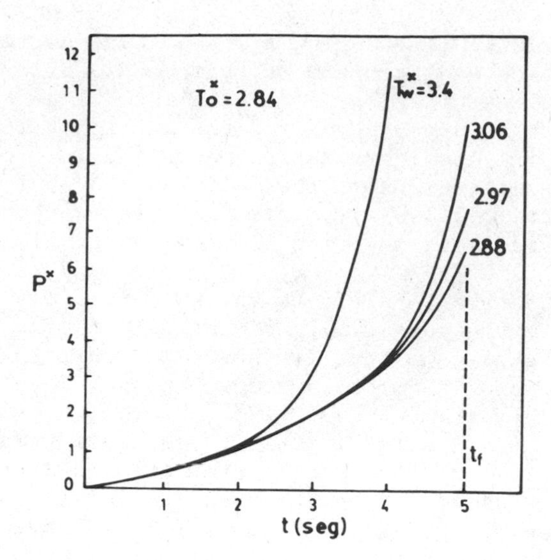

Fig. 2 . Pressure rise during filling for different wall temperatures.

initial material temperature of 55°C ($T_0^* = 2.84$). If we calculate G (0.05) and G_z^{-1} (0.15) for this condition we find that they correspond to the point marked with an x in figure 1. This means that as long as the wall temperature is below 120°C ($T_W^* = 3.4$) no premature gelling occurs. The pressure rise for this filling condition for wall temperatures of 60° C (2.88), 70° C (2.97), 80° C (3.06), and 120° C (3.40) are shown in figure 2. Notice the fact as said before, that pressures are low unless premature gelling occurs. This format was choocen since most times, the initial material temperatures is fixed from good mixing requirements.

From figure 1, we can see that the critical gelling number decreases as we increase G_z^{-1} for larger values of T_W^* but increases for lower values of T_W^*. This depends on whether the mold wall on the average acts as a heat source or a heat sink, since increasing G_z^{-1} (decreasing G_z) means increasing the effect of heat conduction. To understand this, let us first discuss what happens for the cases when $T_W^* \leqslant T_0^*$. For this cases, increasing G_z^{-1} allways increases the critical gelling number since the mold wall acts as a heat sink for all the time. For cases where $T_W^* > T_0^*$, initially the mold wall acts as a heat source until the material becomes hot enough due to the heat liberated by the chemical reaction. For this cases, weather increasing G_z^{-1} decreases or increases the critical gelling number depends of the overall effect: in other words, weather most of the time the mold wall is a heat source or a heat sink.

Finally, based on the experimental results of Castro[7], we think that this graphs could also be used to predict weather or not knit line problems may occur. To do this, one should calculate the local gelling number (G_L) using the time needed by the material in the average to reach the knit line instead of the filling time to compute G. That is

$$G_L = \frac{\text{time to fill the mold up to } L_o (t_{fo})}{\text{isothermal gel time at } T_o} \tag{9}$$

where

$$t_{fo} = \frac{HWL_o}{Q} \tag{10}$$

CONCLUSIONS AND SIGNIFICANCE

We have shown how using the dimensionless groups relevant to the RIM process, we can correlate guidelines for premature gelling. We do not pretend here to give a complete set of dimensionless graphs but rather to show their usefullness. We give here graphs wich are applicable to the experimental RIM type chemical system discussed by Castro and Macosko[5], to generate graphs for any other chemical system is an easy matter. What we need to know are the kinetics and rheology of the chemical system. Methods for obtaining them are discussed in references (5, 6, 8 and 9).

A comment should be made as to the applicability of guidelines developed for a rectangular mold to real parts. The filling stage is of course, dependent of the mold geometry but, we think that the rectangular geometry, of the simple geometries, is the closest to real RIM parts. To apply the premature gelling graphs, when calculating the gelling number what we need to do is to use the mold volume and in calculating the Graetz number, besides the mold volume we need to identifie the mold thickness. Strictly speaking, the guidelines developed here are applicable only for the specific chemical system used here, but we believe that as long as the shape of the viscosity rise is similar, one could use the same graphs for a different polyurethane chemical system.

NOMENCLATURE

c^* = extend of reaction (conversion): $(C_o-C)/C_o$
c^*_g = solidification (gel) point
H = mold thickness
$\hat{k}$ = thermal conductivity
k_{T_o} = kinetic rate constant evaluated at T_o
k^* = dimensionless reaction rate: $(k/k_{T_o})(c^*_g/(1-c^*_g))$
L = mold length
p = pressure
T_o = initial material temperature
T_w = mold wall temperature
T^* = dimensionless temperature = $(T-T_o)/\Delta T_{ad}$

t^* = dimensionless time: t/t_f
v_x^* = dimensionless velocity in the x-direction: $v_x/\bar{v}_x$
$\bar{v}_x$ = average velocity in the x-direction
W = mold width
x^* = dimensionless axial variable = x/L
z^* = dimensionless transverse direction: z/(H/2)
$(\frac{\partial p}{\partial x})^*$ = dimensionless pressure gradient: $(\partial p/\partial x)/(4\ Q\eta_{T_o}/W\ H^3)$
η^* = dimensionless viscosity = η/η_{T_o}

DIMENSIONLESS NUMBERS

G_z = Graetz number = $\dfrac{(\bar{v}_x/L)\ \rho\ C_p}{(\hat{k}/(H/2))^2}$

G = Gelling number = t_f/t_{g_o}

E^* = dimensionless activation energy = $\dfrac{E}{R\Delta T_{ad}}$

E_η^* = dimensionless viscosity activation energy = $\dfrac{E\eta}{R\Delta T_{ad}}$

T_o^* = initial dimensionless temperature = $T_o/\Delta T_{ad}$

T_w^* = wall dimensionless temperature = $T_w/\Delta T_{ad}$

REFERENCES

1. E. Broyer and C. W. Macosko, AIChE J., 22, 268 (1976)
2. J. M. Castro, C. W. Macosko, F. E. Critchfield, E. C. Steinle, and L. P. Tackett, Soc. Plast. Eng. Tech. Papers, 26, 423 (1980).
3. C. W. Macosko, Proceedings of the First International Congress on Reactive Polymer Processing, Pittsburgh, October 1980.
4. J. M. Castro and C. W. Macosko, AIChE J., 28: 2, 250 (1982)
5. J. M. Castro and C. W. Macosko, Soc. Plast. Eng. Tech. Papers 26, 434, (1980).
6. S. J. Perry, J. M. Castro and C. W. Makosko, 52° Anual Congress of the Society of Rheology, February 23-25, 1981. Williamsburg, Virginia.
7. J. M. Castro, Ph.D. Thesis, Dept. of Chem. Eng. and Mat. Sci., University of Minnesota (1980),
8. S. D. Lipshitz and C. W. Macosko, J. Appl. Polymer. Sci., 21, 2029.
9. E. B. Richter, M. S. Thesis, Dept. of Chem. Eng. and Mat. Sci., University of Minnesota (1977)

DESIGN PRINCIPLES FOR BARRIER-SCREW EXTRUDERS (Abstract)

J. T. Lindt and B. Elbirli

Dept. of Metall. and Materials Engineering

University of Pittsburgh, Pittsburgh, PA, USA

As early as in 1959 the first barrier screws were introduced to make the melting process in single screw extruders more efficient. The operating principles of most barrier screws are based on a separation of the molten and unmolten polymer by an auxiliary flight gradually transversing the main screw channel.

Only recently the melting performance of barrier screws has been analyzed systematically; first, a kinematic model for isothermal Newtonian melts has been proposed, and later a more complete dynamic analysis has become available. The latter model allows the melting length, the axial pressure profile,and the solid bed acceleration to be predicted accurately.

This paper presents results of an extensive study on the effects of various barrier screw concepts on the melting of crystalline polymers.A quantitative optimization procedure is carried out using a HDPE extrusion employing a 90 mm Barr as an example. This screw is also analyzed for a case where the melting is completed well before the end of the transition zone.

Not only does this paper give the necessary insight into the operation of the widely used barrier screws, but it also provides practical procedures applicable to design,trouble-shooting and optimization problems.

CONTINUOUS PROCESSING OF POLYMERIZING FLUIDS IV: CONTINUOUS MECHANISM IN A CHANNEL OF A SCREW-EXTRUDER

J.T. Lindt and B. Elbirli

Metallurgical and Materials Engineering
University of Pittsburgh
Pittsburgh, PA 15261

ABSTRACT AND CONCLUSIONS

Reactive extrusion of polymers (REX) is an integrated engineering operation that may combine, completely or partially, the traditionally separated steps of the polymer macromolecule formation, compounding, mixing, devolatilization, shaping and the product structuring into a single engineering operation carried out in a screw extruder as developed over the years for conventional polymer processing. The advantages associated with REX are in substantial reduction of the manufacturing time, energy savings as typically the remelting and often the solvent recovery steps are eliminated, in lower capital investment and in the potential direct feed-back between the polymer mechanical properties and the process conditions.

The present theoretical simulations have suggested that significant interactions exist between the extruder conveying mechanism and polymerization reaction in REX. Our results further suggest that careful temperature control may be required for a successful REX operation involving possibly screw cooling. An analysis of the present type is capable of determining an optimum strategy to maintain a reasonably uniform distribution of the average molecular weight across the channel to ensure both efficient transport and a uniform product. From the extrusion technology standpoint, there is an additional, very important benefit: if the proper temperature control is maintained, the uniformity in radial distribution of molecular weights causes REX to become similiar to conventional melt extrusion in terms of its conveying mechanism.

INTRODUCTION

Reactive extrusion (REX) - the continuous counterpart of Reaction Injection Molding - can be defined as a simultaneous flow, heat transfer and (bulk) polymerization operation that occurs in a screw extruder. Some applications of this technique are given in ref. (1), both single- and twin-screw extruders are used.

Despite the rapid growth of this field, engineering principles of REX have not yet been fully established. The information generated so far (2,3) suggests that significant differences exist between the conventional and "reactive" use of extruders the working of which relies on the drag conveying mechanism. In contrast to the conventional melt pumping, the screw characteristic for REX can be strongly influenced by the distribution of the average molecular weight within the screw channel.

It is the purpose of this paper to give a quantitative discussion of the differences between REX and the conventional operation along with recommendations concerning stability of the REX operation.

CONVEYING MECHANISM IN REX

Drag forces are responsible for transport in single-screw extruders, non-intermeshing twin-screw and intermeshing corotating twin-screw extruders. In view of the phenomenological similarity between these configurations as demonstrated by Tadmor (4) for the case of conventional melt extrusion, the below discussion applies to extruders where no positive transport exists, most notably to the single-screw and the intermeshing corotating twin-screw exturders commonly used in REX.

Typically, the polymerizing fluid during its passage through the helical channel is subjected to drag forces of varying intensity reflecting the progressively increasing viscosity associated with the polymerization process. It should be born in mind that some stagnation in the neighborhood of the screw surface can be expected in view of the long residence times implied there. Such phenomena can be responsible for significant <u>local</u> changes in the conveying efficiency of the screw. Thus, even for a screw of constant depth a highly non-linear pressure profile can be expected (3) in REX while in a partly filled channel the fillage is bound to vary significantly along the screw axis (2).

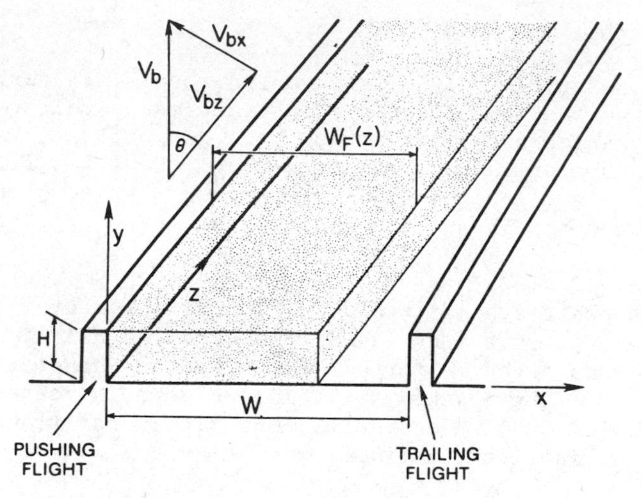

Fig. 1. Drag Flow in Reactive Extrusion.

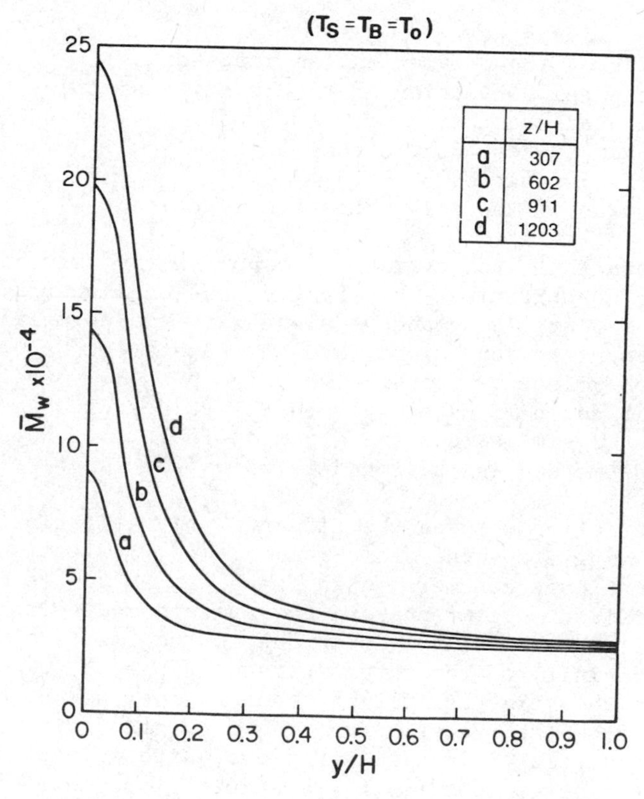

Fig. 2. Average Molecular Weight in the Screw Channel at Various Down-Channel Locations.

TRANSPORT IN PARTLY FILLED CHANNELS

In the pressureless zone of a conventional extruder, the fraction of the channel cross-section occupied by melt remains virtually unchanged along the screw (5). In contrast, in REX, considerable caution is required to control the screw fillage, in particular, when the reaction is combined with devolatilization or, simply, when jamming of the channel should be prevented for other reasons.

Fig. 1 shows a schematic of the flow arrangement. The filled channel width, W_F, as a function of the dwon-channel location, z, has been analyzed recently for REX (2). In the mathematical model of ref. (2), the flow in the partly filled zone is represented by a numerical solution of the simultaneous flow, heat transfer and polymerization kinetics equations.

The following mechanisms are assumed to govern the transport pattern in the screw channel:

- viscous forces
- kinetics of the polymerization reaction
- down-channel convection of the reacting species
- radial heat conduction
- down-channel heat convection
- viscous dissipation
- energetics of the polymerization reaction

It can be shown that the overall transport regime may be characterized quantitatively by eight dimensional groups, namely, L/H, Re, Gr, Da, Br, A, Le and De along with the normalized initial and boundary conditions, and the parameters involved in the conversion-molecular weight-viscosity (p-$\bar{M}_w$-η) correlation (the pertinent definitions along with the values used in the present computer simulations are given in Table 1). Details concerning the method can be found in ref. (2).

The interaction between the flow and polymerization is illustrated using step-wise stoichiometric polymerization of similar bifunctional monomers as an example. As can be seen in Fig. 2, significant differences in the average molecular weight can develop along the channel depth due to the relatively slow flow in the vicinity of the screw surface ($y/H = 0$). The non-uniformity in the molecular weight tends to increase with increasing distance from the inlet. From the flow standpoint, the presence of the relatively highly viscous, high molecular weight material at the screw root implies a reduced transport efficiency associated with the molecular weights of Fig. 2 is illustrated in Fig. 3 in terms of the drag flow capacity and channel fillage (see curve b of Fig. 3). It is important to note that the drag flow capacity-volumetric flow rate at a zero down-channel pressure gradient - strongly depends on the z-location. For the operating parameters of Table 1 (corresponding to a 2.5 inch diameter screw, channel depth 1 mm and screw speed 60 rpm) using the proper temperature control can increase the transport efficiency to levels typical of conventional melt extrusion. For this specific case, maintaining $T_B/T_S \sim 1.16$ allows one to adjust the polymerization rates in the channel such that only small differences in molecular weight and viscosity are permitted to exist along the channel depth and, thus, the flow pattern tends to approach the conventional case.

Fig. 3 is not only important by giving some insight into the flow mechanics of REX but it also suggests that in order to maintain partly filled channels when processing a reactive melt, a careful control of the temperature distribution in the fluid is required. This may call for fairly shallow channels, and in many cases, a cooled screw arrangement.

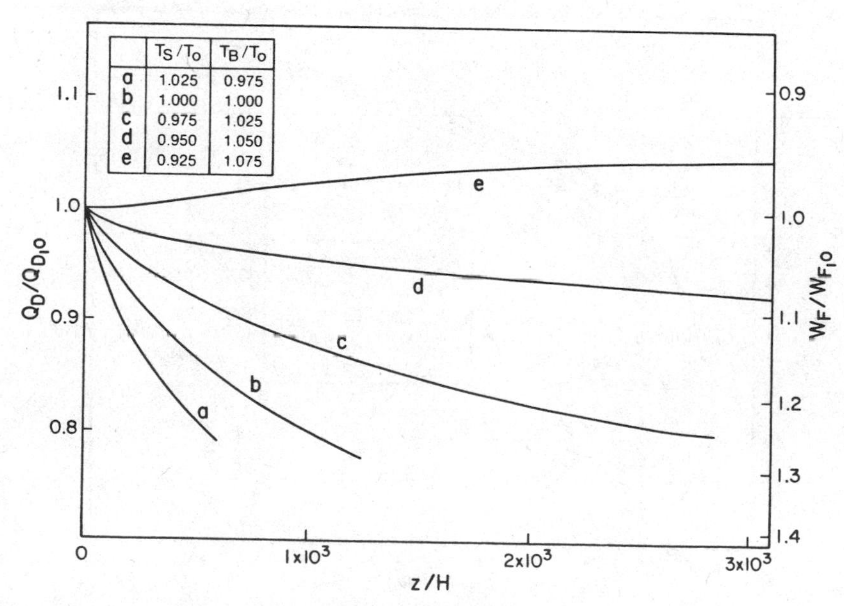

Fig. 3. Drag Flow Capacity and Channel Fillage as Functions of Helical Distance.

PRESSURE BUILD-UP

In completely filled channels, the velocity profile undergoes a similar development to that discussed above. Using data of Table 1, a simulation of REX has been done. This time, a more complete model (3) has been used. It is based on assumptions similar to those given above with the additional feature, however, to represent the cross-channel circulation in the extruder. The solution has been obtained by tracking simultaneously a number of discrete streamlines making use of arguments proposed originally by Tadmor (4) for extrusion of isoviscous fluids. The streamlines are treated as microsegregated plug reactors each of them assocaited with distinct temperature and conversion distributions along its length. The shapes of the streamlines are determined as a part of the analysis thus they depend on the thermal and polymerization histories of the particles travelling along the streamlines. In addition to descriptions of the velocity, temperature and conversion fields the analysis yields predictions of the axial pressure profiles, stresses and heat fluxes at the screw and barrel surfaces, MWD within the channel and some additional quantities derived from the above for a given set of material, geometrical and operational parameters.

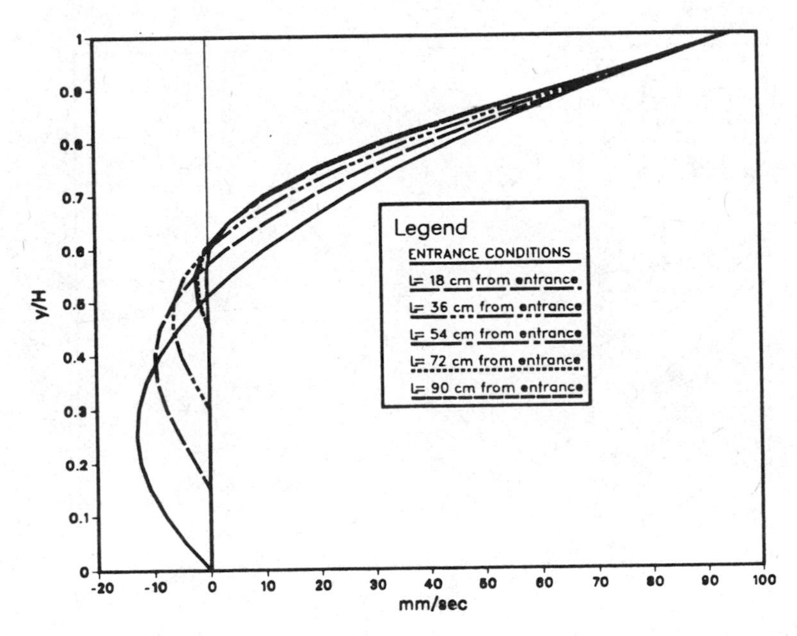

Fig. 4. Down Channel Velocity Profiles in the Extruder for Reactive Processing. Ts = Tb = To = 315 K.

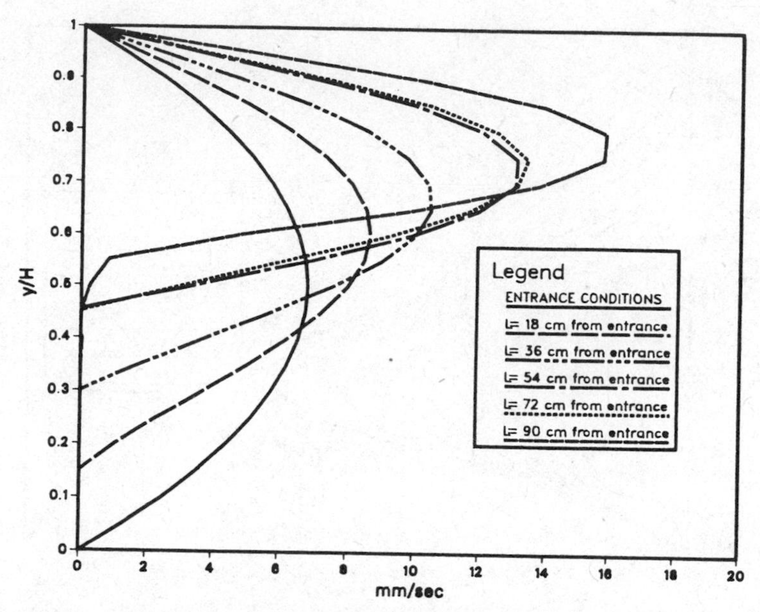

Fig. 5. Cross-Channel Velocity Profiles in the Extruder Channel for Reactive Processing. Ts = Tb = To = 315 K.

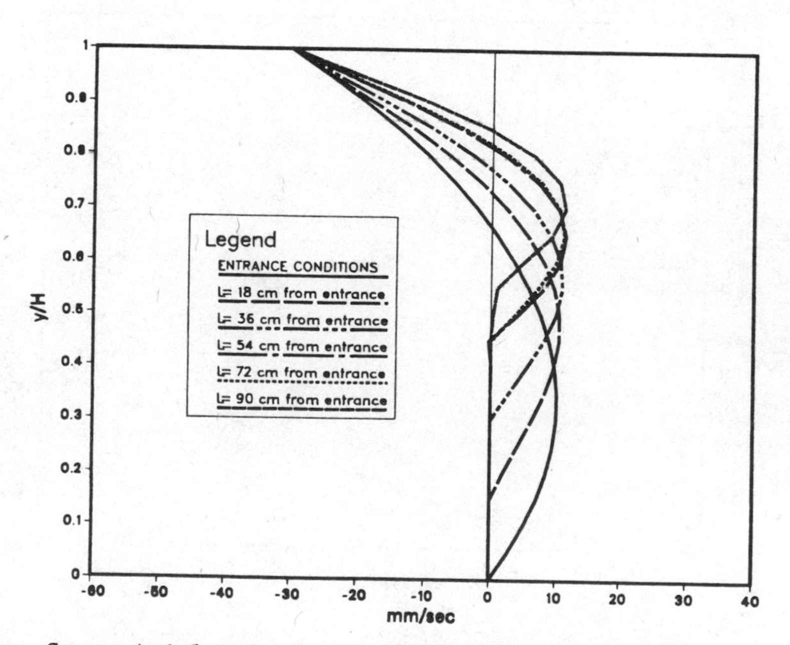

Fig. 6. Screw Axial Velocity Profiles in the Extruder Channel for Reactive Processing. Ts = Tb = To = 315 K.

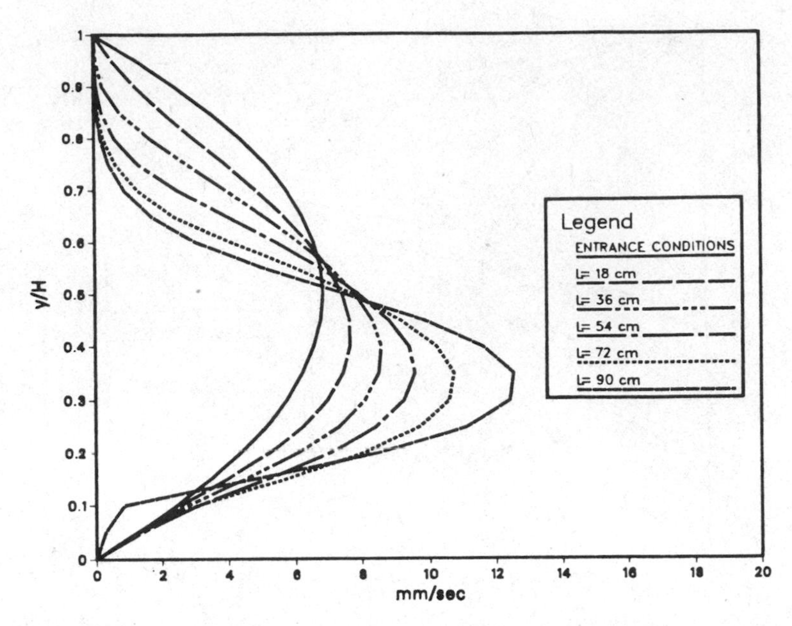

Fig. 7. Screw Axial Velocity Profiles in the Extruder Channel for Reactive Processing. Ts = 298 K, Tb = 333 K, To = 315 K.

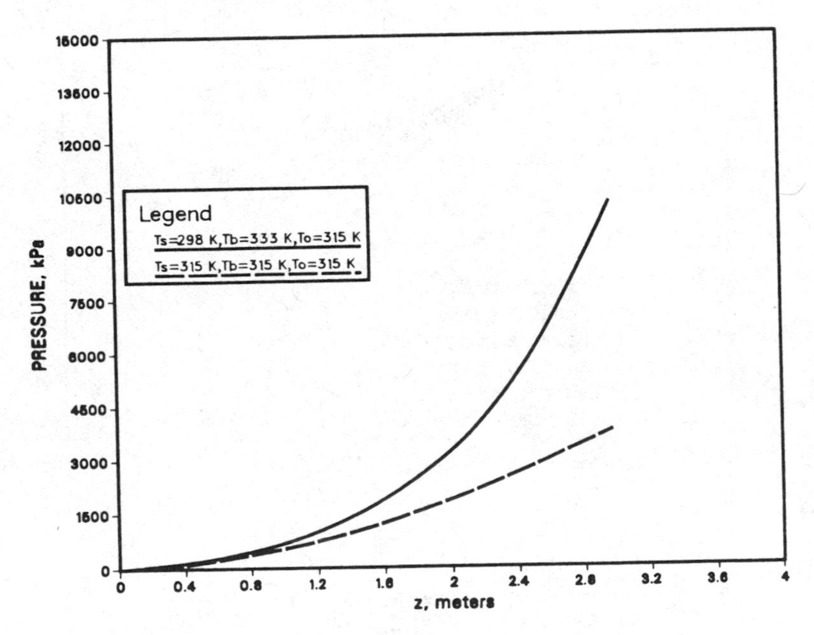

Fig. 8. Pressure Development in the Extruder Channel for Reactive Processing.

Figs. 4-6 show results on the closely related down-channel, cross-channel and axial velocity profiles, respectively. As expected, for $T_S = T_B = T_0$ progressive flow stagnation tends to develop in the channel also in the presence of pressure gradients. The qualitative arguments explaining the existence of the stagnant layer, and the means of suppressing it are similar to those given for a partly filled channel. Again, by promoting the rate of polymerization at the barrel surface the stagnating layer of polymer at the screw root can be eliminated; see Fig. 7. In contrast to partly filled channels, however, the change in the thermal regime to control the molecular weight distribution in the channel is associated with a change in the axial pressure profile as shown in Fig. 8.

REFERENCES

1. U.S. Patent 3,503,944, Polymerization of Ethylene, BASF, 1970.

   U.S. Patent 3,642,964, Continuous Process for Polyurethane, the Upjohn Co., 1972.

   U.S. Patent 3,960,802, Process for the Manufacture of a Vulcanizable Silicon Composition, General Electric Co., 1976.

   German Patent DOS 2,232,877, Continuous Polycondensation of Polyarylesters, Dynamit Nobel, 1966.

   Belgian Patent 740,219, Production of Polycaprolactone-Based Urethane Elastomers, Union Carbide Corp., 1970.

2. J. T. Lindt, Int. Journ. Polym. Proc. Eng. (1983) (in press)

3. B. Elbirli and J. T. Lindt, Continuous Processing of Polymerizing Fluids III, 2nd International Conference on Reactive Processing of Polymers, Pittsburgh, 1982.

4. A. Kaplan and Z. Tadmor, Polym. Eng. Sci., 14 58 (1974).

5. P. H. Squires, SPE Journal, 14, 24 (May 1958).

ACKNOWLEDGEMENTS

This work has been supported by a grant from the National Science Foundation (CPE-8007184).

PROCESSING OF THERMOSETS

J.C. Halpin,* A. Apicella and L. Nicolais

Polymer Engineering Laboratory
University of Naples
Piazzale Tecchio, 80125 Naples, Italy

ABSTRACT

The cure behaviour of commercial grade TGDDM-DDS mixtures is analysed by means of Differential Scanning Calorimetry (DSC), High Performance Liquid Chromatography (HPLC) and Steady Shear and Dynamic viscosity measurements. The kinetic analysis of the DSC thermal scans suggested that DDS primary amine and epoxide etherification dominate the cure reactions. The P.A.-Epoxide addition has an overall heat of reaction, referred to the weight of the epoxy component, of 250 cal/g and an activation energy of 16.6 Kcal/mole, while the corresponding values for the etherification are 170 cal/g and 41 kcal/mole.

The Flory approach has been used to describe the molecular characteristics of the non linear polymerization of the TGDDM-DDS systems. The gelation limits in isothermal tests, calculated by considering the copolymerization of a tetrafunctional TGDDM with a bifunctional DDS, were in good agreement with the values determined in the rheological tests.

The gelation theory was succesfully combined with the WLF theory to describe the viscosity profiles of the reacting melt during the cure.

INTRODUCTION

Carbon fiber reinforced composites are currently made from preimpregnated materials containing high performance epoxy polymers and aromatic amines as matrices. Prepregs are generally slowly heated at constat rate up to the final cure temperature in order

* Wright-Patterson Air Force Base, Dayton, Ohio, USA

to remove the excess resin and to produce a void free composite. The viscosity of the resin initially decreases with increasing temperature but, as the cure reactions occur, its value reachs a minimum and then starts to grow. The magnitude and shape of the viscosity-temperature-time profiles are crucial in the early stages of the process where adequate flow properties of the resin are requested. A proper design of the cure cycle depends on the understanding of the critical phenomena associated with the thermokinetic of the polymerization, chemorheology and flow of the resin during the phase of the cure preeceding the gelation of the system, which marks the transition from a liquid to a rubbery material.

Thermokinetic

The TGDDM-DDS system has been recognized to be characterized by differently activated cure reactions (1-5). The principal reactions which dominate the cure behaviour are the primary and secondary amine addition and the etherification of the epoxide (homopolymerization). The secondary amine addition, however, has been found much more activated than the etherification, and hence involved in the postcuring only at temperatures above 200°C. The low reactivity of the former may be attributed to its inaccessibility due to the steric hinderence of the DDS aromatic ring.

The activation energy and heat of reaction as well as the relative concentration of the active sites determine the rate of consumption of the epoxides in the different reactions. In order to assign a characteristic exothermicity to the epoxide ring opening in the primary amine and homopolymerization, the heats of reaction, determined by DSC, were referred to the weight of the epoxy component. The etherification was separated by thermal polymerization

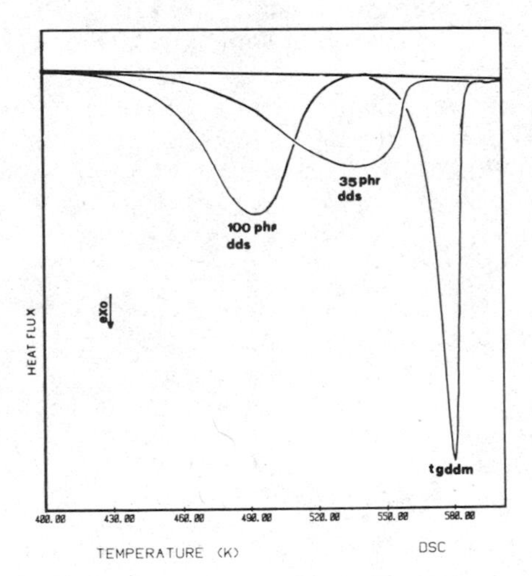

Reaction	Heat of Reaction, cal/g	Activation Energy, Kcal/mole
P.A. add.	250	16.6
Homopolym.	170	41.0

Figure 1: DSC thermograms, heats of reaction and activation energies of TGDDM-DDS systems of different composition

of a TGDDM sample, while the primary amine addition was isolated by scanning a mixture containing more than the stoichiometric composition of DDS primary amine (>95 PHR). The DSC thermograms of the two limiting systems are compared in figure 1 with the scan relative to a mixture containing 35 PHR of hardener. The three scans show well distinct characteristics associated to the different mechanisms of reaction. The 35 PHR system resumes the characteristics of the two limiting systems. Other investigations (5-7) suggested that the epoxides are first consumed in the primary amine addition, while the etherification principally occurs at high temperatures, when all the primary amines are exhausted. The heat of reaction of the former is significantly higher than that of the etherification, i.e. 250 and 170 cal/ grams of TGDDM, respectively. The activation energies derived from a kinetic analysis of the thermograms of the two limiting systems were, respectively, 16.6 and 41.0 Kcal/mole for the two reactions.

The theoretical values of the of the heats of reaction for the TGDDM-DDS mixtures of composition ranging from 10 to 60 PHR of hardener, calculated assuming that the primary amine are first consumed, favorably compare with the values experimentally determined in the DSC scans, Table 1. Notice that in the event that primary and secondary amine additions occured, the heat of reaction referred to the weight of epoxy should be indipendent on the DDS composition.

Table 1: Comparison between the theoretical and experimental values of the heats of reaction of TGDDM-DDS mixtures

DDS, PHR	0	10	20	30	40	49	60	95
H_{exp}, cal/g	170	184	194	205	215	218	230	255
H_{theo}, cal/g	--	179	188	197	206	214	224	--

The relationships between the DDS composition, fractional heat of reaction directly determined from the DSC scan and actual epoxy conversion for this system characterized by multiple reactions has been presented in previous publications (6,7).

Isothermal cure of the mixture containing 35 PHR of DDS in the range of temperatures ranging from 140° to 205°C indicated that the reaction was characterized by a maximum rate, which was attributed to the autocatalytic overall mechanism of reaction, see the curve "a" of figure 2. The comparison between a first order kinetic, and the advancement of the reaction directly derived from the DSC thermogram and from the epoxy conversion (6,7) is also reported on the same figure 2.

The isothermally cured samples are not completely cured and present a residual heat of reaction since the vitrification leads to a quenching of the reactions due to the control of the diffusion on the postcure. Gelation and vitrification, that in ordinary

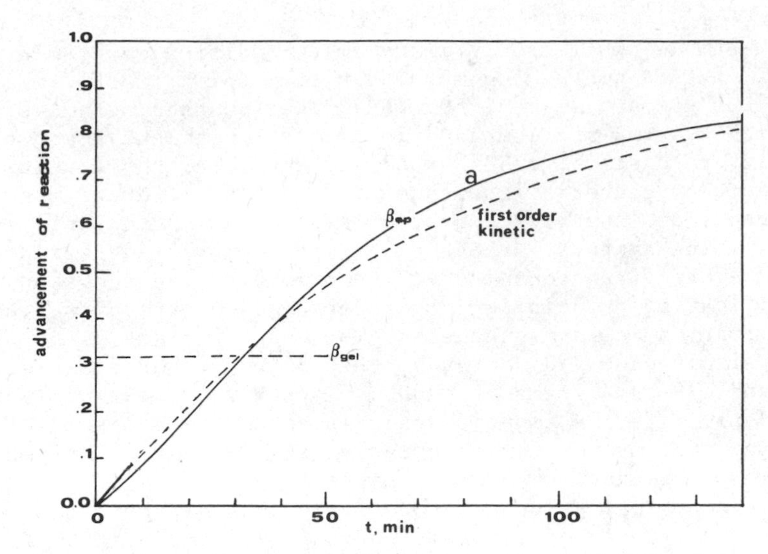

Figure 2: Isothermal DSC cure and kinetic of the advancement of the reaction (180°C) for a 35 PHR of DDS mix.

conditions occur in two distinct phases of the cure process, strongly influence the thermoset molecular structure and hence material properties. We are first interested to correlate the processing parameters to the molecular characteristics of the reacting system up to the gel point. The basic relationships between the extent of reaction and the molecular structure were first introduced by Flory (8), and are summarized in the following.

Since the DDS molecule is essentially acting as a bifunctional curing agent, while the TGDDM is tetrafunctional, the relation between epoxy and primary amine conversion is:

$$P_a = P_{ep} m_{ep}/m_a = P_{ep} 4\, m_{To} /\, 2\, m_{Do} = r\, P_{ep} \qquad (1),$$

where m_{To} and m_{Do} are the initial number of moles of TGDDM and DDS. The weight average molecular weight at a fixed conversion is:

$$\overline{M}_w = \frac{(2r/f)(1+rP_{ep}^2)N_T^2 + (1+(f-1)rP_{ep}^2)N_D^2 + 4rP_{ep} N_T N_D}{(2rN_T/f+N_D)(1-r(f-1)P_{ep}^2)} \qquad (2)$$

where N_T and N_D are the molecular weight of the TGDDM and DDS. The branching coefficient introduced by Flory is related with the epoxy and amine conversion by:

$$\alpha = P_a P_{ep} = r\, P_{ep}^2 \qquad (3).$$

Gelation will occur at a value of the branching coefficient and of the epoxy conversion, P_g, where the system develops an infinite network and the expression 2 of $\overline{M}_w$ diverges:

$$\alpha_g = 1/(f-1) \quad \text{and} \quad P_g = \sqrt{1/(f-1)(g-1)r} \qquad (4),$$

where f and g are the functionalities of the TGDDM and the DDS.

Chemorheology

The rheological behaviour is particularly sensitive to the molecular modification of the reacting melt. The increase of the molecular weight during the cure is reflected on a macroscopic level in a progressive increase of the viscosity. The High Performance Liquid Chromatography (HPLC) has been succesfully used to determine the changes of the molecular characteristics and hence the molecular parameters, previously introduced, during the cure. In particular, the isothermal cure at 180°C has been investigated.

The intensity of the peacks in the HPLC chromatograms is proportional to the concentration of each specie present in the reacting melt, but is not indicative of the relative concentrations when the UV detector adsorbance of the reaction products are not determined. Figure 3 shows the HPLC chromatograms for TGDDM-DDS mixtures containing 35 PHR of hardener, isothermally cured at 180°C for times ranging from 1 to 35 minutes. The concentrations expressed in weight fractions of the TGDDM and DDS monomers has been taken as a function of time assuming that$[A(t)/A_o]w_o=w(t)$. The initial weight fractions of monomers in the TGDDM-DDS mixture were $w_{oT}=0.74$ and $w_{oD}=0.26$. The variation of the weight fractions of the single monomers is reported in figure 4a.

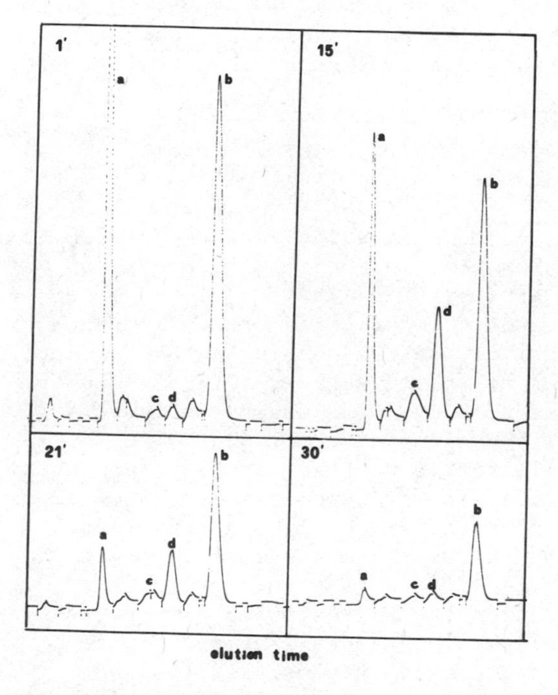

a) DDS

b) TGDDM

d) TGDDM-DDS

c) DDS-TGDDM-DDS

Figure 3: HPLC chromatograms during isothermal cure at 180°C of a TGDDM-DDS system.

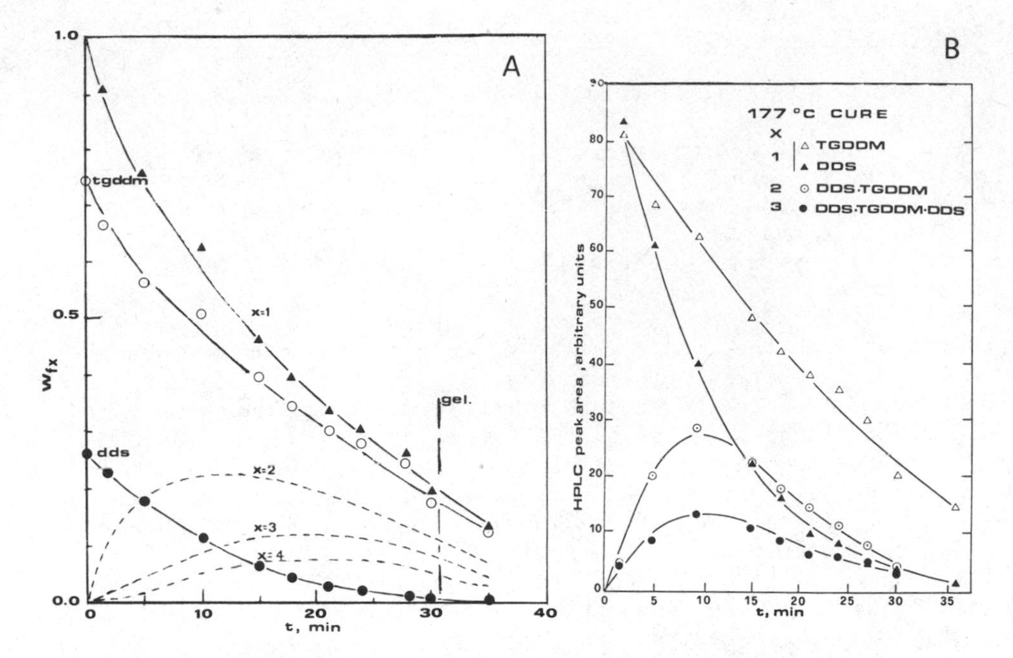

Figure 4: a) Wheight fractions of the monomers and reaction products and b) HPLC peack areas during the isothermal cure at 180°C of a TGDDM-DDS system

The relationships between the epoxy conversion, branching coefficient, weight fractions of x-mers, reagent functionalities and stoichiometric imbalance are:

$$w_x = ((1-\alpha^2)/\alpha) x \omega'_x \beta^x, \text{ where } \omega'_x = f(fx-x)!/(fx-2x+2)!x! \text{ and } \beta = (1-\alpha)^{f-2}. \quad (5)$$

It is possible from equations 5's to derive the variations of $\alpha(t)$ from the experimental values of $w_1(t)$. The weight fractions for x=2, 3 and 4, calculated from $\alpha(t)$ by using the 5's, are also reported in figure 4a. The shapes of the curves favorably compares in figure 4 with those qualitatively determined in the HPLC (figure 4b). The value of w_1 at the gel point, where α_g=0.33, is about 0.20, which correspont to a gel time of 30-32 minûtes (figure 4a).

The complete rheological behaviour of the TGDDM-DDS system has been characterized in oscillatory and steady state shear tests. Whereas the steady state shearing flow measurements describe the rheological properties of the material in the liquid state, the oscillatory shearing is used to characterize the material even when entering the gelation stages.

Gel times were extimated from the viscosity vs time plots relative to both types of rheological characterization. Figure 5 compares the steady shear (●) and complex viscosities (▲) and reports the tanδ relative to the isothermal cure at 180°C of the mixture containing 35 PHR of DDS. In the early stages of the cure

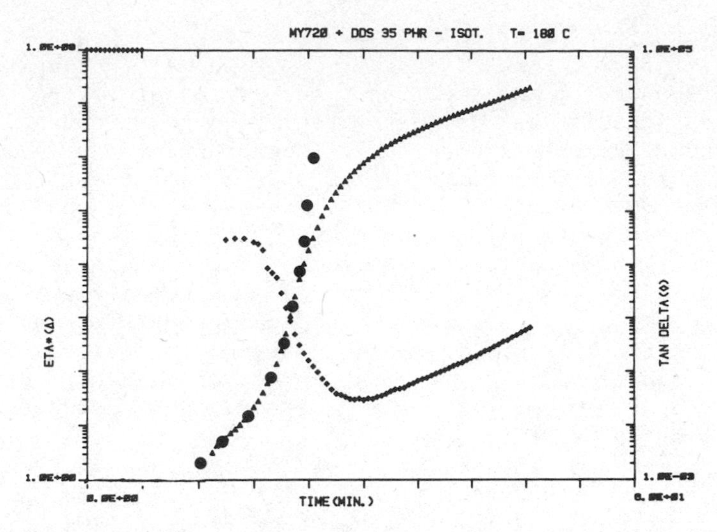

Figure 5: Steady shear (●) and oscillatory (▲) viscosity measurements of the TGDDM-DDS system (35 PHR)

the viscosity slowly increase with the time up to a critical point where, probably, extensive branching and /or entanglements are occurring, and then rapidly increases approaching very large values. The vertical portion of the constant shear rate curve and the minimum in tanδ have been used to extimate the gel times. These values are in good agreement with the value previously determined from the analysis of the molecular characteristics (HPLC). Moreover, a DSC calorimetric extimate of the gel time was also possible from the reaction kinetic curve reported in figure 2, how the time necessary to reach the theoretical value of the epoxy conversion for gelation calculated from equation 4, P_g=0.315. The values of the gel times determined from HPLC, rheological and calorimetric analyses are compared in table 2.

Table 2: Gel times determined from different techniques in the isothermal cure at 180° of a TGDDM-DDS system

Gel times from	HPLC	Steady	Dynamic	DSC
min.	32	30	30	32

The use of the DSC, HPLC and viscosity analyses gives us the capability of fully characterize the rheological, molecular and kinetic properties of this system.

A commercial formulation, used as matrix for carbon fiber prepregs, has been characterized at different degrees of cure preeceding gelation by using these tecniques. In particular, the dynamic viscosities between 35° and 135°C of samples characterized by values of the branching coefficient, determined by HPLC, ranging from 0.025 to 0.19 have been determined at 1.0 and 10.0 Hertz. The viscoelastic properties of a thermosetting melt during the cure are influenced by two phenomena; the variation of the average molecular characteristics and glass transition temperatures. Both processes induce changes of the resin flow behaviour, however, while the system is not reacting, a progressive increase of the temperature always results in a decrease of the viscosity, a more complex behaviour is observed for systems in which the increase of the temperature activate the polymerization of the components. Generally, a minimum in the viscosity is observed. The two basic phenomena associated to the changes of the rheological properties may be analyzed by using the Flory approach and the classical WLF theory, once the modifications of the average molecular characteristics and glass transition temperatures are determined as a function of the advancement of the cure.

The influence on the viscosity of the increase of the temperature for systems of constant T_g and α is determined by using the WLF equation, if the system is not reacting (i.e. $T < 120\text{-}125^{\circ}C$):

$$\ln \frac{\eta(T,\alpha)}{\eta_r(\alpha)} = - \frac{C_1(T-T_g(\alpha))}{C_2+T-T_g(\alpha)} + \frac{C_1(T_r-T_g(\alpha))}{C_2+T_r-T_g(\alpha)} \qquad (6),$$

where $\eta(T,\alpha)$ is the viscosity for a given value of the branching coefficient, $T_g(\alpha)$ the corresponding glass transition temperature, while the subscript r refers to the value at a reference temperature, T_r. Figure 6 compares the viscosity profiles experimentally determined and theoretically calculated by using the values of T_g reported on the curves and the constants $C_1=40.5$ and $C_2=52^{\circ}C$. The branching coefficients determined by HPLC are also reported.

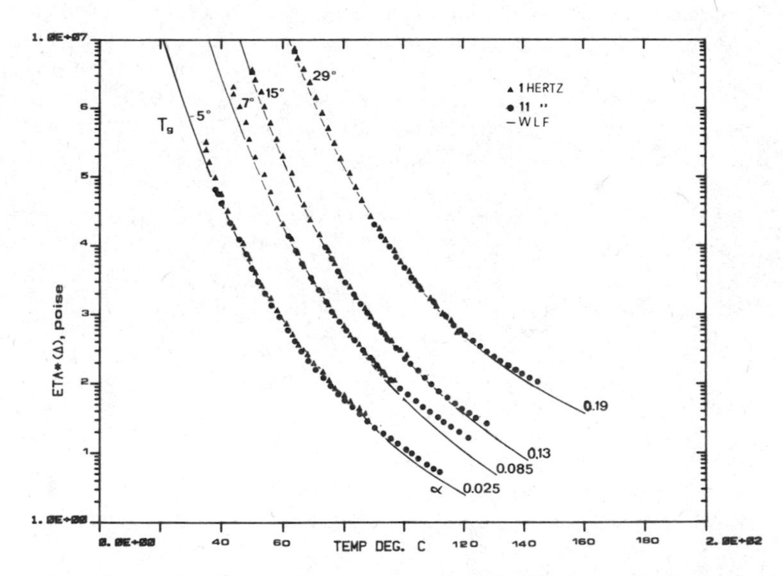

Figure 6: Viscosity profiles experimentally determined (● ▲) and theoretically calculated from the WLF equation 6 (—)

On the other hand, the variation of the viscosity at a fixed temperature, determined by the changes of the molecular characteristics, may be derived as follows; while for a linear polymer $\eta = KM_w^a$ with $1 < a < 2.5$ for $M_w <$ of a critical value and $a = 3.4$ for $M_w >$ of the critical value, a similar relationship has been suggested (9) for the non linear polymerization, $\eta = K(gM_w)^a$, with a =3.4 and g, the ratio of the square of the radii of giration of a branched to the linear chain of the same molecular weight, defined as:

$$g = \frac{1-\alpha(f-1)}{2\alpha} \ln\left(\frac{1-\alpha}{1-\alpha(f-1)}\right) \quad (7).$$

Referring the viscosity to its initial value when $\alpha = 0$,

$$\eta(\alpha)/\eta_o = (g\bar{M}_w(\alpha)/\bar{M}_{wo})^a = (g\ \bar{X}_w)^a \quad (8),$$

where $\bar{X}_w$ is the number average molecular weight and is function of α:

$$\bar{X}_w = (1-\alpha)/(1-\alpha(f-1)) \quad (9).$$

Combining equations 7, 8 and 9:

$$\eta(\alpha) = \eta_o\left(\frac{1-\alpha}{2\alpha} \ln\left(\frac{1-\alpha}{1-\alpha(f-1)}\right)\right)^{3.4} \quad (10).$$

However, while in the phenomenological expression relative to the linear polymers the variation of the exponent "a" was probably taking into account the changes of the glass transition with M_w, the choise for the non linear polymer of a constant exponent should be balanced by explicity considering the functionality of T_g with $\bar{M}_w$. We have, in fact, found good agreement between the experimental values of figure 6 and the theoretical values of the viscosity at different temperatures and degree of cure calculated by using:

$$\eta(T,\alpha) = \eta_{ro}\left(\frac{1-\alpha}{2\alpha}\ln\left(\frac{1-\alpha}{1-\alpha(f-1)}\right)\right)^{3.4} \exp\left(\frac{C_1(T_r-T_{go})/C_2+T_r-T_{go}}{C_1(T-T_g(\alpha))/C_2+T-T_g(\alpha)}\right) \quad (11)$$

where η_{ro} is the viscosity at the reference temperature and T_{go} the glass transition temperature of the unreacted system.

The variation of the glass transition temperature at different levels of cure has been experimentally determined by DSC both for the commercial formulation and the mixture containing 35 PHR of DDS. In particular, the thermograms relative to the latter samples isothermally cured at 180°C for times ranging from 1 to 200 minutes were used to determine the glass transition temperatures and residual heats of reaction. The epoxy conversions, determined according to reff. 6 and 7 from the residual heats of reactions, and the T_g's are reported in figure 7 (right side). The plot shows a change in the slope in the range of epoxy conversions which corresponds to both the exhaustion of the DDS primary amines, $P_e = .308$, gelation, $P_e = 0.315$ On the left side of the same figure are reported the viscosity profiles measured during the cure described in the figure and theoretically calculated according to equation 11.

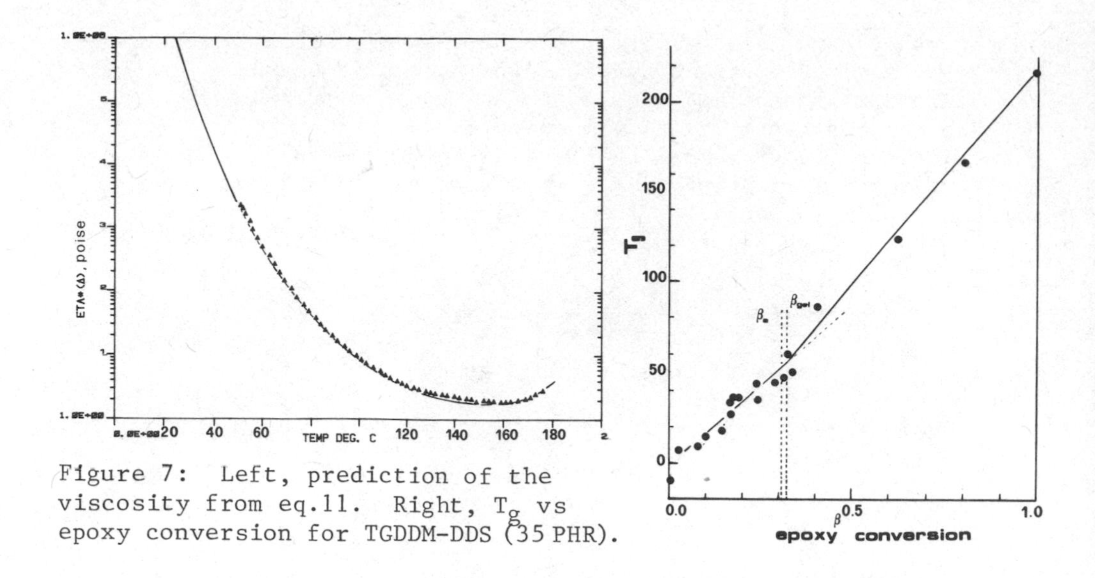

Figure 7: Left, prediction of the viscosity from eq.11. Right, T_g vs epoxy conversion for TGDDM-DDS (35 PHR).

In conclusion, the cure in the DSC gives the capability of continuosly monitor the advancement of the reaction in a system undergoing a selected cure cycle and hence to predict its rheological behaviour before entering the gelation. The extimates of the gel times and viscosities are, in fact, possible from the gelation and WLF theories once the epoxy conversions and T_g's are determined as a function of the time.

REFERENCES

1)S.Lunak, J.Vladika and K.Dusek, Polymer, 19, 913 (1978).
2)P.Perser, and W.Bascom, J.Appl.Polym.Sci, 21, 2359 (1977).
3)M.Cizmeicioglu and A.Gupta, SAMPE Quart. 13, No3, 16 (1982).
4)K.Dusek, M.Bleka, S.Lunak, J.Polym.Sci., Polym.Chem., 15,2393 (1977).
5)E.T.Mones, R.J.Morgan, ACS Polym.Prep., 22, No2, 249 (1981).
6)"Thermokinetics and Chemorheology of the Cure Reactions of TGDDM DDS Epoxy Systems", A.Apicella, L.Nicolais, M.Iannone and P.Passerini, sub. to J.Appl.Polym.Sci.
7)"The Role of the Processing Chemo-Rheology on the Ageing of High Performance Epoxy Matrices", A.Apicella, L.Nicolais, J.C.Halpin, 28th National SAMPE Meet., Anaheim, USA (1983).
8)P.J.Flory, "Principles of Polymer Chemistry", Cornell University Press, Ithaca, N.Y., 1953.
9)E.M.Walles, and C.W.Makosko, Macromolecules, 22, 521 (1979).

ACKNOWLEDGEMENTS

This work has been partially supported by the Aeritalia aircraft company and the CNR progetto finalizzato Chimica fine e Sec. and NSF Program, Grant No DMR-820-1228.

ADAPTING A MICROPROCESSOR SYSTEM TO AUTOCLAVE PROCESSING

Robert E. Coulehan

OMO CORPORATION
13 Allen Street
Dobbs Ferry, NY 10522
USA

INTRODUCTION

The processing of advanced composites has taken great strides in the last six years. The early interests in the control and quality assurance of resin systems and prepregs soon led to a broad development of techniques used to follow the processing history.This included analysis of materials from resins to fibers by the prepregger, through the fabrication of the individual composite part and into the service for performance data. The results of these interests have given the processing industry a broad range of instrumental techniques that furnish a wide variety of process-property data such as viscosities, chemical species distribution, reaction energies, etc, that can be used to judge optimum cure profiles.[1] The implementation of these results are seen in the many material specifications that have been written in the past four years.

Today when a new part is to be fabricated, there are a variety of evaluation steps available to the processor to ensure that the cure cycle will fit the processing facilities. The specific area of interest in this paper is the autoclave process and ways to upgrade and introduce flexibility into existing units.

The autoclave has been one of the dominant factors in putting advanced composites into the market place. These units have been called upon to handle a wide variety of resin-fiber systems in all shapes and sizes. Yet, there is the constant complaint by the material supplier about the lack of flexibility with the autoclave cure cycle. In spite of these comments, new applications are constantly being added to the success of the autoclave system. It is in these successes that the flexibility of the autoclave processing has grown. Yet, there is a great need to further improve this processing potential for many autoclaves especially the larger, older style units that are in many shops. With this fact in mind, we will outline a development microprocessor system that can bring a new dimension to the operations of many autoclaves.

BACKGROUND

The very large array chips that have proliferated the digital electronics market since the mid seventies has changed the way we think and act in processing data. What formerly required special rooms to house a computer installation now can be put on a desk, in a lab or in a factory. Now we can go much further in our interfacing with computers. We can take it apart, reconfigure it to do specialized operations in process control problems, and at a cost that is within everyones budget. However, it would be misleading to leave the impression that such changes can be done by a simple substitution for the present process control system for the autoclave. There is much more to it. Such a microprocessor system will not only be an integral part of the control system, but will allow the operator to expand the whole processing window and begin to treat each composite piece on an individual basis. This is done by regulating certain parameters independently of the demands of other parts in the autoclave. At the same time process modifications are easily programmed for changes and updating that is needed for different materials.

BUILDING THE MICROPROCESSOR SYSTEM

The system that will be used as our model is based on the 6502 microprocessor. This is an 8-bit microprocessor that has been very successful in a wide variety of applications from the personal computer in the home to the harsh environment of the production floor. The advantage of this particular microprocessor is the ability to interface with input-output devices for a wide variety of problems.

Let us consider a practical example in building such a system for an autoclave. The first step in the procedure is to fix the precision level of the data base. If temperature information is needed to a value of one degree Celius, we then know the minimum voltage level required in setting up the specifications for the analog-to-digital (A/D) converter. If now the temperature range is 200°C, then the A/D converter must read to one part in 400. Such a range and precision indicates that the A/D converter be at least a 10-bit unit. With such a unit there will be adequate range for future expansion. Similar considerations are used for pressure and vacuum information as well as other special sensing devices that are used in the system.

It is important to choose the range and precision of the sub components of the microprocessor control system so as to cover all possible variations that might arise in actual use. This will also help in the next level of consideration, that of the memory of the system both type and amount. This decision will be dependent upon the two levels of memory, program and data. For the programmed memory, one chooses from a number of options such as ROM (Read Only Memor), EPROM (Erasable programmable ROM), etc. that permits the cure algorithm to be made inaccessable to outside changes unles there is a specific need to put such functions in the program. This memory space can now be used exclusively to direct the process and cannot be interrupted or changed.

Data memory can also be scaled to reflect some flexibility to the particular process. Although nothing was said about the number of data points that one needs for an autoclave cycle, it is in this area that the power of the microprocessor system lies. Consider a very simple case of ten thermocouples on a part and the need to evaluate a temperature distribution over this part at each step in the process. During the initial heat up, the average temperature of the part can be computed at all times. The program interrogates each thermocouple in an order set down by the algorithm and stores that value in memory. The rate at which each thermocouple is read can be adjusted through interval timers so that the actual part heat up rate can be phased to the desired temperature cycle. The algorithm set up the number of readings for each thermocouple, let us say 10, in the time interval. These 10 values for each of the 10 thermocouples are averaged and compared to the desired temperature profile at each time interval. These data can now be stored in a memory stack to be retreived at regular intervals as a record of the process. Suc a conclusion is rather obvious.

However, let us say that we have two such parts in the autoclave and for some reason want to apply vacuum at different temperatures or one part reaches the temperature at an earlier time. The microprocessor system now inputs two sets of temperature scans into two independent parts of the program and nests a routine to operate the action of the vacuum for each as is required. The microprocessor will not distinguish between the two parts. It views the data from each as data from some memory location whether it is internal to the microprocessor system or from the outside world. Hence, if we can put two composite parts in that may need different cure schedules then there is really no reason why we could not put in three or four or any number if there is adequate memory in the microprocessor system. This is exactly the direction that this work will take to give the flexibility to this type of processing.

Now what is the memory size that is needed for such an operation? Is it 8K, 16K, 32K or what? The choice of memory size must be made during the project development phase and with the particular type of microprocessor selected for the job. There are many compatible memory boards that plug in directly to the microprocessor system with a minimum of firmware modifications such as memory address locations. In selecting the type of memory, it is important to know if a static RAM (Radom Access Memory) is needed since the power requirements can be considerably different from dynamic RAM.

We can now begin to see some of the potential of the microprocessor system and how older autoclaves can be modified to give updated performance. An important concern in all these microprocessor applications is the language needed for the different interactive levels. Most higher level languages are too slow to permit direct, on-line process control of an autoclave system. The compilers for these languages will take up much needed memory space that would make them impractical for most shops to operate. The two levels of language found most appropriate for this type of control is assembly language and BASIC. With assembly language one can interface directly with the input-output commands of the microprocessor and still remain transparent to any attempts to interrupt it from outside commands except for the controlled reset type command.BASIC is a compatible language that canbe used to monitor and retreive data for record keeping purposes. There are many innovative techniques that have been developed using these two language levels to add operational capabilities to these control systems. One of the more innovative is the nesting of process control loops- in some instances up to 20 nested loops in one control program - and still have memory to keep data for records.

Finally, there is an important consideration that gives such a microprocessor a new role in process control. This is the ability to operate under harsh conditions[2] of temperature, humidity, electrical and other environmental factors where stability and long term control is needed. With proper design of the microprocessor system there are many ways to put these systems into the production area that will give excellent reliability and performance. The added benefit of the low cost of these systems - under $2000 - permits duplication for increased flexibility for multiple process control.

REFERENCES

1. May, C. A. et al ,"Process Automation - A Rheological and Chemical Overview of Thermoset Curing", IUPAC Athens, 1982.

2. Coulehan, R. E., "Introduction to Microprocessor Control in Hostile Environments", To be published by Carnegie Press during 1983.

PART III

RELATIONSHIPS BETWEEN PROCESSING AND PROPERTIES

HEAT TRANSFER AND KINETICS OF FLOW INDUCED CRYSTALLIZATION AS INTERTWINED PROBLEMS IN POLYMER MELT PROCESSING

H. Janeschitz-Kriegl[x)] and F. Kügler[xx)]

x) Johannes-Kepler-University, A-4040 Linz, Austria

xx) Chemie Linz AG., A-4021 Linz, Austria

Dedicated to Prof. O. Kratky on the occasion of his acceptance of an honorary doctor's degree at Vienna University on Nov. 3rd, 1982.

ABSTRACT

It has been known for a rather long time that texture (and quality) of an article processed from a semicrystalline polymer is dependent on a number of process variables like temperature settings or speed of processing. Type of polymer (molecular mass distribution), type and concentration of additives are of importance as well. A survey will be tried of efforts made in order to elucidate mechanisms being influenced by the mentioned variables. Emphasis will be laid on more fundamental approaches rather than on purely empirical rules.

Several important points are discussed as there are: The influence of the finite speed of crystal growth of layers solidifying along the mould wall, the reasons for the high degree of orientation of these layers, the implications emerging from the coupling of energy and momentum equations, the outlook for a quantitative description of the mould filling process of semicrystalline polymers.

INTRODUCTION

The present authors do not maintain to be grown-up experts in the field of crystallization kinetics as indicated in the title. However, partly as a logic step to a further extension of previous research, partly as a consequence of immediate industrial interest, they started to reconnoitre this field, not without discovering that a great deal of the necessary research work in this field has still to be done. In particular, this holds for the combined influences of a steep temperature gradient and a shear flow, as prevailing in many processes like injection moulding. In contrast to the processing of amorphous polymers, where birefringence patterns in cuts of moulded articles can readily be understood[1,2,3,4,5,6,7,8], no clear picture seems to exist for the mechanisms which cause the texture of cuts obtained from moulded articles of semicrystalline polymers. Apparently, a new phenomenon is encountered with these polymers, viz a delay in solidification due to supercooling. With rapid cooling, in amorphous polymers the glass transition occurs at a higher temperature and without delay[9], whereas in crystallizable materials the crystallization temperature experiences a cooling rate dependent shift to a lower value with some counteraction by shear nucleation. Surprisingly enough, this phenomenon has so far been ignored in pertinent engineering calculations. Only recently, one of the present authors - together with his coworkers - has proposed an experimental technique for obtaining the necessary kinetic data[10]. A paper presenting preliminary results has been prepared as well[11]. While doing this work, they realized that our knowledge in the field of crystallization kinetics of polymers is insufficient. The present paper can be considered as the result of an effort to select those findings from the huge literature on polymer crystallization, which seem of primary relevance to the indicated research aims. In this connection the authors wish to make their apology, if in some respect this presentation comprises not only what has been done but also what has to be done, which is considered a justifiable ingredient of a review lecture.

SOME CONCEPTIONAL CONSIDERATIONS

Referring to the comment made with respect to the usual engineering calculations, we feel that this type of calculations deserves a more detailed scrutiny. For the purpose the usual treatment of the heat penetration problem is recalled. This problem bears realistic features of usual polymer processing techniques where the relatively hot polymer melt touches the relatively cold wall of a machine part (mould in injection moulding). In this case the "cold" penetrates into the melt. This is described with the one-dimensional Fourier equation

$$\rho c \frac{\partial T}{\partial t} = \lambda \frac{\partial^2 T}{\partial x^2} \qquad \ldots(1)$$

where T is the temperature, t the time, x the distance from the mentioned plane, ρ is the density, c the heat capacity and λ the heat conductivity of the medium filling the half space. In a well-known text book of recent date[12] two recommendations are given for the case that a phase transition (crystallization) occurs in the course of the cooling of the medium, the latter initially being at a temperature well above the transition temperature.

Recommendation a) For the two homogeneous phases the validity of eq. (1) (with the corresponding material constants) is assumed. At the moving boundary the boundary condition

$$\lambda_c \left(\frac{\partial T}{\partial x}\right)_c = \rho_c H \frac{dx_c}{dt} + \lambda_\ell \left(\frac{\partial T}{\partial x}\right)_\ell \qquad \ldots(2)$$

is assumed, where subscripts c and ℓ indicate the crystalline and liquid media, respectively, H is the latent heat of crystallization and (dx_c/dt) is the speed of the moving boundary. As will be shown below, this boundary condition implies, however, that the process of crystallization is "heat diffusion controlled" under all circumstances. In other words, the speed (dx_c/dt) must be capable of assuming an arbitrarily high value, if the solution of this boundary value problem requires such a value. In fact, according to this boundary value problem the required (dx_c/dt) turns out to be infinite in the very beginning of the experiment, when the temperature gradient at $x = 0$ is infinitely high as well[13]. Such a crystallizing system does not exist in reality. It has recently been shown[11] that, with polymer melts the deviation from the results of the sketched procedure are particularly large. This should be obvious in view of the well-known slow crystallization of polymers. In fact, even if the speed of crystallization of polymers is enhanced by shear nucleation, the just made remarks remain valid in principle.

In Fig. 1 a well-known unified presentation of the radial growth speeds of spherulites[14] of a rather large number of polymers is reproduced. From this figure we learn that the curve representing the speed of spherulite growth as a function of temperature is bell-shaped. Somewhere in the middle between glass transition temperature T_g and (thermodynamic) melting point T_m this curve possesses a maximum. The theoretical explanation for this shape can be found in the literature[33,34]. For our purpose we are interested in the fact that the "nucleation rate controlled" speed of crystal growth is finite everywhere in the interesting temperature range. It is practically zero at the (thermodynamic) melting point (due to a lack of primary and secondary nuclei) and is also practically zero at the glass transition temperature (due to the high viscosity of the system). As will be reviewed in the next section, with the application of shear the spherulitic growth of crystals disappears.

At the same time shape, location and maximum height of the curve

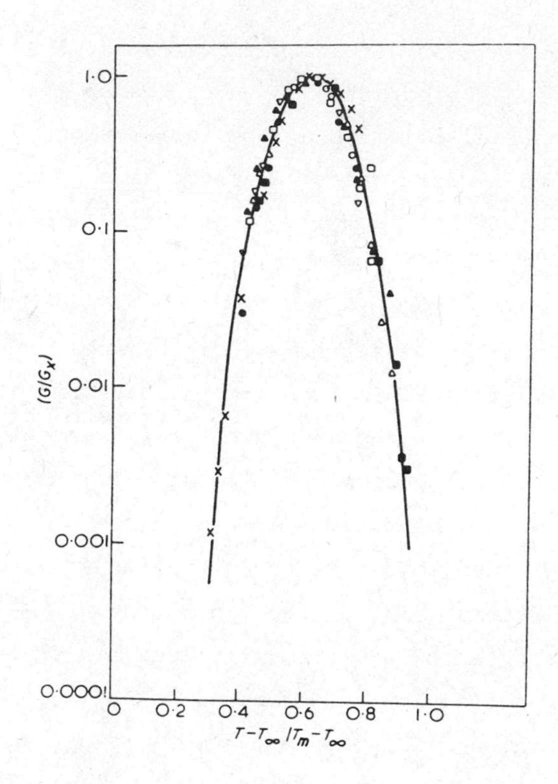

Fig. 1: Corresponding states plot of reduced sherulite growth speed (with respect to maximum speed) vs. reduced temperature $(T-T_{\infty})/(T_m-T_{\infty})$, where T_m is the melting point and T_{∞} is the temperature of zero free volume[9], according to ref. 14.

describing growth as a function of temperature, will certainly change, but the essential remarks made in this section will remain valid, devaluating method a). However, also the second method usually recommended will turn out to be incorrect.

Recommendation b) The assumption is made that the heat of crystallization H can be incorporated into the specific heat c which assumes, as a function of temperature, a bell-shaped appearance similar to the curve in Fig. 1. In this way, one hopes to take into account the fact that crystallization does not occur at the thermodynamic melting point but in a temperature range below this point. A moving boundary is no longer necessary in this treatment. One simply uses eq.(1) with c as a proper function of temperature. The polymer is treated like a mixture of polymers of varying melting points, the curve of c vs. T reflecting the composition of this mixture. However, it goes almost without saying that such a model is incapable of describing the effect of a quick quench, after which for many known industrial polymers a supercooled amorphous phase can be obtained. Such a model material cannot be cooled down to a temperature in the vicinity of the glass transition without complete crystallization.

Also in treatment b) one overlooks the profound difference between heat diffusion controlled crystallization and nucleation rate controlled crystallization. In fact, according to this treatment progress of crystallization must be comparatively slow in the vicinity of the temperature at which the maximum occurs in the "apparent" specific heat, since at those temperatures too much heat must be removed by heat diffusion. If, on the other hand, crystallization occurs into the supercooled melt, which can easily absorb the heat released during nucleation rate controlled crystallization, one may just obtain, according to Fig. 1, a relatively high crystallization speed in this temperature range.

It is hoped that these preliminary remarks can be substantiated by examples given later in this review. However, before this can be done, one has to look after reasons for the observed high orientation occurring in the so-called shear zones, as found in cuts of processed articles. It will turn out that both parameters envisaged, viz. shear flow and temperature gradient, can cause high orientation independently.

SHEAR INDUCED CRYSTALLIZATION

An important progress in the investigation of crystallization under shear stress was achieved by Haas and Maxwell[15]. These authors used an apparatus which consisted of two parallel glass plates. Between these plates a thin layer of the polymer was placed. This system was quenched from a temperature well above the melting point of the polymer to a chosen crystallization temperature below the melting point T_m of the unloaded melt. When a previously selected shear stress was suddenly applied, shear flow started in the polymer. After some time this creep flow came to a halt because of the induced crystallization. One interesting quantity was the total shear achieved at various crystallization temperatures and shear stresses. For our present purpose the most important quantity, however, is the induction time. For low shear stresses the time was determined, at which spherulites formed a bridge from one glass plate to the other. For high shear stresses the time at the inflexion point of the creep curve was chosen. When going from low to high shear stresses the authors observed a transition from spherulitic crystallization to a "structureless" crystallization characterized by the appearance of a large birefringence.

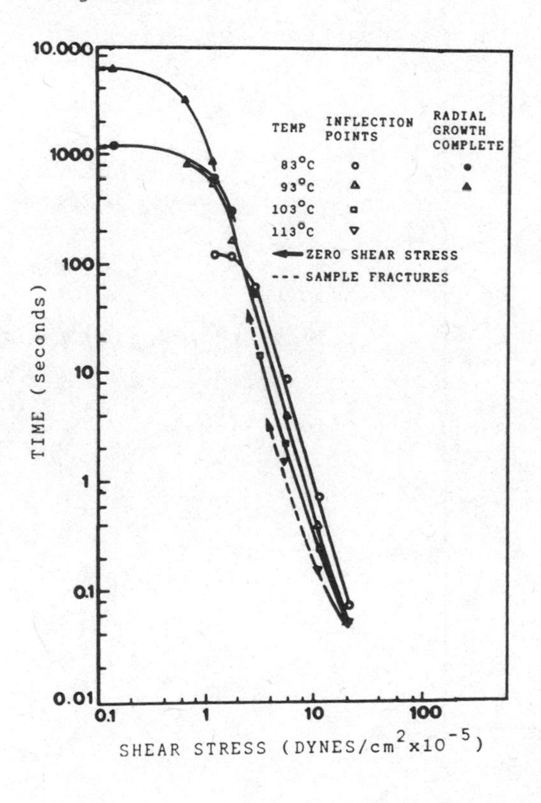

Fig. 2: Crystallization times vs. shear stress for high molecular mass polybutene-1 at various temperatures, according to ref. 15.

In Fig. 2 results as obtained for a high molecular weight (HMW) polybutene-1 are shown. Samples were quenched from 190° C to several temperatures below T_m = 113° C. From Fig. 2 one notices that an increase of the shear stress by a little more than one decade caused an increase of the crystallization speed by as much as four decades. A closer inspection of Fig. 2 reveals a cross-over of the curves obtained at 83 and 93° C. From this experience one can derive the conclusion that the shape of the curve describing the crystallization speed as a function of temperature must depend on the shear stress applied. This conclusion is confirmed by a look on Fig. 3, which is also taken from the work of Haas and Maxwell. This figure shows the temperature dependences of the rate constants as obtained at various shear stresses. With low shear stresses, where spherulitic crystallization prevails, the rate constants decrease, when the temperature approaches the melting point T_m. This is in accordance with Fig. 1. With high shear stress, however, the rate constants show a tendency to increase with the temperature.

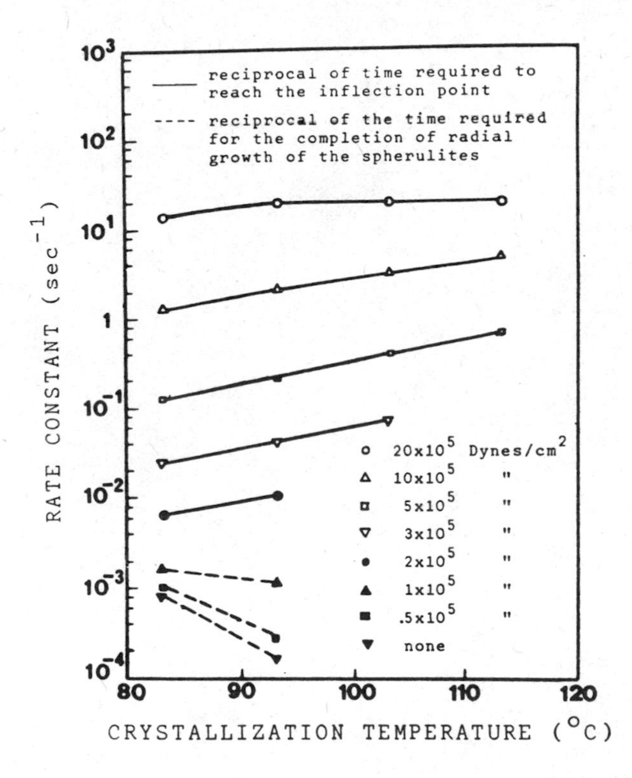

Fig. 3: Rate constants (i.e. inverse crystallization times) vs. crystallization temperature for high molecular mass polybutene-1 at various shear stresses, according to ref. 15.

Several other authors continued to carry out work as lined out by this pioneering paper[16,17,18]. However, as we were interested only in the principal features of the crystallization phenomena, we decided to switch to the next problem, which was ignored by this isothermal crystallization work, viz. the influence of steep temperature gradients. This subject will be surveyed independently in the next section.

ZONE CRYSTALLIZATION[19,20]

For these experiments a thin rod of the polymeric material is contained in a thin walled glass tube. This glass tube is slowly moved along its axis through an oven, where the polymer is melted. At the exit of the oven the polymer is exposed to a rather steep temperature gradient, in which the polymer crystallizes. If the speed of the tube is comparable with the speed of growth of the spherulites, one observes intriguing phenomena. In this case the polymer crystallizes against a rather steep temperature gradient. As a consequence of the high temperatures prevailing ahead of the crystallization front no nuclei enter this front, as they did not get the chance to be formed. In the steady state the speed of the tube is equal to that of the crystallization front so that from an observer outside of the system the crystallization front is stationary. If the speed of the tube is continuously increased from extremely low values, say of the order of microns per second, the temperature of the crystallization front decreases from the thermodynamic melting point to values in the vicinity of the temperature where the maximum occurs in Fig. 1.

The mechanism of the crystallization appears to be the same as with spherulitic growth (no shear), but the appearance of the crystallized area is quite different. Since from the time when the tip of the glass tube leaves the oven and is cooled to a temperature, where nucleation occurs, no new nuclei are formed anywhere in front of these first nuclei, no spherulites appear in the field at some distance from the tip. One obtains an almost ideally oriented system in which the macromolecules are oriented perpendicularly to the direction of the movement of the tube.

If the speed of the glass tube is further increased, however, the crystallization front shifts to a temperature below the temperature of the maximum speed of crystal growth. From that moment nuclei are formed ahead of this front. As a consequence this front quickly disappears.

Fig. 4: Longitudinal section through sample of poly(ethylene oxide) crystallized at a heater temperature of 100° to 110° C, a cooler temperature of -5 to -10° C, a temperature gradient of 3° C/mm and a zone velocity of 15 µm/min, according to ref. 21. Right end quenched at room temperature.

In Fig. 4 results, as obtained by Lovinger and Gryte[21] for a polyethylene oxide, are shown. This picture is obtained from a longitudinal cut through a sample treated at a speed of 15 µm/min (oven temperature between 100 and 110° C). On the left side of the picture the oriented field is shown. The right side crystallized after a quench to room temperature. The advantage of polyethylene oxide for the present purpose is that it does not develop more than one crystal modification. Another situation is encountered with polypropylene. With this polymorphic polymer one obtains the interesting situation shown in Fig. 5, as taken from another paper by Lovinger, Chua and Gryte[22]. In order to obtain this picture the authors chose a speed at which the crystallization front of the faster crystallizing β-modification was stationary. On the left side of the picture this front is seen. It is an intriguing feature of the polypropylene crystallization that the slower crystallizing α-spherulites have a higher nucleation tendency[23]. As a consequence, their nucleation happens already ahead of the crystallization front of the β-crystals. As a consequence of their lower growth speed, the α-spherulites are encircled by the β-field and assume a teardrop-like shape. The magnitude of these sperulites depends on the distance from the crystallization front, where their nucleus is formed.

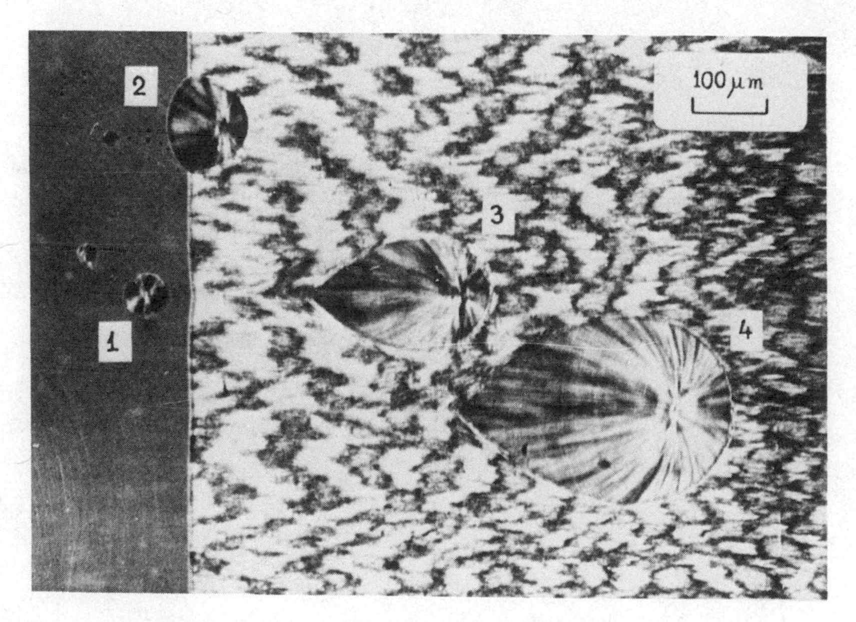

Fig. 5: Longitudinal section through a zone crystallized isotactic polypropylene. Growth front at the left side with α-spherulites nucleated ahead of this front, according to ref. 22.

A MODEL EXPERIMENT FOR A MORE GENERAL CASE

As this experiment has been designed only recently [10,11], the pertinent apparatus is depicted schematically in Fig. 6. The hot polymer melt (at temperature T_i) is extruded from a slit die onto the surface of a roller which can be cooled down to a temperature T_w well below the thermodynamic melting point of the polymer. In order to accomplish the shearing of the fluid layer deposited on the roller surface, a (concave) coaxial cylinder segment (segment of an outer viscometer cylinder) is placed on the roller in front of the slit die. The temperature of this segment is kept at the die temperature T_i with all experiments. The fluid layer moves freely into the gap between the roller and the said segment. This is the best way to satisfy the condition of continuity which means that the rate of flow must be the same in the slit die and in the gap. If this condition is not fulfilled, a surplus of the fluid, which

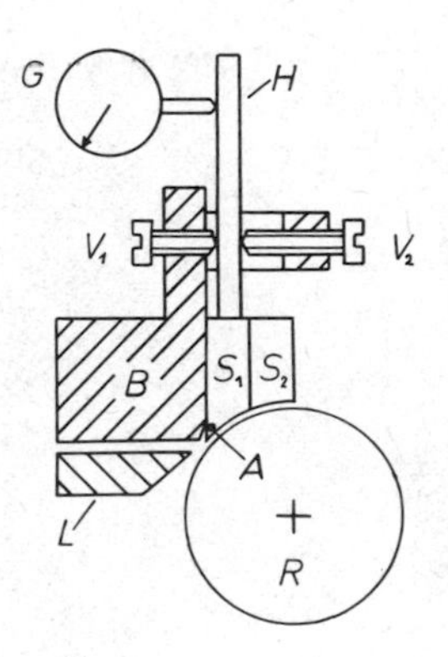

Fig. 6: Schematic drawing of the apparatus for the on-line measurement of the solidified layer thickness: L...lip, B...block (together forming the slit-die), r...roller, S_1... segment, S_2... additional segment, A...axle, H...handle, V_1, V_2 ... screws for adjustment of segment position, G... dial gauge.

cannot be carried away by the drag flow in the gap, piles up at the entrance or, contrariwise, air is entrained into the gap. This is a simple but sharp criterion in view of the proposed method to measure the thickness of the layer ultimately solidified on the surface of the roller at the end of the gap, in case the roller temperature is kept below the melting point of the polymer. In fact, the thickness of such a layer increases continuously from the entrance to the exit of the gap. Since with this solidification process the higher speed of the roller surface is transferred into the polymer, the effectiveness of the drag flow increases from the entrance to the exit, seemingly violating the continuity condition inside the gap. In reality the polymer film contracts on the hot side in a lateral direction. This can easily be observed, since the hot side is on the side of the said segment, which can be viewed from above. If the segment is properly tilted around an axle mounted at the edge of the entrance, the gap width can be reduced near the exit, until the contraction of the polymer film is undone. It can easily be shown for a solidification caused by a phase transition that the downward displacement of the edge at the exit is practically equal to the thickness of the ultimately solidified layer. In this way the coworkers of one of the present authors not only developed a convenient on-line method for the measurement of the thickness of the solidified layer. They were also able to give evidence that only the "structureless" layer, as found on a microscopic picture of a cut made from the ejected film, had really been formed during the period of combined action of shear and temperature gradient.

Another point of interest is found by the fact that in this system the solution for the energy equation is practically independent of that for the momentum equation, on condition that a solidified layer of a thickness of no more than one tenth of the gap width is considered. In fact, one always has

$$v_w \geq v(x_c) \geq v_w(1-x_c/D) \;, \qquad \ldots(3)$$

where v_w is the speed of the roller wall, $v(x_c)$ is the speed at distance x_c from this wall and D is the gap width at the exit. In other words, at a distance x_c from the roller wall the speed of the polymer is at most by a factor $(1-x_c/D)$ smaller than at the surface of the wall. The penetration concept, as formulated in Section 2, can be applied in these practical cases. The time t for the contact of the material under shear is given by

$$t = s/v_w \qquad \ldots(4)$$

where s is the length of the segment.

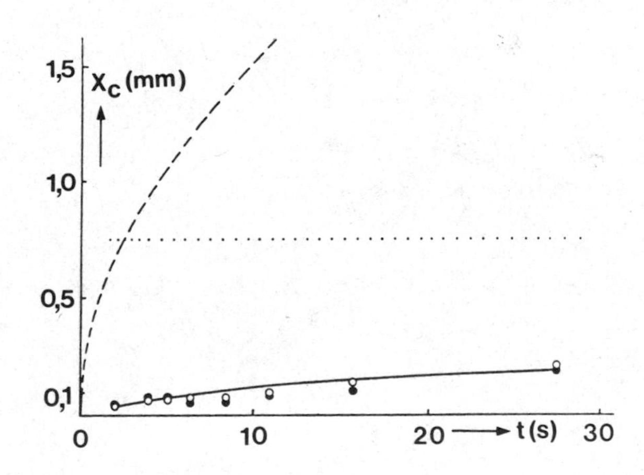

Fig. 7: Crystallized layer thickness vs. contact time for a polypropylene grade at $T_i = 245^o$ C, $T_m = 190^o$ C, $T_W = 90^o$ C and $\dot{\gamma} \propto 1/t\ s^{-1}$. Dashed line calculated according to recommendation a), dotted line steady state situation at D = 1 mm, open circles from on-line measurement, closed circles from microscopic observation[11].

For the thermal data of polypropylene[24] in Fig. 7 some simple theoretical and experimental results are gathered for $T_i = 245^\circ$ C and $T_w = 90^\circ$ C. The distance x_m, at which during the cooling process the thermodynamic melting point T_m is reached, is given as a function of contact time t according to the penetration theory with infinitely fast crystallization speed (boundary condition of eq.(2)). One may observe that this line has an infinite slope at zero time. By a horizontal dashed line the distance is indicated where T_m is located in the stationary situation of simple heat conduction through a polymer layer of thickness D, with the segment at T_i and the (stationary) roller at T_w (taking into account the different heat conductivities of melt and crystallized area). With moving roller this distance is reached at a sufficiently long contact time t (at a sufficient distance s from the entrance to the gap). The figure suggests that the more complicated mathematical solution for the transition from the initial penetration situation to the steady situation may be confined to the corner near the intersection of the two lines. The experimental points indicated by open circles are obtained at shear rates $\dot{\gamma} \propto 1/t$ with the aid of the discussed on-line method. The closed circles give microscopically determined thicknesses of the "structureless" layers, as obtained from cuts of the corresponding ejected films. The agreement of the two types of measurement is very reasonable. On the other hand, these experimental results are obviously in disaccord with the theoretical prediction on the basis of eq.(2).

In Fig. 8 a photographic picture is given of such a "structureless" layer together with the spherulitically crystallized part of the film, from which we now know that it crystallizes after the moment that the film has left the gap. From the sharp boundary between this layer and the interior of the film one may conclude that the nucleation of this layer happens at the roller surface, from where the crystallization front proceeds. From the initial slope of the line, as drawn through the origin and the experimental

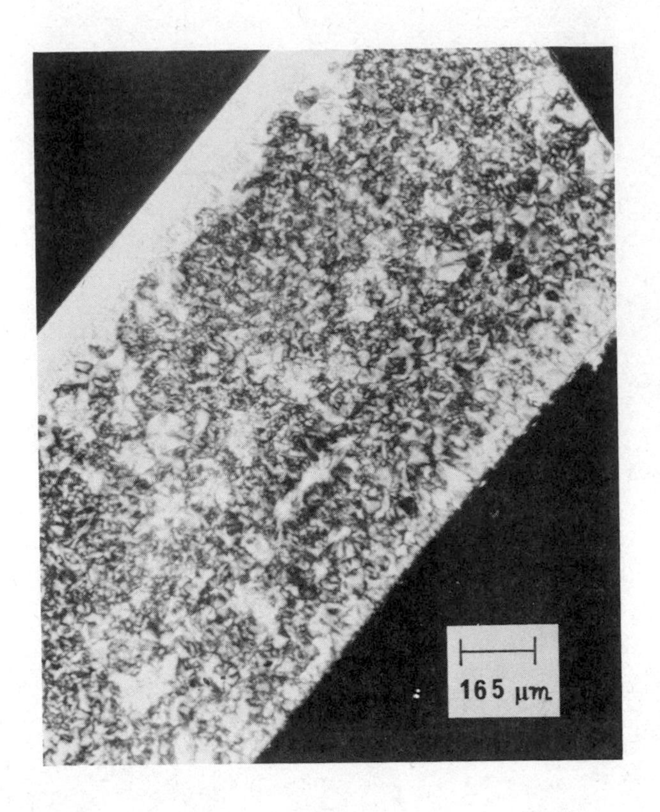

Fig. 8: Microphotograph of a 35 m microtom cut of ejected film in the 1,2-plane of flow (1...flow direction, 2...direction perpendicular to shearing planes). Enlargement 32:1, v_w = 5mm s^{-1} T_i = 245° C, T_w = 100°C.

points of Fig. 7, one derives a speed for the crystallizing front which is a factor 2000 larger than that of the β-spherulites in a quiescent melt[25]. This is in accord with the findings by Haas and Maxwell[15]. The comparison with the growth speed of β-spherulites is justified by the observation that the few interspersed bright β-spherulites of high birefringence[26] (the picture is taken between crossed polarizer and analyser) are interconnected with the row of bright crystallites apparently nucleated after the cessation of flow at the boundary of the "structureless" layer. In fact, this layer must be highly crystalline. Its density has been measured and found to correspond to 70% β-crystallinity[27]. Its degree of crystalline order is very high as well[28]. With heating its birefringence starts to relax at about 120° C[27]. This birefringence is positive with respect to the tangential direction of the roller surface pointing to the fact that the polypropylene molecules are aligned in this direction[29].

At this point relationships to the results obtained with zone crystallization should probably be stressed. As with zone crystallization the macromolecules in the crystallization front are oriented parallel to this front. Moreover, this front moves into a polymer melt, in which the temperature increases sharply in the direction of the propagation of the crystallization front. The argument for this statement will be given below. Its correctness anticipating one concocts that the probability must be small that nuclei are formed ahead of the crystallization front. This may be an explanation for the homogeneity of the "structureless" layer and for its sharp boundary. As soon as shear flow stops at the exit of the gap, the speed of growth of the layer is assumed to be reduced by a tremendous factor so that nucleation, leading to spherulite growth, gets its chance (compare the right side of Fig. 4).

Finally, the explanation for an assumed increase of crystallization temperature in the direction of the propagation of the crystallization front must be given. This explanation will elucidate once more the profound difference between heat diffusion controlled and nucleation rate controlled crystal growth. In case that eq. (2) would hold, the temperature of the crystallization front should invariably be constant and equal to T_m. With always finite speed of propagation, however, the polymer close to the roller surface will quickly be cooled down to the roller temperature before any crystallization has happened. With increasing contact time the crystallization front moves into the bulk of the polymer with a speed which depends on the varying crystallization temperature T_c at this front. The fact that this temperature T_c increases with distance x_c of the crystallization front from the roller surface, can easily be shown by means of the sketch given in Fig. 9.

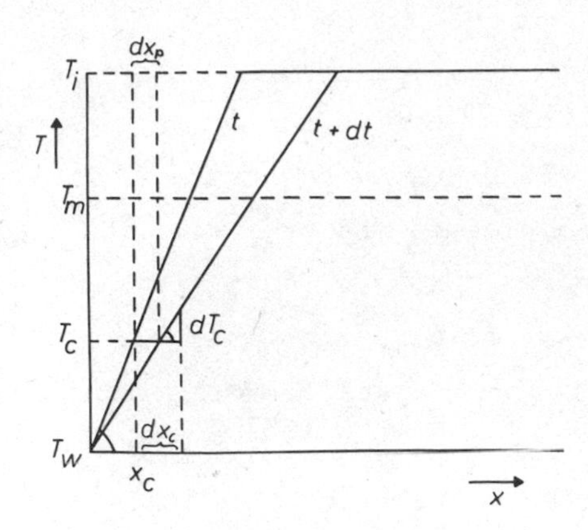

Fig.9: Illustration of the construction of a differential equation for the crystallization temperature T_c.

From Fig. 9 it follows that the differential equation for T_c as a function of x_c must approximately read:

$$\frac{x_c}{T_c - T_w} \frac{dT_c}{dx_c} = 1 - \frac{dx_p}{dx_c} , \qquad \ldots(5)$$

where T_i, T_m, T_c, T_w are the temperatures used earlier in this review, x_c is the distance of the crystallization front from the wall, dx_c is its propagation during the time interval dt and dx_p is the corresponding propagation of the penetration at level T_c. For simplicity the temperature profiles are indicated by straight lines. Since the differentials are infinitesimal, the two angles indicated by arcs are identical for our construction. At time zero dx_p is infinite at T_i, due to the fact that the pertinent initial slope in Fig. 7 is infinite. This means that dT_c/dx_c is negative and infinite. As soon as T_c is in the vicinity of T_w, the value of dx_p becomes much smaller than that of dx_c. This means that T_c starts to increase with x_c. The larger dx_c becomes, the smaller dx_p will be, since the released latent heat of crystallization slows down the penetration at level T_c. If T_w is chosen at a level where the speed of growth of the crystals is not too small, the initial slope of the line connecting the experimental points in Fig. 7 gives a reliable measure of the speed of crystal growth at temperature T_w and at the indicated shear rate. It goes without saying that this speed is nucleation rate controlled. A systematic exploitation of this method is in progress.

INJECTION MOULDING EXPERIENCES

Sandwich structures in injection moulded semicrystalline polypropylenes have been reported by Kantz, Newman und Stigale[30] as early as in 1972. These authors discerned three different layers or zones, as shown in Fig. 10 being taken from their paper. The outer layer or "skin" is "structureless", highly birefringent and shows a very high crystallite orientation as revealed by the X-ray wide angle diagram. The next zone or "shear zone" contains relatively small α-spherulites and some bright β-spherulites. According to the experiences reported in Section 5 there must be some doubt whether the name "shear zone" is correct. May be that this zone is the consequence of the packing stage. In fact, this zone contains clear traces of row nucleation. This type of nucleation has first been described by Keller and Machim[31] for crystallization under shear. Such a shear may accompany the additional filling necessary because of the setting of the core. Almost simultaneously Fitchmun and Mencik[29] have reported a similar but more extended investigation into the morphology of injection moulded polypropylene. These authors found even more layers. Their layer 3 shows a very high

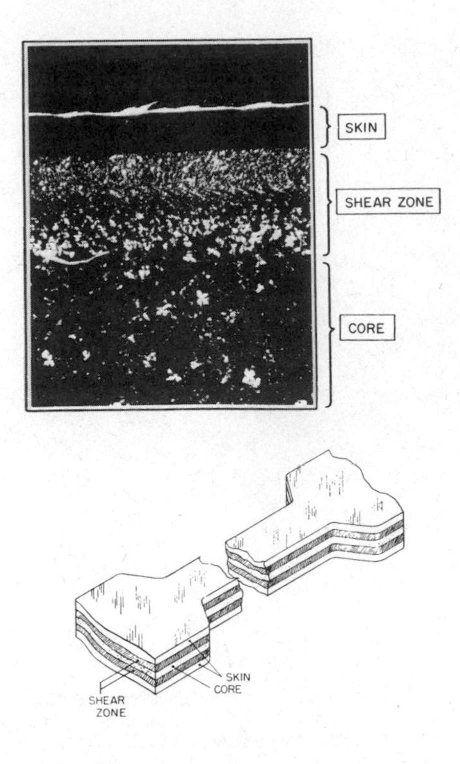

Fig. 10: Tensile bar (schematic) moulded from isotactic polypropylene. Microphotograph shows morphology of the three zones obtained from longitudinal microtom cut, according to ref. 30.

crystallite orientation[32], whereas their layer 4 and their core are practically unoriented. Similar but much less pronounced textures have been revealed in cuts from extruded tubes[35-38].

DISCUSSION

At the present stage of our knowledge it will certainly be difficult to give a conclusive interpretation for the morphology of injection moulded parts of semicrystalline polymers. However, some ways of describing the mould filling process can certainly be excluded. At the same time, some hope rises that we are not too far from understanding the basic features of this process.

In Section 2 arguments are brought forward for dismissing recommendation b) as erroneous. These arguments are based on the fact that by incorporating the latent heat of crystallization into a temperature dependent heat capacity one spoils every chance to describe the consequences of supercooling. On the basis of the experiences gathered in Sections 5 and 6 we can add another negative argument. A sharp boundary of the crystalline layer formed during the mould filling can never be explained on the basis of this concept.

On the other hand, if it is true that only the "structureless" layer or "skin" is formed during the primary filling process, there will be a good chance for the formulation of a mouldability criterion. In fact, during primary mould filling only this layer would reduce the passage for the polymer melt aiming at the flow front. Possibly, the growth of this layer will be calculable on the basis of the experiences gathered with the experiment sketched in Section 5. There is a chance that usual nucleation agents will have no influence on the growth of this layer, as this layer is clearly nucleated at the mould wall.

ACKNOWLEDGEMENTS

The authors are indebted to Mr. G. Krobath, Linz University, for providing some of his still unpublished results and to Dr. H. Dragaun, Vienna University of Technology and Plastics Technology Institute LKT-TGM, for his cooperation with respect to the X-ray measurements. The authors are grateful to American Chemical Society, Butterworth, Society of Plastics Engineers and John Wiley for granting permission to reproduce figures from their journals.

REFERENCES

1. H. Janeschitz-Kriegl, "Polymer Melt Rheology and Flow Birefringence", Polymers/Properties and Applications, Vol. 6, p.424, Springer 1983.

2. J.L.S. Wales, J. van Leeuwen and R. van der Vijgh, "Some Aspects of Orientation in Injection Moulded Objects", Polym.Eng.Sci. 12, 358(1972).
3. H. Janeschitz-Kriegl, "Injection Moulding of Plastics: Some Ideas about the Relationship between Mould Filling and Birefringence", Rheol.Acta 16, 327(1977).
4. H. Janeschitz-Kriegl, "Injection Moulding of Plastics II. Analytical Solution of Heat Transfer Problem", Rheol. Acta 18, 693(1979).
5. A.I. Isayev and C.A. Hieber, "Towards a Viscoelastic Modelling of the Injection Molding of Polymers", Rheol. Acta 19, 168(1980).
6. A.I. Isayev, C.A. Hieber and D.L. Crouthamel, "Orientation and Residual Stresses in the Injection Molding of Amorphous Polymers", SPE Techn. Papers 27, 110(1981).
7. J.L. White and W. Dietz, "Considerations of the Freezing-In of Flow-Induced Orientation in Polymer Melts by Vitrification with Application to Processing", J.Non-Newtonian Fluid Mech. 4, 299(1979).
8. H. van Wijngaarden, J.F. Dijksman and P. Wesseling, "Non-Isothermal Flow of a Molten Polymer in a Narrow Rectangular Cavity", J.Non-Newtonian Fluid Mech. 11, 175(1982).
9. J.D. Ferry, "Viscoelastic Properties of Polymers", John Wiley, 3rd Ed., New York 1980.
10. H. Janeschitz-Kriegl, M. Lipp, W. Roth and A. Schausberger, "A Model Experiment for Injection Moulding of Plastics: On-Line Measurement of Solidified Layer Thickness", Proceedings 3rd Austrian-Italian-Yugoslav Chem.Eng.Conf., Vol.II, p. 213, Graz 1982.
11. H. Janeschitz-Kriegl, G. Krobath, W. Roth and A. Schausberger, "On the Kinetics of Polymer Crystallization under Shear", European Polymer J., in press.
12. Z. Tadmor and C.G. Gogos, "Principles of Polymer Processing" pp. 288, 295, 607, John Wiley, New York 1979.
13. H.S. Carslaw und J.C. Jaeger, "Conduction of Heat in Solids", 2nd Ed., p. 282, Clarendon, Oxford 1959.
14. A. Gandica and J.H. Magill, "A Universal Relationship for the Crystallization Kinetics of Polymeric Materials", Polymer 13, 595(1972).
15. T.W. Haas and B. Maxwell, "Effects of Shear Stress on the Crystallization of Linear Polyethylene and Polybutene-1", Polym. Eng. Sci. 9, 225(1969).
16. A. Wereta and C.G. Gogos, "Crystallization Studies on Deformed Polybutene-1 Melts", Polym.Eng.Sci. 11, 19(1971).
17. K. Kobayashi and T.Nagasawa, "Crystallization of Sheared Polymer Melts", J.Macromol.Sci.-Phys., B4, 331(1970).
18. K. Katayama, Sh. Murakami and K. Kobayashi, "An Apparatus for Measuring Flow-Induced Crystallization of Polymers", Bull. Inst.Chem.Res., Kyoto Univ. 54, 81(1976).

19. Y. Fujiwara, "Über die Sphärolithstruktur von isotaktischem Polypropylen durch orientierte Kristallisation aus der Schmelze", Kolloid-Z. 226, 135(1968).
20. J.M.Crissman and E. Passaglia, "Mechanical Relaxation in Polyethylene Crystallized with Various Degrees of Lamellar Orientation", J.Res.Natl.Bur.Std. A 70, 225(1966).
21. A.J. Lovinger and C.C. Gryte, "The Morphology of Directionally Solidified Poly(ethylene Oxide) Spherulites", Macromolecules 9, 247(1976).
22. A.J. Lovinger, J.O. Chua and C.C. Gryte, "Studies on the α and β Forms of Isotactic Polypropylene by Crystallization in a Temperature Gradient", J.Polym.Sci., Phys.Ed., 15, 641(1977).
23. E.J. Addink and J. Beintema, "Polymorphism of Crystalline Polypropylene", Polymer 2, 185(1961).
24. D.W. van Krevelen, "Properties of Polymers", 2nd Ed., Elsevier, Amsterdam 1976.
25. H.J. Leugering and G. Kirsch, "Beeinflussung der Kristallstruktur von isotaktischem Polypropylen durch Kristallisation aus orientierten Schmelzen", Angew.Makromol.Chemie 33, 17(1973).
26. F.L. Binsbergen and B.G.M. de Lange, "Morphology of Polypropylene Crystallized from the Melt", Polymer 9, 23(1968).
27. Measurements by G. Krobath.
28. Measurements by H. Dragaun.
29. D.R. Fitchmun and Z. Mencik, "Morphology of Injection-Moulded Polypropylene", J. Polym.Sci., Phys.Ed., 11, 951(1973).
30. M.R. Kantz, H.D. Newman and F.H. Stigale, "The Skin-Core Morphology and Structure-Property Relationship in Injection-Moulded Polypropylene", J.Appl.Polym.Sci. 16, 1249(1972).
31. A. Keller and M.J. Machin, "Oriented Crystallization in Polymers", J. Macromol. Sci. (Phys.), B 1, 41(1967).
32. Z. Mencik and D.R. Fitchmun, "Texture of Injection-Molded Polypropylene", J.Polym.Sci., Phys.Ed., 11, 973(1973).
33. B. Wunderlich, "Macromolecular Physics", Vol. 2, p. 161, Academic Press, New York 1976.
34. D.W. van Krevelen, "Crystallinity of Polymers and the Means to Influence the Crystallization Process", Chimia 32, 279(1978).
35. H. Dragaun, H. Hubeny and H. Muschik, "Shear Induced β-Form Crystallization in Isotactic Polypropylene", J.Polym.Sci., Phys.Ed. 15, 1779(1977).
36. H. Dragaun, "Feinstrukturuntersuchungen am Schichtaufbau von extrudiertem Polypropylen", Prog.Colloid & Polym.Sci. 62, 59(1977).
37. H. Dragaun, H. Hubeny, H. Muschik und G. Detter, "Zur Technologie und Morphologie von Polypropylen-Druckrohren", Kunststoffe 65, 311(1975).
38. H. Muschik und H. Dragaun, "Zum Schereinfluß auf die Kristallisation und die mechanischen Eigenschaften von isotaktischem Polypropylen",Prog.Colloid & Polym.Sci. 66, 319(1979).

ON THE DEFORMATION OF POLYCARBONATE IN THE GLASSY STATE

K. Bielefeldt*, B.-J. Jungnickel,
and J. H. Wendorff

Deutsches Kunststoff-Institut
Schlossgartenstrasse 6 R
D - 6100 Darmstadt, West Germany

INTRODUCTION

Thermoplastics are usually processed in the molten state either by extrusion or by injection moulding techniques (1). Although an additional drawing step, for instance for a film or a fiber, may be performed in the solid state in order to achieve a preferential orientation of the chain molecules, this will happen at temperatures above the glass transition temperature. In principal it is of course also possible to process thermoplastics in the solid state at temperatures below the glass transition temperature of amorphous materials and at temperatures below the melting temperature of partially crystalline materials (2-7). Cold working is a well known technique in the area of metallurgy. It has, however, been applied to thermoplastics only in a limited number of special cases. The deformation to which the samples are subjected are well outside the range of the linear elastic response. This range is characterized by a spontaneous total recovery of the sample after the stress release. Cold working techniques include cold drawing of films and fibers, rolling of thicker samples as well as other forging and pressing techniques (8, 9).

Cold working offers several advantages over extrusion and injection moulding (10-13). The energy con-

*Visitor from the Technical University Zielona Gora, Poland

sumption, for instance, is lower. It works well even for very high molecular weight materials. Often it results in favorable changes of the bulk and surface mechanical properties, such as the impact strength or the surface hardness. The disadvantages of cold working include a limited range of deformation, which can be achieved, and a tendency towards shrinkage of cold-worked parts.

In principal all polymers, which behave ductile rather than brittle, should be appropriate for cold-working. The properties " ductile " and " brittle " are, however, not absolute quantities but depend on the structure of the material, on the mode of deformation, to which the material is subjected, on the temperature at which the deformation takes place and finally on the deformation rate (14).

The deformation of the thermoplastic material, independent of the mode of deformation, consists of several characteristic components. We are able to differentiate, for instance, between reversible and irreversible deformation, keeping in mind that the relative contributions of these components depend on the temperature as well as on the deformation time. The total deformation may also be decomposed into an elastic and a viscoelastic component, which in turn consists of an anelastic and a plastic component. In general the various components are related to each other. This relation can be expressed in terms of suitable models. In the case of materials composed of chain molecules we will often observe that the anelastic deformation is much larger than the truely plastic deformation.

The deformation components, introduced above, will take place also during cold-working. They may be attributed to elastic, anelastic and plastic deformation processes on a molecular and on a supermolecular scale. The exact nature of these processes, which determine the capability of a material to be cold-worked to a great extent, are still open to many questions. A few papers have been published which were concerned with uniaxial compression and rolling in the solid state (12, 15-17). From a technical point of view the main points of interest were the upper deformation, which can be achieved and the accuracy of the dimensions of technical parts obtained by cold-working techniques. In addition the changes induced in the mechanical properties were of interest.

It was the aim of our investigations to understand the relation a) between the different methods of cold working and the technical parameters chosen for this purpose, b) the molecular processes which are induced in the material and c) the changes of the bulk and surface properties of the material which are controlled by the local deformation processes. This paper is specifically concerned with the cold-working properties of amorphous polycarbonate, which is known to be ductile at temperatures far below the glass transition temperature. We studied the case of an uniaxial compression at room temperature.

EXPERIMENTAL

The experiments were performed on a bisphenyl-A polycarbonate (Makrolon 3100, Bayer AG), which is reported to be amorphous, in agreement with our X-ray and calorimetric investigations. Cylindrical samples of different heights were placed between two parallel plates of a press. The plates were carefully polished in order to reduce the friction. The cylindrical samples were compressed in the direction of the cylinder axis. The deformation was characterized in terms of the ratio λ of the height after (h) and (h_o) before the compression, $\lambda = h/h_o$. The deformation rate was about $\dot{\lambda} = \lambda^{-1}\ min^{-1} \approx {}^{o}2\ min^{-1}$.

The largest deformation λ_G which we were able to obtain, turned out to be a function of the ratio of the height to the diameter of the samples. We therefore chose the dimensions of the samples in such a way that the final height after cold compression was 4 mm whereas the diameter was 40 mm, independent of the total deformation. If the critical value of the deformation was surpassed, brittle fracture took place. The samples usually failed in a manner similar to an explosion of the material.

The cold worked samples were decomposed into microtome cuts along the radial, the tangential as well as along the normal direction of the cylinder. The cuts were taken from different positions in order to determine the homogeneity of the sample. The structural parameters, which we characterized, were the orientation, as determined by means of birefringence and X-ray flat camera studies and the density of the material, as obtained in a density gradient column.

Our investigations were furthermore concerned

with the relaxation of the sample in the stress free state after cold-working. We studied the spontaneous recovery of the elastic deformation component, of the anelastic component at room temperature as well as the recovery of the sample induced by a slow increase of the temperature up to temperatures above the glass transition temperature. In addition we determined the variations of the recovery stress of the compressed samples at constant height as the temperature was increased up to the range above the glass transition temperature. For this purpose we placed the samples between the parallel plates of the press, the distance between which was kept constant. The force acting on the plates due to the recovery of the samples was monitored as a function of the temperature and the original deformation.

RESULTS

Relaxation and retardation

The true stress strain curves were obtained both for the case of the uniaxial tensile as well as for the uniaxial compression at room temperature. For this purpose the experimentally obtained deformation forces were reduced to the true diameters of the samples for each deformation interval. They are characterized by a large elastic modulus (Fig. 1a). In the case of the compression the stress was found to decrease slightly above the yield point and to increase again with increasing compression for λ larger than 0.7, as the critical deformation is approached at which failure takes place. The critical values turned out to be dependent on the sample for experimental reasons. This observation is in agreement with quantitative results reported in the literature (18, 19). If the stress was allowed te relax at constant elongation (this being equivalent the limiting case $\dot{\lambda} \longrightarrow 0$), we observe a shift of the region at which yielding takes place to smaller values (Fig. 1b).

We found that the largest compression obtainable for polycarbonate is about $\lambda_G \approx 2$ for the case of uniaxial tension and about $\lambda_G \approx 0.13$ for the case of the uniaxial compression. The molecular deformation processes which occur in the material for these two types of deformation seem to differ, since the stress-strain curves differ for uniaxial tension and compression.

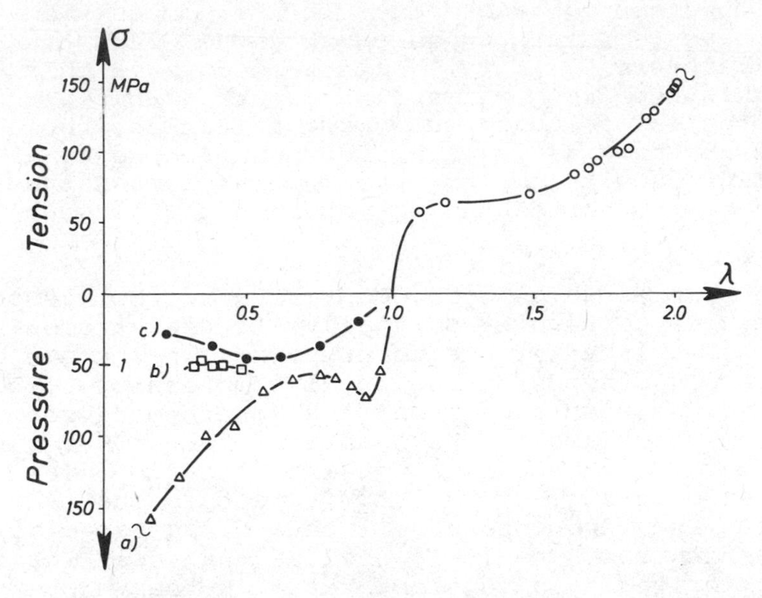

Fig. 1. Stress-strain curves of polycarbonate

(a) $\dot{\lambda} = \lambda^{-1}$ min^{-1} at 295 K, left scale of ordinate
(xxx = compression
ooo = tension)

(b) $\dot{\lambda} \rightarrow 0$ at 259 K right scale of ordinate

(c) $\dot{\lambda} = \lambda^{-1}$ min^{-1} at 432 K T_g + 7 K)

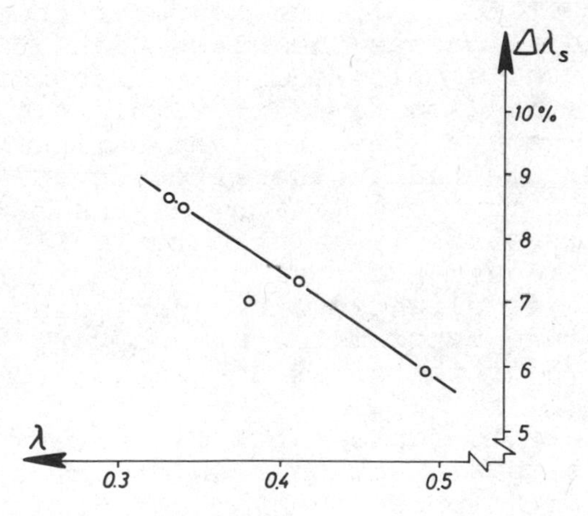

Fig. 2. Relative spontaneous elastic retardation of polycarbonate as a function of the compression

No strain hardening takes place if the compression is performed at temperatures surpassing the glass transition temperature (Fig. 1c). Failure occurs in this case within the flow regime. The upper value for the compression is of course much larger than for the compression in the glassy state. Furthermore the compression stress is lower by two orders of magnitude as compared to the corresponding value in the glassy state.

The samples which were compressed at room temperature are able to recover partially, if the stress is removed directly after the compression. The extent of this recovery depends on the total compression to which the sample had been subjected. The spontaneous recovery, as characterized by the quantity $\Delta\lambda_s = (h' - h)/h_0$ is a linear function of the total deformation. This is demonstrated in Fig. 2. The requirement is that the sample had been compressed well beyond the range of linear elastic response. It is obvious that cold working is characterized by a constant ratio of the recoverable versus the total deformation.

Besides the elastic recovery, which takes place within a time scale of 1 s, no anelastic recovery was observed at room temperature. A temperature increase, on the other hand, resulted in an additional partial recovery within a time interval of several hours (Fig. 3). The extent of this recovery depends on the temperature. It is small at temperatures far below the glass transition temperature. If the temperature of the compressed samples surpasses the glass transition temperature, however, the sample is able to recover totally within a short time. In fact the sample displays exactly its original shape after recovery. This behavior has already been documented in the literature (15). Our studies indicate that the maximum of the relaxation time spectrum is strongly shifted to smaller time scales with increasing temperature. The relaxation strength increases slightly with increasing temperature in the temperature range below T_g and very strongly in the neighborhood of the glass transition temperature.

We were able to represent the recovery of the compressed samples as a function both of the temperature and of the total compression by means of a master curve. This required an appropriate reduction of the recovery data. Obviously the processes which are responsible for the recovery on a molecular as well as on a supermolecular scale are independent of the total deformation.

Their time dependence may vary however.

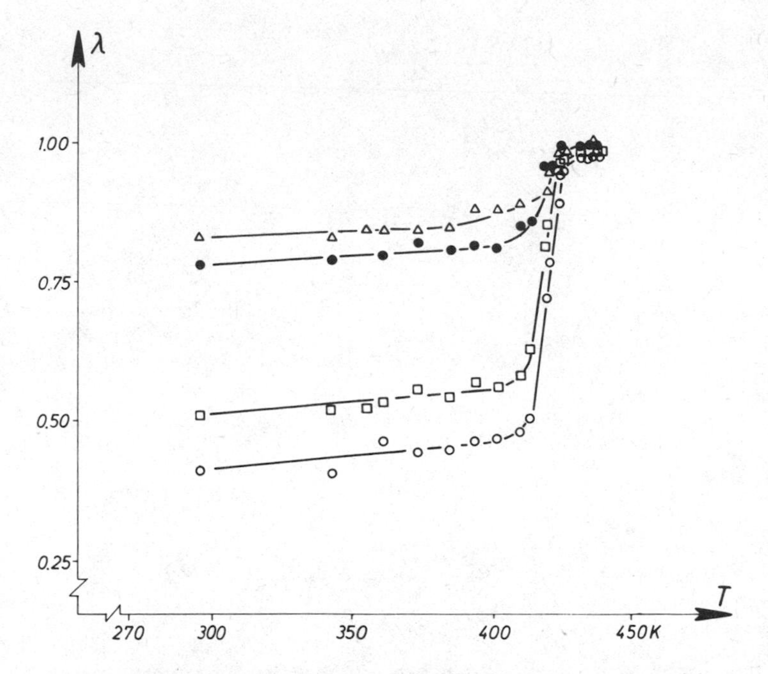

Fig. 3. Equilibrium retardation as a function of the temperature

ooo $\lambda' = \lambda_0 + \Delta\lambda_s = 0.41$

□□□ $\lambda' = 0.51$

xxx $\lambda' = 0.78$

△△△ $\lambda' = 0.83$

The relaxation stresses, which were determined as described earlier, are nearly independent of the temperature in the temperature range below 90° C. They increase very steeply with increasing temperature in the neighborhood of the glass transition temperature. The largest value σ_M is obtained at temperatures above T_g. The relaxation stresses decrease again with increasing temperature at still higher temperatures. It is obvious that this happens because of the onset of flow processes. The largest values of the relaxation stresses σ_M obtained in this manner have to be looked upon as the lower limit of the true relaxation stresses. The dependence of this stress on the total compression is displayed in Fig. 4. Here the total compression is represented by the quantity $\lambda^2 - 1/\lambda$, for reasons which will become apparent later on. The re-

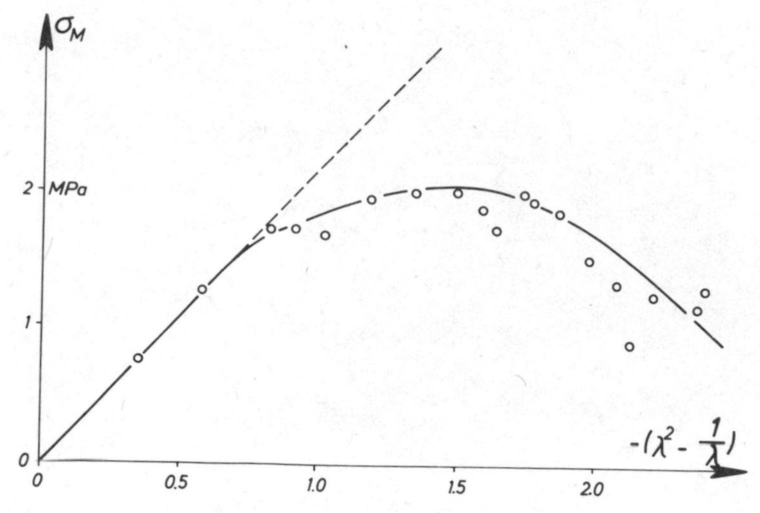

Fig. 4. Maximal retardation stress σ_M as a function of the compression

covery stress increases linearly with increasing values of $\lambda^2 - 1/\lambda$. Deviations from this behavior occur at larger compressions, probably due again to the onset of flow processes. In fact, a sample cannot sustain recovery stresses which are larger than the yield stress. A comparison of Figs. 1 and 4 reveals at once that the recovery stress and the yield stress are of the same order of magnitude. Deviations may occur because of the different rates of flow in the two cases. The observation that the temperature T_M at which the maximum value of the relaxation stress is observed increases with increasing total compression (Fig. 5) is in qualitative agreement with results of Arshakov et al. (20). They studied the recovery of the different thermoplastics as a function of the temperature and of the deformation. They attributed this effect to varying contribution of different relaxation processes to the total recovery.

Our studies on the relaxation behavior of different parts of macroscopical samples revealed that the compression occurs homogeneously throughout the sample.

Studies on the structure of the deformed samples

The deformation of the material at room temperature did not result in changes of the segmental

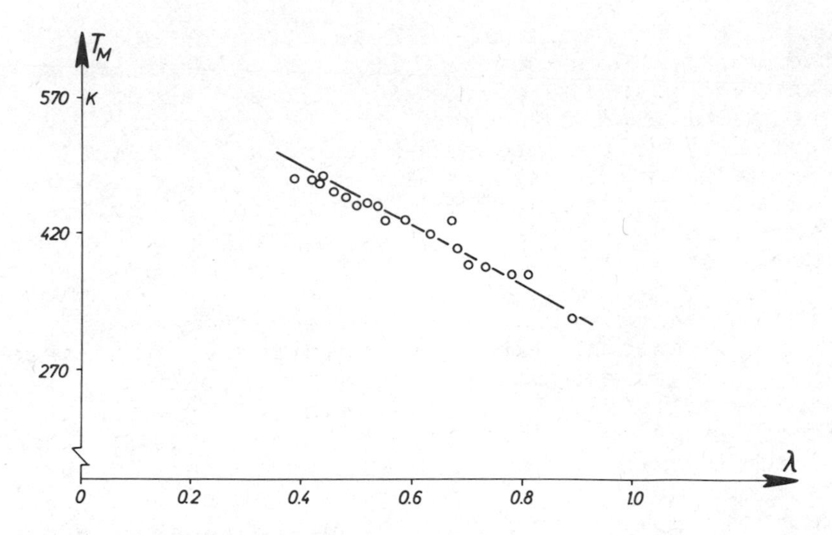

Fig. 5. Temperature at which the relaxation stress obtains its largest value, as a function of the compression

distance distribution as becomes obvious from X-ray scattering studies, dilatometric and calorimetric investigations. This includes changes of the degree of crystallinity due to cold-working. Small angle X-ray studies on thermal density fluctuations and their dependence on the deformation were performed in order to look for possible changes of the thermodynamic state of the glassy polymer, which might happen due to deformation-induced variations of the free volume. No indications were obtained for changes in thermal density fluctuation and thus for the thermodynamic state (21).

Changes resulting from cold working were, however, observed with respect to the density of the material. The density was found to decrease up to a deformation of $\lambda = 0.85$ and to increase continuously with increasing deformation above this value. It is obvious that two contributions having different effects on the density occur simultaneously. At smaller values of the deformation apparently cavities are introduced into the material. Their existence was proved by means of electronmicroscopical studies. Their sizes turned out to be of the order of several um. Their number is small enough in order to have a neglible effect on the total mechanical properties. This point will be discussed in more detail below. Small angle X-ray studies showed that cold working does not result in the formation of small microvoids in the size range between about 1.0 nm and 100.0 nm. The formation of such microvoids is a characteristic failure process in partially crystalline or amorphous materials if subjected to a tensile deformation (22). In addition to the formation of cavities, cold-working seems to lead to a compression of the matrix of polycarbonate.

X-ray wide angle flat camera diagrams revealed the existence of a planar texture in cold worked polycarbonate, the plane of compression being the texture plane. The chain segments are located within this plane, which is oriented perpendicular to the direction of the uniaxial compression. The chain segments are distributed isotropically in this plane. As far as we know the texture is homogeneous throughout the sample.

The dependence of the birefringence Δn on the deformation parameter $\lambda^2 - 1/\lambda$ is shown in Fig. 6. A li-

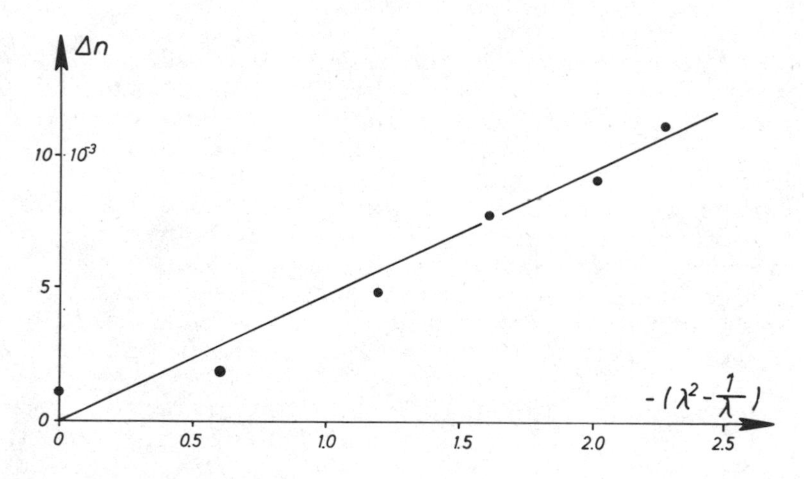

Fig. 6. Birefringence as a function of the compression

near relation is observed. The total orientation and consequently the total birefringence, which is obtainable, is small compared with the values corresponding to a fully oriented state of the chain molecules (Δn_0 = 0.145).

DISCUSSIONS

The concept which will be used in order to account for the experimental results on the cold working of polycarbonate is based on the following observations:

(a) The recovery is total at temperature above T_g
(b) The relaxation stress is a linear function of the deformation, as characterized by $\lambda^2 - 1/\lambda$, for small values of the deformation
(c) The orientation $f_\theta = \Delta n / \Delta n_0$, occuring as a consequence of the deformation is also a linear function of the deformation parameter $\lambda^2 - 1/\lambda$
(d) No crystallization takes place during cold working

These observations suggest that the compression of polycarbonate may be described in terms of the deformation of an entropy-elastic Gaussian network (23). The experimental results on the recovery behavior and on the orientation yield two sets of data, which allow the calculation of the relevant parameters of the Gaussian network. The birefringence is given by (23)

$$\Delta n = \frac{N}{15 \varepsilon_0} \frac{(n^2+2)^2}{6n} \Delta\alpha_s (\lambda^2 - 1/\lambda)$$

where N is the crosslink density, i.e. the number of network chains per cm , and $\Delta \alpha_s$ the anisotropy of polarisation of a statistical chain element. The stress is given by

$$\sigma = NkT(\lambda^2 - 1/\lambda)$$

where σ is the stress necessary to sustain the compression λ. The factor NkT may be extracted from the linear part of the curve displayed in Fig. 4. The value obtained is $N = 4.3 \cdot 10^{26}\,m^{-3}$ and thus $\Delta \alpha_s = 6.6 \cdot 10^{-40}\,Fm^2$. The value obtained for the chain element density refers of course to the neighborhood of the glass transition temperature. The value of the anisotropy of the polarisibility, on the other hand, was determined from the relationship between Δn and λ at room temperature, assuming N to be temperature invariant, to a first approximation.

From the data we have to conclude that the chain elements between crosslinks are characterized by a mass of 1680 g Mol^{-1} which corresponds to about seven monomer units per chain element on the avergage, and thus to a contour length of about 8 nm. A comparison between the anisotropy of the polarizibility of the monomer units and of the statistical chain element shows them to be identical. The values obtained so far for the network model are surprising but not unusual. Ward for instance reported a value of about 4230 $gMol^{-1}$ for the network chain (which equals about 25 monomer units) and 3.3 monomer units per statistical chain element for the case of poly(ethylene terephthalate) drawn at 80° C (24).

The network discussed so far will probably restrict the upper deformation, if no disentanglement processes take place.

The rather large crosslink density can hardly be attributed directly to structural features, such as chemical crosslinks etc. In addition, the large crosslink density does not seem to be dependent on the particular chemical structure of the chains, since similar effects have been reported for polystyrene, poly(methylmethacrylate) and polyvinylchloride.

A formal description of the kinetics of relaxation and its temperature dependence can be given on the basis of a model by Bonart and coworkers (25). In this model the real structure is represented by a coupled system of stress elements and fixation elements. The stress elements are responsible for the elastic recovery of the deformed material, whereas the fixation elements control the temperature at which the stress elements are able to relax. The stress elements are characterized by different length, elongation and coil constants. In the case of cold working we have to conclude in terms of this model, that a considerable part of the fixation elements are uneffective already at room temperature. This allows a partial recovery of the deformation. In addition, we can conclude, that the structure of the network, consisting of stress elements and fixation elements is independent of the total deformation. This gives rise to a master curve representation of the relaxation behavior and to a deformation and recovery behavior similar to that of an ideal entropy elastic rubber.

ACKNOWLEDGEMENTS

The authors gratefully acknowledge the financial support of the Bavarian Department of Commerce and Transport. One of the authors (K. Bielefeldt) expresses his gratefulness for award of a fellowship of the Alexander von Humboldt - Stiftung.

REFERENCES

1. E. C. Bernhardt, "Processing of Thermoplastic Materials", Reinhold Publ. Corp., New York (1959)
2. L. J. Broutman and S. Kalpakjian, SPE Techn. Pap. 25:46 (1969)
3. H. Bühler and E. v. Finkenstein, WTZ Ind. Fert. 59:569 (1969)
4. P. M. Coffman, SPE Techn. Pap. 25:50 (1969)
5. T. Maeda, Jap. Plast. Age 10:7 (1972)
6. K. M. Kulkarin, Polym. Eng. Sci. 19:474 (1979)
7. W. Ziegler and K. H. Scholl, Kunststoffe 63:19 (1973)
8. E. v. Finkenstein, Metall 24:340 (1970)
9. H. Ebeneth, K. Heidenreich and H. Röhr, Kunststoffe 60:15 (1970)
10. P. M. Coffman, Mat. Eng. 67:74 (1968
11. H. Käufer and G. Arnold, Kunststoffe 67:457 (1977)
12. J. Baldrian, Faserforsch. Textilt. 26:255 (1975)
13. B. S. Thankar, L. J. Broutman and S. Kalpakjian, SPE Techn. Pap. 24: 295 (1978)
14. I. M. Ward, "Mechanical Properties of Solid Polymers", Wiley Interscience, London (1971)
15. K. C. Rusch, Polym. Eng. Sci. 12:288 (1972)
16. M. J. Miles and N. J. Mills, J. Mat. Sci. 10:2092 (1975)
17. N. D. Crisp, M. J. Miles and N. J. Mills, J. Mat. Sci. 12:1625 (1977)
18. G. Titomanlio and G. Rizzo, Polymer 19:1335 (1978)
19. G. Peilstöcker, Struktur und Eigenschaften aromatischer Polycarbonate, in: "Kunststoff-Handbuch, Bd. VIII: Polyester" (Editors: R. Vieweg and L. Goerden), Hanser-Verlag, München (1973)
20. S. A. Arzhakov, N. F. Bakeyev and V. A. Kabanov, Polymer Sci. USSR 15:1296 (1973)
21. J. H. Wendorff and E. W. Fischer, Kolloid-Z./Z. Polymere 251:876 (1973)
22. S. N. Zhurkov and V. S. Kuksenko, Int. J. Fract. 11:629 (1975)
23. L. R. G. Treolar, "The Physics of Rubber Elasticity", Clarendon Press, Oxford (1975)
24. F. Rietsch, R. A. Duckett and J. M. Ward: Polymer 20:1133 (1979)
25. R. Bonart, L. Morbitzer and F. Schultze-Gebhardt, Kolloid-Z./Z. Polymere 251:1015 (1973)

THE DEVELPMENT OF SUPER-STRONG POLYMERS IN PLANAR DIRECTIONS (Abstract)

Anagnostis E. Zachariades

IBM Research Division

San Jose, California, U.S.A.

Over the last decade there has been considerable interest toward the development of polymer morphologies with high tensile modulus and strength by inducing a high degree of chain extension and orientation using various techniques. In general, the high modulus products from all the above techniques are uniaxially drawn filaments or films which fibrillate readily because of their lateral weakness. As a result, the development of polymer compositions with ultra-high mechanical performance in more than one direction has been achieved only by the construction of fiber reinforced composite structures.

In recent studies in our laboratory, we have investigated the formation of homogeneous polymer compositions with ultra-high mechanical performance in more than one direction. The polymer is processed at a temperature below or near its isotropic melting point under curvilinear flow conditions which are generated by the combined effects of compression and rotation. The morphologies obtained are multiaxially oriented and have high planar tensile properties and impact strength. In this paper results of our studies with thermoplastic and thermotropic polymers using simple torsional flow are discussed.

RELATIONSHIP BETWEEN PROCESSING INDUCED MORPHOLOGY, MOLECULAR ORIENTATIONS AND MECHANICAL BEHAVIOUR OF AMORPHOUS POLYMERS

K. P. Großkurth

Institut für Baustoffe, Massivbau und Brandschutz
Technische Universität Braunschweig,Hopfengarten 20
D-33 Braunschweig, Federal Republic of Germany

INTRODUCTION

Although HOUWINK [1] suggested already in 1936 the existence of colloidal superstructures in amorphous polymers newly developed methods of electron microscopic preparation techniques allow in the recent years for the first time to obtain direct evidence of well defined morphologies in many high polymers. So for example it was possible to show in the case of atactic polystyrene [2], styrene/ acrylonitrile copolymer [3], polyethylene terephthalate [4], polycarbonate [4] and polymethyl methacrylate [5] that the isotropic and slightly oriented states are characterized by globular network structures; the highly oriented materials possess a line by line structure. Geometrical shape and regularity of the morphology only depend on the degree of molecular orientations, but not on the way producing themselves by varying the hotstretching parameters. Therefore similar relationships can be assumed between morphological changes and processing induced molecular orientations in injection moulded polymers.

EXPERIMENTAL

The material used for the experiments was commercial bulk polystyrene with a molecular weight of $\overline{M}_n$ = 111 000. In order to mould specimens with significant different molecular orientations in the cross section a low injection temperature of only 180°C and a mould cavity wall temperature of 40°C have been chosen. The dimensions of the longitudinally moulded prismatic specimen were 120 x 15 x 4 mm^3.

To get informations about the processing induced internal structure due to the wanted distance to the sample surface thin layers have been cut by a microtome knife. Now the morphology of the remained uncovered internal surface could be identified by means of electron microscopy and one-step-replica method after oxygen ion etching. The selective etching process is based on different degradation rates of regions with locally differing densities and/or differing packing types of the chain molecules.

Fig. 1 shows the scheme of the used etching apparatus type GEA 004- S. The bulk specimen is treated on the watercooled sample plate. Oxygen ions are produced in the high-frequency field (27.12 MHz) of the annular electrode at a partial pressure of about 10^{-4} ... 10^{-3} mbar. Superimposed low dc voltages (max. 500 V) between the electrode at the top of the recipient and the sample plate accelerate the oxygen ions onto the sample surface. In addition to the selective chemical etching effect of the activated oxygen a mechanical microerosion process based on the kinetic energy of the accelerated ions is degradating the polymer surface. By means of mass

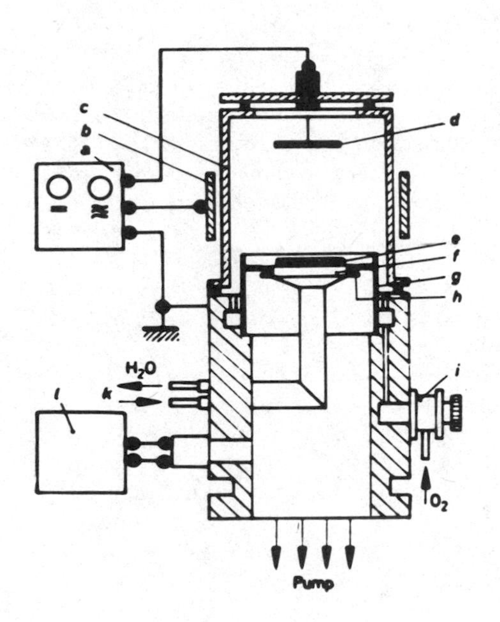

a electronic control unit

b Hf annular electrode

c glasrecipient

d dc electrode

e sample

f watercooled sample plate and opposite electrode

g cylinder

h deflector

i proportionating valve

k cooling water

l vacuum gauge

Fig. 1 : Scheme of the etching apparatus GEA 004- S

spectroscopy it was possible to point out that the degradation products mostly are gaseous and removed over the diffusion or turbomolecular pump [6]. The etching time of normally some hours strongly depends on the plastic deformation during microtome pretreatment in the surface layers. Otherwise artefacts would be expected.

RESULTS

Fig. 2 represents a characteristic series of processing induced supermolecular structures in moulded polystyrene at the sprue distant end of the specimen. Qualitatively, however, the demonstrated morphological spectrum is also typical for the other sample sections. The depth values include the microtome abrasion and etching degradation.

For the investigation of the direct surface structure a microtome cutting was not necessary. Thus, the original sample surface was etched for only 30 minutes. In a depth of nearly 1 µm a superlattice structure ordered perpendicular to the injection direction can be observed (Fig. 2a). With increasing distance to the surface the morphology at first becomes globular (Fig. 2b). An alignment of the globules normal to the injection direction again is clearly visible. The morphological regularity reaches its maximum in a depth of 20 ... 40 µm, where a line by line structure is realized (Fig. 2c). Then the structure becomes more and more irregularly. At a distance of 300 µm to the surface the change from the fine globular morphology into a coarse globular one is indicated (Fig. 2d); in a depth of 500 µm the change has been finished already (Fig. 2e) and now the structure remains nearly constant (Fig. 2f).

DISCUSSION

It is well known, that the injection moulding process effects molecular orientation distributions in the cross section. Therefore the observation of rather similar superstructures in polystyrene subjected to uniaxial hotstretching is a significant factor for interpreting these phenomena.

The morphology of the isotropic and oriented state also was detected by means of electron microscopy and the same preparation techniques [2]. Fig. 3a shows the typical structure of the isotropic material. It consists of an irregularly formed network similar to grain boundaries with a mesh size of some µm. After stretching of only 25% a globular morphology can be observed (Fig. 3b). An orientation of the globules in the transverse direction is indicated. With further drawing the globules become smaller and change after stretching of 50% to a line by line structure (Fig. 3c). The lines are ordered perpendicular to the direction of molecular orientations. Their regularity increases with the stretching ratio (Fig. 3d).

The observed superstructures are based on locally discontinuities of deformation processes, which produce density differences [3],[4]. Previous conclusions obtained from structural studies on amorphous polystyrene [2] indicate that the line boundaries possess a higher density than the intermediate material. So a nearly parallel alignment of the macromolecules within the boundaries was suggested.

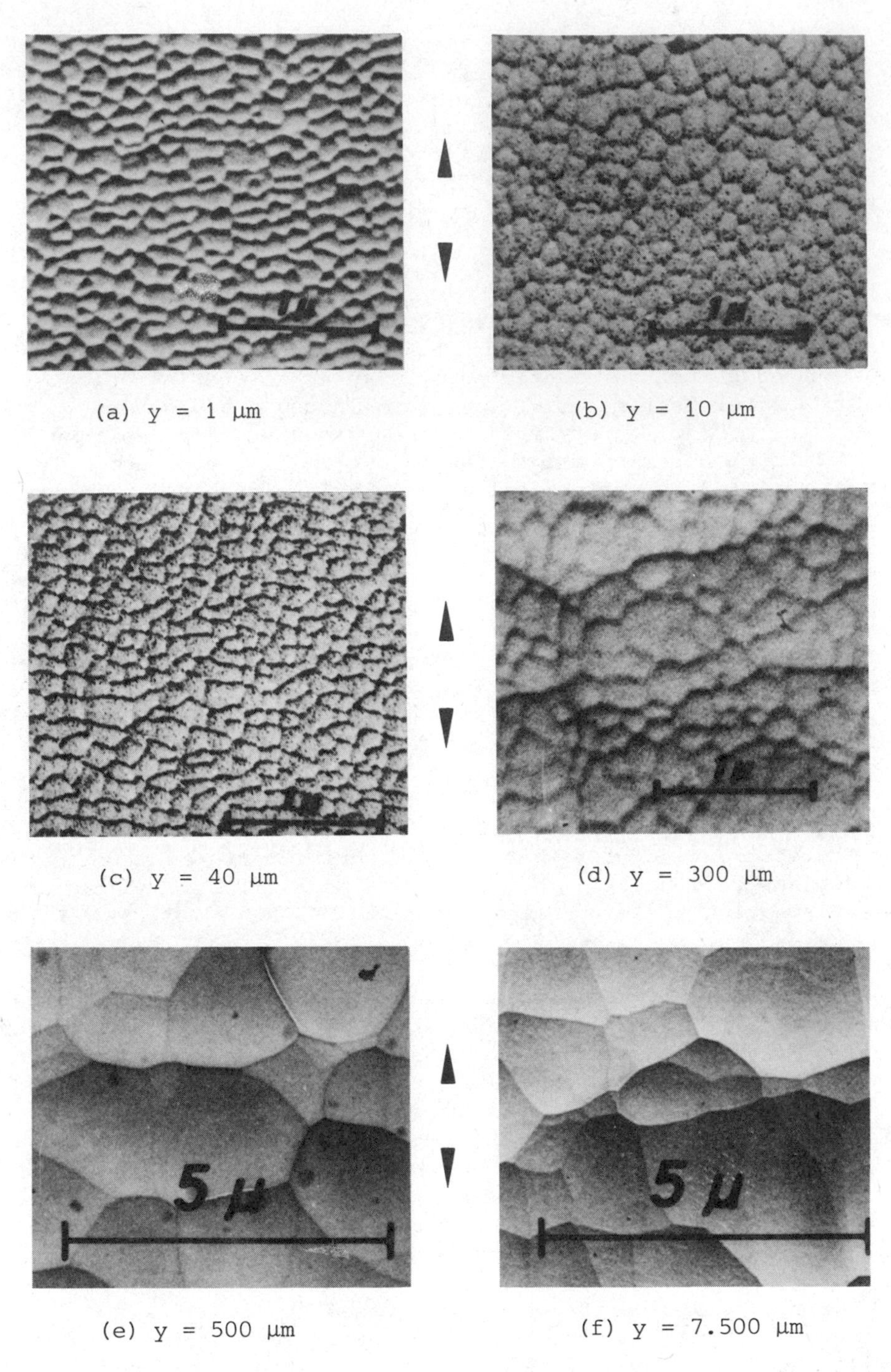

(a) y = 1 µm (b) y = 10 µm

(c) y = 40 µm (d) y = 300 µm

(e) y = 500 µm (f) y = 7.500 µm

Fig. 2: Morphology of moulded PS. Injection temperature 180°C. Depth y. Arrow marks the direction of injection.

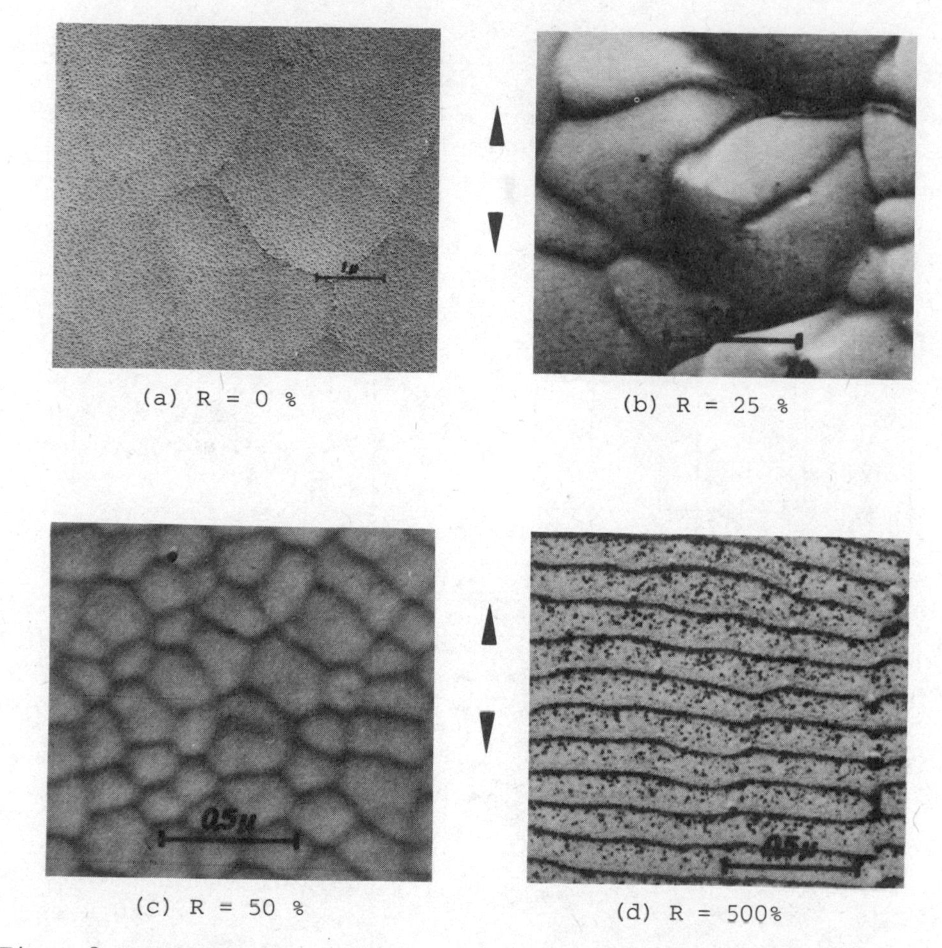

(a) R = 0 % (b) R = 25 %

(c) R = 50 % (d) R = 500%

Fig. 3: Morphology of isotropic and hotstretched PS. Stretching parameters: time 15 s , temperature 115°C, degree R. Arrow marks the direction of stretching.

By quantitative analysis of the electron micrographs mean spacing of structure details - boundaries of globules or lines - and corresponding standard deviation were determined. Both parameters characterize geometrical shape and regularity of the supermolecular structures. In Fig. 4 the standard deviation as a size for the reciprocal regularity is plotted against the chosen stretching conditions. The three-dimensional diagram illustrates that the formation of structure becomes more regularly with increasing degree of orientation and decreasing temperature of stretching. The most significant changes result from drawing up to about 200% and stretching temperatures just beyond the glass transition.

The relationships between stretching parameters and mean spacing of structure details are rather similar. A minimum line mean spacing of about 100 nm has been found. It is nearly independent of further drawing.

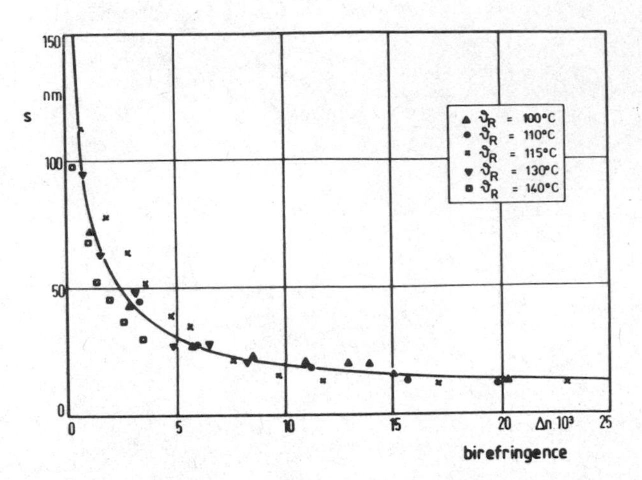

Fig. 4 :

Standard deviation s of structure details mean spacing as function of stretching degree R and temperature. Stretching time 15 s .

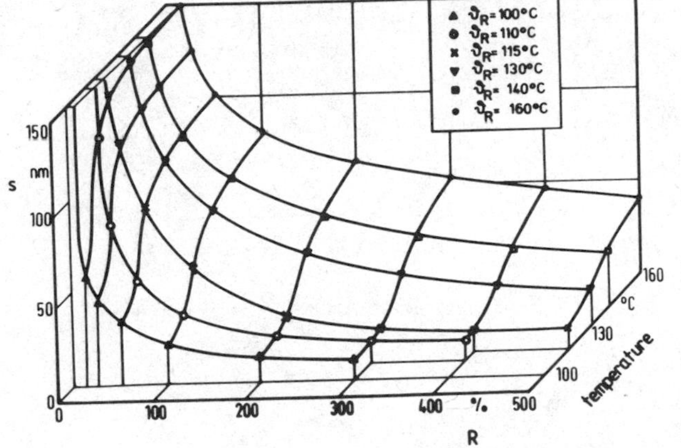

Fig. 5 :

Like fig. 4, plotted against birefringence Δn.

Using the birefringence as a measure for molecular orientation instead of the stretching parameters the standard deviation reduces to a very simple function as shown in Fig. 5. Thus, it must be pointed out, that geometrical shape and regularity of the morphology only depend on the degree of molecular orientations, but not on the way producing themselves.

This is why the morphological results found at the moulded material readily can be understood. Frozen-in molecular orientations produced by melt shearing are the reason for changes in morphology. Owing to the strongly oriented chain molecules the area near the surface possesses a well-defined line by line structure; contrary to this region internal a globular structure due to a low degree of molecular orientation is evident.

A quantitative analysis comparison of the processing induced and hotstretching effected morphology now is suitable to detect the local degree of molecular orientation in microregions of mouldings. The results plotted in Fig. 6 show extremely differing orientations in the cross section of moulded polystyrene. These conclusions are of high technical significance, because the local mechanical and fracture behaviour mainly depends on the orientation distribution function in the moulding section.

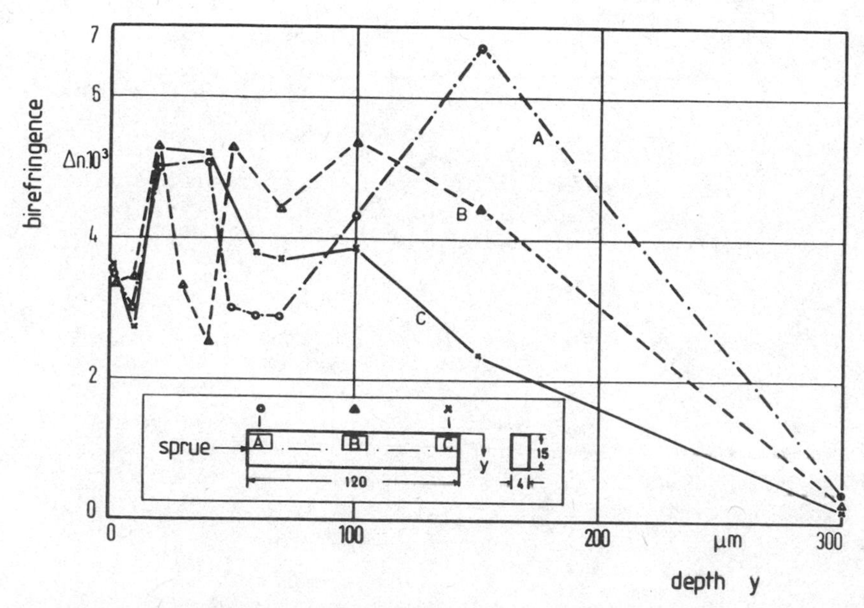

Fig. 6 : Birefringence calculated from the morphological data as a function of depth. Moulded PS. Injection temperature 180 °C.

The connections between morphology, mechanical properties and molecular orientations, as measured by birefringence, quantitatively can be studied on hotstretched samples. As shown in Fig. 7 tensile stress (σ_B) and strain at maximum load (ε_B) reach their greatest values at the beginning of the line by line structure area and then remain approximately constant. Better connections exist between tensile impact strength (a_{zn}) and the structural parameters mean line concentration $1/d_z$ and regularity 1/s. Obviously morphological defects are representing internal notches or weak points and therefore effect an enhanced decrease of tensile impact strength in the area of lower molecular orientations.

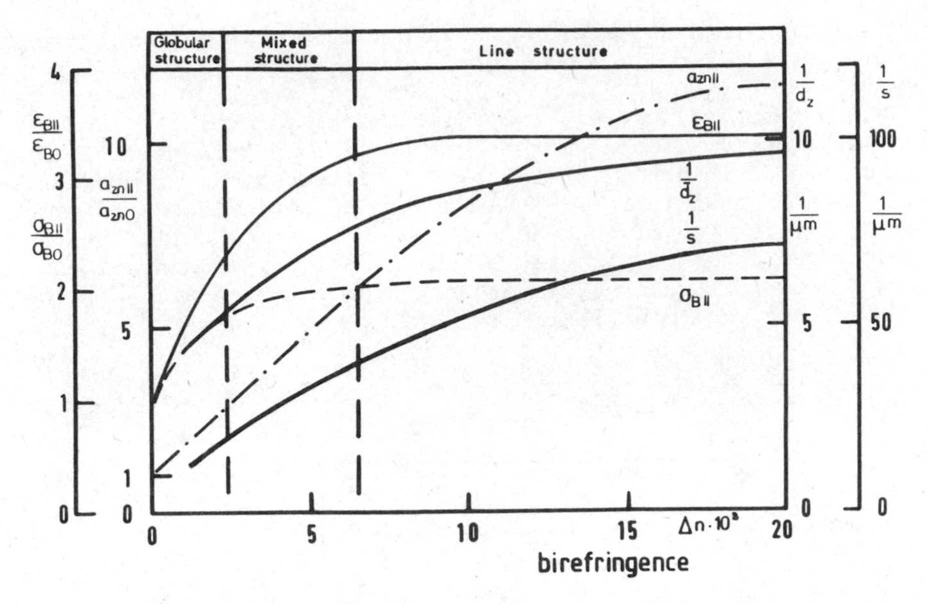

Fig. 7 : Effect of molecular orientations on morphology and mechanical properties of hotstretched PS in the longitudinal direction

With respect to these results in the case of injection moulded polymeric material the local mechanical and fracture behaviour strongly depends on the orientation distribution function in the moulding section. In the injection direction both the tensile and impact strength increase with the degree of molecular orientation, in the case of atactic polystyrene the ductility arises too. Therefore mouldings mainly craze internally while the highly oriented material near the surface remains still uncrazed.

REFERENCES

[1] Houwink, R., Trans. Farad. Soc. 32:122 (1936)
[2] Großkurth, K.P., Colloid & Polymer Sci. 255:120 (1977)
[3] Großkurth, K.P.; Gummi Asbest Kunststoffe 25:1159 (1972)
[4] Kämpf, G. and H. Orth, J. Macromol. Sci. - Phys. B 11 (2):151 (1975)
[5] Großkurth, K.P., Progr. Colloid & Polymer Sci. 66:281 (1979)
[6] Großkurth, K.P., Gummi Asbest Kunststoffe 30:848 (1977)

EFFECTS OF PROCESSING ON THE STRUCTURE AND PROPERTIES OF PVC

Björn Terselius and Jan-Fredrik Jansson

Royal Institute of Technology
Department of Polymer Technology
S-100 44 Stockholm, Sweden

ABSTRACT

This work includes an extensive study of the effect of the gelation or fusion stage of processing on the ultimate mechanical properties of PVC pipes by variation of the mass temperature at extrusion. Moreover, the influence of gelation on the supermolecular structure and on some physical properties was studied.

INTRODUCTION

The sensitivity to thermal degradation in combination with enhanced melt viscosity due to a low, but significant amount of crystallinity makes processing of PVC difficult. In addition, virgin PVC is built up as a particle hierarchy (1) difficult to break down in a controlled way adapted to the combination of properties wanted. Balancing the properties requires detailed knowledge of the compaction and gelation (fusion) routes, i.e. of the preparation for and the actual break-down of part of the supermolecular structure and the build up of a strong load-bearing network of coherent molecular entanglements.

The discovery by Berens and Folt (2), 15 years ago of primary particles persisting in the melt during processing made the crucial importance of gelation for the quality of PVC pipes clear. Today the challenge is to increase the knowledge of the gelation process enough to make sure that pipes may stand careless handling as well as at least 50 years of internal water pressure.

In this work we have carried out an extensive study of the ultimate mechanical properties of PVC pipes extruded at different mass temperatures. We have also tried to explain the gelation controlled development of mechanical strength in terms of structural changes.

EXPERIMENTAL

Materials

A set of PVC pipes (110 x 4.5 mm) was processed in a Krauss Maffei KMD90 twin-screw extruder at mass temperatures ranging from 176 to 205°C. A lead-stabilized compound was used based on a skin-free suspension PVC of K-value 68, corresponding to a weight average molecular weight of $\overline{M}_w$=172,000 and a number average of $\overline{M}_n$= 60,000. The resin is manufactured by Kema Nord under the trade name Pevikon S688. By gel permeation chromatography (Waters Associate Inc., PGC M6000A) was confirmed that the thermal degradation of the pipes extruded was insignificant.
From the same dry blend sheets were roll milled for 10 min (after formation of a coherent sheet) at temperature settings from 120 to 210°C. The average temperature was determined by a Heimann KT15 IR pyrometer with a target spot located in the vincinity of the rolling bank. The temperature variation of the sheet along the rolling mill was about 1°C.

Methods

The gelation level was estimated by capillary rheometry (3) measuring the capillary entrance loss in a zero-length die at 135°C using an expression derived by Gonze (4).

Internal water pressure resistance tests were carried out at 60°C (5). 15 to 20 data points were collected for each failure curve. The circumferential strain was determined for each pipe by measuring the outer diameter before loading and after unloading (at the maximum position).

Uniaxial tensile testing was carried out at rates from 0.01 to 500 mm/min at temperatures from -10 to +180°C in an Instron 1122 machine and at $5.8 \cdot 10^{-4}$ mm/min at +60°C in a homemade apparatus (6). The test specimens were cut out axially from the pipes (50 x 15 x 4.5 mm) and provided with a central hole (Ø 9 mm).

Charpy impact tests were carried out on specimens cut out axially from the pipes to the dimensions 50 x 6 x 4.5 mm (7). Part of the specimens were provided with a 45°-notch with 1/4 mm tip radius. The depth below the notch was kept constant, 2.95 mm. The test temperature was varied between -190 to +72°C. Data points represent average values from tenfold or more tests.

Instrumented falling weight tests were performed on curved plates (100 x 115 x 4.5 mm) from the pipe wall (7). The plates were supported by a curved anvil with a central hole (Ø 40 mm) allowing penetration of the specimen by a flat-headed dart (Ø 10 mm).

Sorption of p-xylene. The weight gain of pipes and curved plates immersed in p-xylene at 30 ± 0.2°C was followed and plotted vs. reduced sorption time (8).

Density measurements were carried out at 23 ± 0.1°C in a gradient column filled with an aquous solution of calcium nitrate (8).

Differential Scanning Calorimetry (DSC) thermograms were recorded by a Perkin Elmer DSC-2 for 13 mg samples at 10 K/min and 1 mcal/sek (8).

Transmission electron microscopy (TEM) were performed on samples embedded in methyl/butyl-methacrylate (30/70) and ultrasectioned to about 80 nm thickness (8).

Residual grain content of pipes were determined from photomicrographs of polished pipe cross-sections by the line intercept method (8).

RESULTS AND DISCUSSION

By *capillary rheometry* the G-levels of the pipes extruded at mass temperatures from 176 to 205°C were seen to range between 20 and 60% when related to a roll milling standard processing (see Fig. 1).

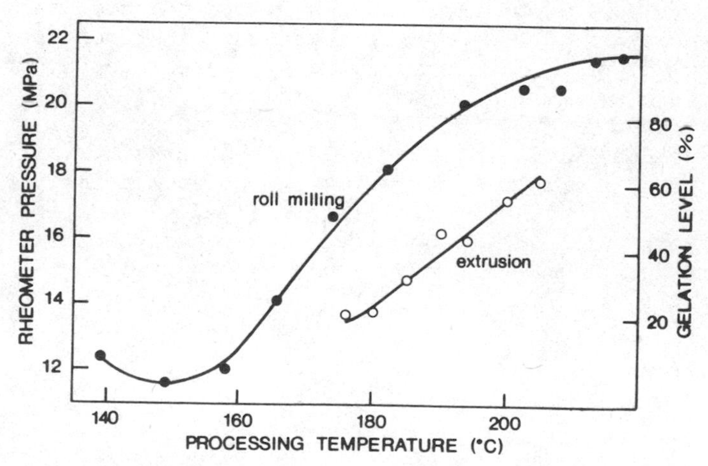

Figure 1. Estimation of gelation level of PVC-pipes by capillary rheometry.

At increasing G-level the *internal pressure resistance* at 60°C for the PVC pipes of K=68 increased although not unambiguously. The character of the curves changed from linear to bent upwards showing a levelling off tendency at long failure times. Premature failure due to a knee in the curve was not observed in any case, but a transition from ductile to creep crack mode of failure shifting to longer failure times at increasing G-level was found.

Extensive *uniaxial tensile testing* showed that the yield parameters are insensitive to variations in G-level irrespective of drawing rate and temperature. Contrary to this the ability of post-yield deformation (cold drawing) is very sensitive to G-level. A ductile-brittle transition similar to the one showed in Fig. 2 is observed shifting to longer times to break at higher G-level.

Charpy-UNIS at low temperatures, reflecting resistance to crack initiation, showed a weak maximum at moderate G-levels (see Fig. 4) and at room temperature the *falling weight-IS* showed a very distinct maximum. However, *Charpy-NIS* at various temperatures, reflecting resistance to crack propagation, increased to a higher level at moderate G-levels (see Fig. 5).

The experiments carried out clearly demonstrates that it is the ability of post-yield deformation (cold drawing) that controls the effect of G-level on the ultimate mechanical properties. Post-yield deformation proceeds by large-scale yielding in internal pressure resistance, notched impact strength and tensile tests but in unnotched impact tests by local yielding and void formation. Void formation is likely to be favored in the presence of a network of badly entangled zones at moderate G-levels.

DSC thermograms and TEM on methacrylate embedded samples showed a correlation between disintegration of primary particles and development of cooling crystallinity represented by the melting endotherm A (see Fig. 6). Increased *density values* (see Fig. 7) and decreased p-xylene sorption rate is explained by the presence of residual grain inclusions of higher crystallinity. In conclusion, it is though that at increasing G-level primary particles are broken down, primary crystallinity melted, a coherent entanglement network established and fixed by secondary imperfect crystallites formed during cooling.

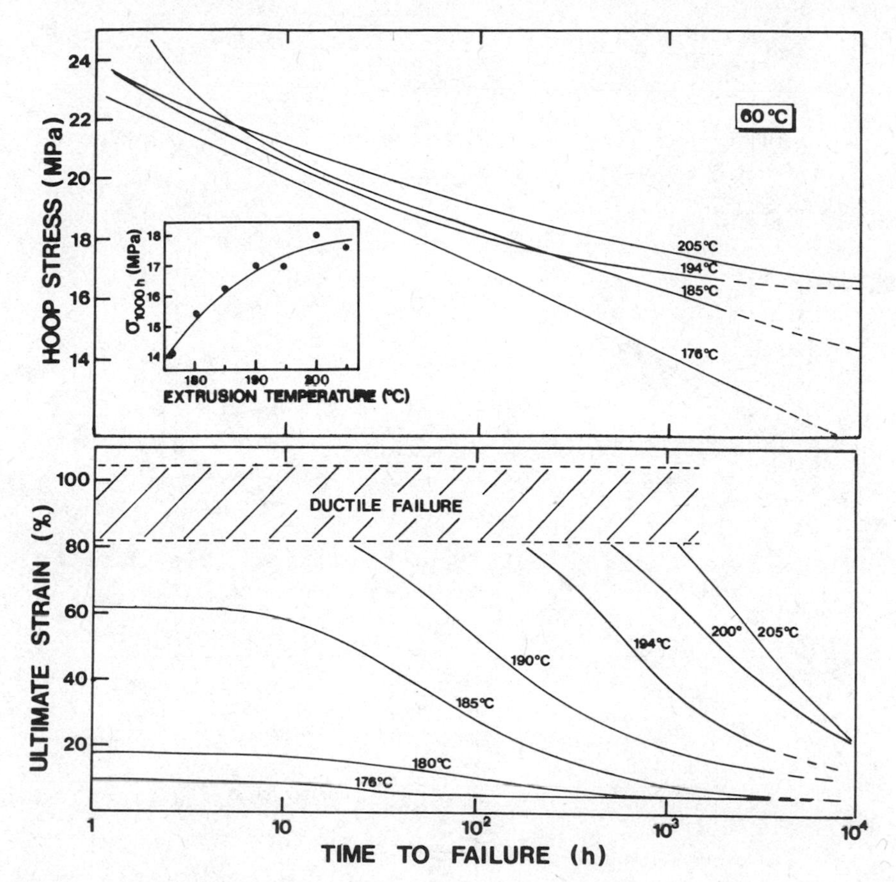

Figure 2. Internal water pressure resistance at 60°C for PVC pipes extruded at temperatures from 176 to 205°C. Ultimate hoop stress (above) and ultimate circumferential strain (below) vs. time to failure.

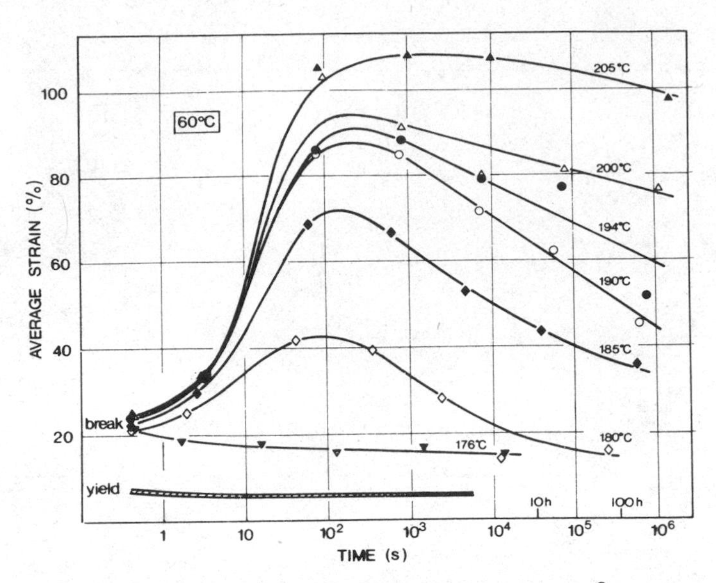

Figure 3. Average strain at yield and break at 60°C vs. time to break for PVC pipes of different extrusion temperature.

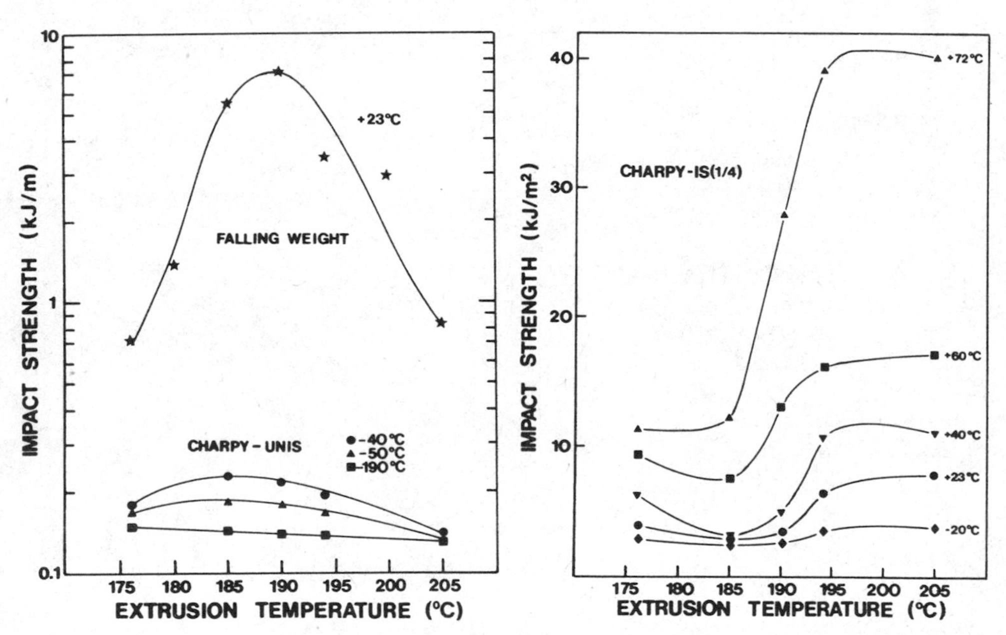

Figure 4. Effect of extrusion temperature on falling weight-IS and Charpy-IS for unnotched specimens at different temperatures.

Figure 5. Effect of extrusion temperature on Charpy-NIS at different temperatures.

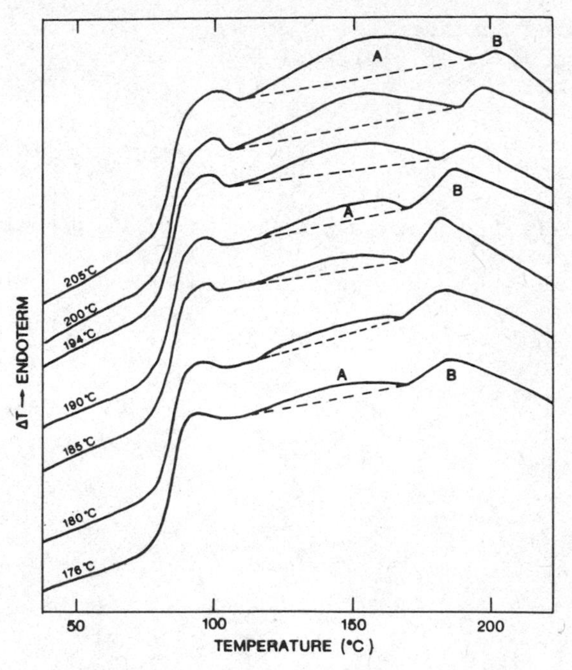

Figure 6. DSC traces for samples of PVC pipes extruded at mass temperatures from 176 to 205°C.

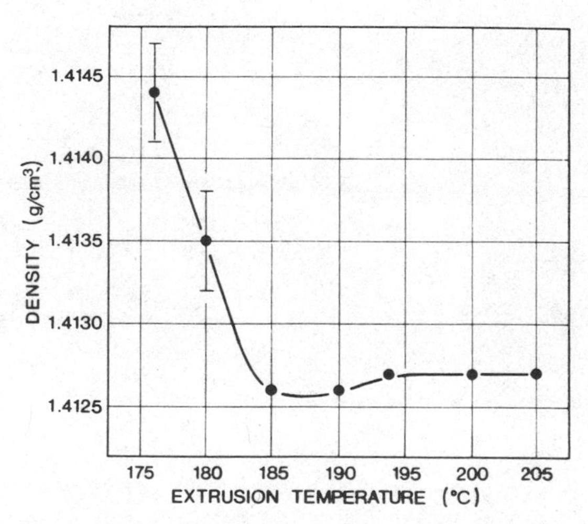

Figure 7. Effect of mass temperature at extrusion on the density of pipes.

REFERENCES

1. M.W. Allsopp in "Manufacture and Processing of PVC", Ed. R.H. Burgess, Applied Science Publ., London (1982), Ch. 8.
2. A.R. Berens and V.L. Folt, Trans. Soc. Rheol., 11, 95 (1967).
3. B. Terselius, J-F. Jansson and J. Bysted, J. Macromol. Sci.-Phys., B20, 403 (1981).
4. A. Gonze, Chim. Ind.-Gén. Chim., 104, 1 (1971).
5. B. Terselius and J-F. Jansson, Gelation of PVC, part 2, to be published.
6. B. Terselius and J-F. Jansson, Gelation of PVC, part 3, to be published.
7. B. Terselius and J-F. Jansson, Gelation of PVC, part 4, to be published.
8. B. Terselius and J-F. Jansson, Gelation of PVC, part 5, to be published.

CORRELATION OF NON-LINEAR RHEOLOGICAL PROPERTIES OF POLYMER MELTS WITH WELD-LINE STRENGTH

R. Pisipati and D. G. Baird

Department of Chemical Engineering
Virginia Polytechnic Institute and State University
Blacksburg, VA 24061-6496

INTRODUCTION

Weld-lines are formed either by the direct impingement of two flowing fronts or by the rejoining of a melt stream previously separated by an obstacle, such as an insert. In both cases, the elongational flow field at the interface causes molecular orientation that is predominantly along, rather than across the weld region. In general, the presence of weld-lines reduces mechanical strength and affects surface appearance of the molded part. Although the reduction in strength of a part as the result of a weld-line is a widely recognized problem, a review of the literature reveals that there is very little fundamental understanding of the mechanism which accounts for the formation of a weld-line.

The reduction of part strength at the weld-line has been attributed to three factors:[1] (i) orientation of molecules along the weld-line rather than across it as a result of the deformation of the polymer after the two melt fronts have come together (ii) partial cooling of the melt front at the interface leading to skin formation (iii) the formation of a relatively sharp V-notch at the weld-line caused by the entrapped air which is forced to the walls. As the polymer freezes, it retains this notch which acts as a stress concentrator.

Studies on weld-lines have been limited to determining the effects of processing conditions such as melt and mold temperatures and injection speed on weld-line tensile yield strength[1-13]. In general it was concluded that weld-line strength increases with increasing melt and mold temperature and is not greatly affected by changes in injection speed. Other factors that influence weld-line formation such as the location of the weld relative to the gate, melt rheology and crystallization kinetics have not been systematically studied.

The quality of a weld-line is associated with the flow field which arises as the two fronts meet each other. Tadmor[14] proposed that a rectangular fluid element moving from the bulk of the fluid to the advancing front will decelerate in the axial direction, while at the same time being accelerated in the perpendicular direction and consequently is stretched at a constant rate. Furthermore, when two advancing parabolic fronts meet, the molecules are stretched transverse to the flow direction. The above two flows generate molecular orientation that is predominantly along, rather than across the weld interface. Two processes that most likely precede weld healing are stress relaxation, which results in a loss of this orientation, and molecular re-entanglement, which is a self diffusion phenomenon. Self diffusion has been modeled[15,17,20] using the tube reptation theory of deGennes[21]. In this model, each polymer chain is considered to be confined in a tube which represents the constraints imposed on its motion by other chains in the bulk. The chain executes a random back-and-forth motion which causes it to slip out of a section of the tube in time. While this happens, an additional equal length of new tube is being added at the opposite end. Good agreement with the experimental data[16,17] obtained by Jud et al. and Wool and Oconnor[17] through the use of a crack healing technique in a double cantilever beam. Jud et al.[16] found that a log-log plot of refracture toughness versus healing time was linear over the entire healing period. The value of the exponent obtained for the increase in refracture toughness was 1/4. Most of this work, however, deals with molecular motion in the solid state starting from a randomly oriented condition. Clearly, for weld healing to occur in processing flows, the initial orientation must be taken into account followed by entanglements being reformed in the melt. In this paper, a rheological test is designed to obtain a measure of the relaxation and re-entanglement times for an amorphous polymer melt. It has been shown that shearing a molten polymer alters its response to further deformation and this effect which may last for long times after the initial deformation[22]. In the case of crystallizable polymers with long branches, this could be several hours[26,27]. In sheared flexible chain polymer systems (above the critical molecular weight for entanglements) entanglements are constantly being lost and then reformed. Although the loss of entanglements depends on the total imposed strain, the reformation of structure is governed by thermal motion requiring many segmental motions of the chain. This may occur at a much slower rate than the time required for the disentanglement process. In studying stress growth upon the inception of steady shear, Stratton and Butcher[28] found the magnitude of the maximum stress relative to the steady state value to depend upon the time the material had been at rest after prior shearing. Utilizing this dependence, they devised an interrupted shear test to determine an empirical re-entanglement time. Although this re-entanglement time is in reality characterized by a spectrum of relaxation times, one time constant may be used as only the early part of the relaxations are considered. This interrupted shear test was also used by Dealy and Tsang[29] who found the re-entanglement time for linear and branched polyethylene to be significantly greater than the stress relaxation time.

In this paper, a modified interrupted shear test, referred to as a double-step shear test, is used to evaluate the re-entanglement time, t_e for three polystyrenes of different molecular weight (but nearly the same molecular weight distribution). This value of t_e is correlated with the strength of weld-lines in injection molded tensile bars. Similar tests are used to determine whether this approach may be applied to filled nylon 6,6 systems.

DOUBLE STEP SHEAR

In the double step shear test a polymer melt is subjected to steady shearing deformation for a time t_s, i.e. a constant imposed strain. After time t_s, shearing is stopped and restarted after a rest time t_r. An overshoot in shear stress is obtained after both shearing operations. Using the peak height τ_m as a characateristic feature of the stress growth curve, the effect of rest time on peak height is measured. This is shown in figure 1. The percent recovery in shear stress following the second deformation $\tau_m(\dot{\gamma}, t_r)$ is expressed as:

$$\% \text{ Recovery} = \frac{[\tau_m(\dot{\gamma}, t_r) - \tau_{ss}] / \tau_m(\dot{\gamma}, t_r)}{[\tau_m(\dot{\gamma}, \infty) - \tau_{ss}] / \tau_m(\dot{\gamma}, \infty)} \tag{1}$$

$$\lim_{t_r \to 0} \% \text{ Recovery} = 0 \tag{2}$$

$$\lim_{t_r \to \infty} \% \text{ Recovery} = 1 \tag{3}$$

When a polymer melt is subjected to a shear rate in the non-Newtonian region, the instantaneous response tends to be that of an elastic solid. This is because the establishment of an equilibrium entanglement density lags the stress, as the experimental time scale is too short. This causes an overshoot in the shear

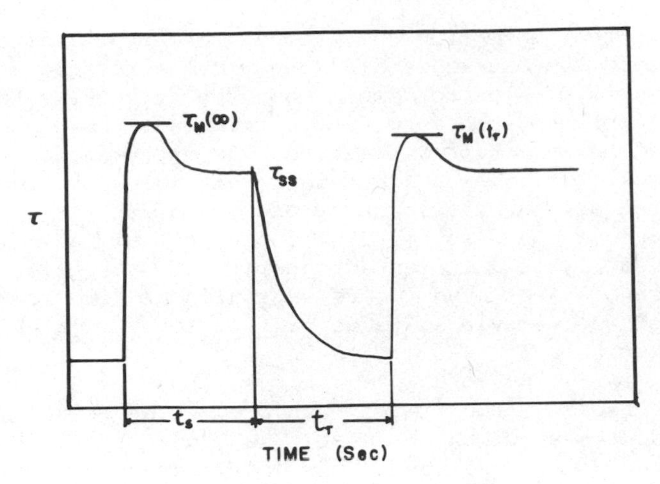

Fig. 1. Double step shear test. $\tau_m(\infty)$ is the peak shear stress after an infinite rest time; τ_{ss} is the steady state value and $\tau_m(t_r)$ is the peak shear stress after a rest time t_r.

stress response. At longer times, the new equilibrium entanglement density corresponding to the shear rate is attained, and the stress relaxes to a steady state value. After a rest time t_r, even though the stress relaxes to a zero value, the entanglement density present prior to shearing is not reached. Hence, at the start up of shear after a rest time t_r, the magnitude of shear stress overshoot decreases. The above experiment is repeated for different rest times, using a fresh sample each time. This ensures that each sample is subject to the same initial strain. The overshoot in shear stress at the start up of shear flow at different rest times is shown in figure 2. The percent recovery is plotted against rest times. The straight line obtained is linearly extrapolated to 100 percent recovery, and the corresponding rest time is empirically referred to as the re-entanglement time. It must however be noted that the percent recovery would approach complete recovery in a sigmoidal manner as the driving force for the formation of re-entanglements decreases with increasing time.

EXPERIMENTAL

The materials used are shown in Table I. The three polystyrene samples have different weight average molecular weights but about the same molecular weight distribution. In the filled nylon systems, the total volume fraction of filler was maintained at 20%. The fillers used were mineral, which is a particulate material having an average diameter of 3-5µm, and glass fiber with an average length of 0.1 mm and an aspect ratio of 10. The relative volume fractions of glass fiber and mineral used are as indicated. Rheological properties were measured by means of a Rheometrics Mechanical Spectrometer using a cone-and-plate geometry with diameter 2.5 cm and cone angle of 0.1 rad.

The polymers were injection molded into a tensile bar cavity which could be gated at both ends or with one of the runners blocked to obtain bars without weld-lines, on an Arburg injection molding unit (model 221-55-250, 1 oz. shot size). Fill time was 6 sec. Mechanical properties were measured using an Instron tester. Melt temperatures of 170°C, 180°C and 190°C were used for PS while the mold was held at room temperature. The filled nylon systems were injection molded at 285°C into a mold at room temperature.

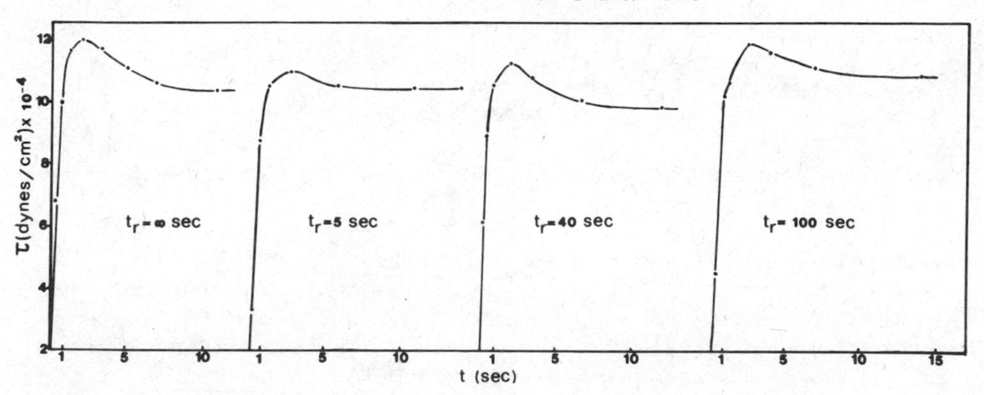

Fig. 2. Double step shear test. Shear stress overshoot after various rest times.

Table I. Materials Used

	Tensile Strength (MPa)	M_w	M_w/M_n	η_o (Pa.S)
Styron 678	37.23	255,000	2.55	480
Styron 666	41.22	240,000	2.50	600
Styron 685	51.85	285,000	2.70	2,200
Nylon 6,6	62.33	30,000	2.00	200
20 Vol % Mineral Filled	83.61	–	–	5,000
20 Vol % Glass Filled	150.00	–	–	6,000

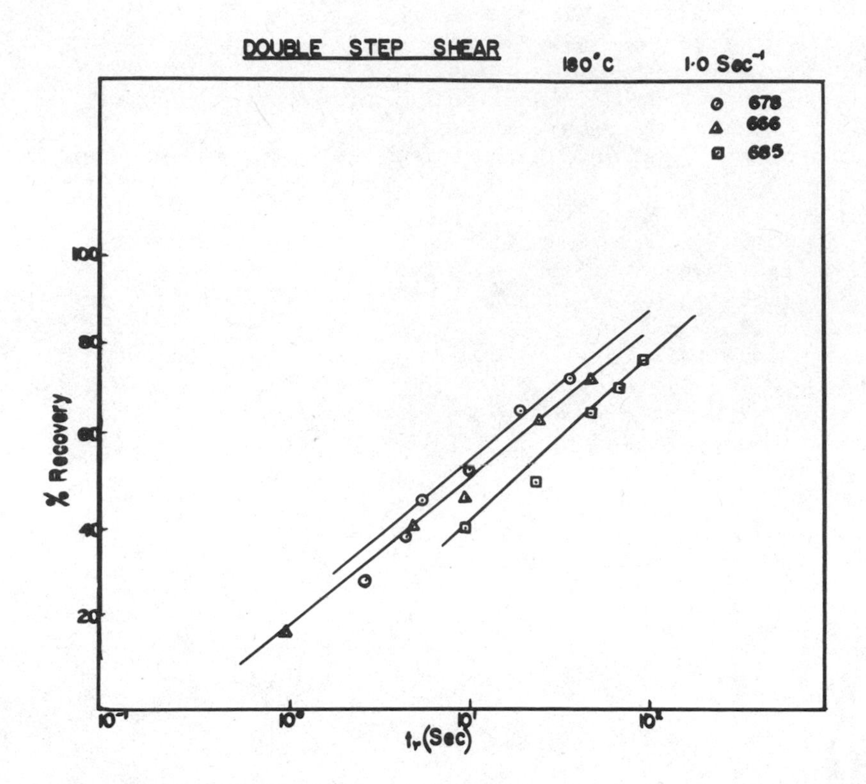

Fig. 3. Percent recovery versus rest times for the three PS samples.

RESULTS AND DISCUSSION

Results of the double step shear test are shown in figure 3. The total imposed strain was chosen such that a steady state value of shear stress was achieved prior to cessation of the flow. It is expected that edge fracture is not occurring since a steady state torque value is obtained. It is seen that 678 has the greatest percent recovery followed by 666 and 685, at the same rest time, i.e. recovery increases with decreasing molecular weight. Temperature dependence of stress recovery is shown in figure 4, with recovery increasing with increasing temperature. It appears that temperature dependence of recovery increases with increasing molecular weight.

Re-entanglement times (t_e) obtained by extrapolating recovery curves are shown in figure 5. t_e decreases with increasing temperature, for all three samples. In Table II values of the time for the stresses to relax to zero after the first shearing t_d and t_e are compared. It can be concluded that the re-entanglement time is greater than the relaxation time by over an order of magnitude. This means that stress relaxation occurs when the entanglement density is at the value existing in steady state flow.

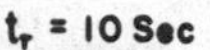

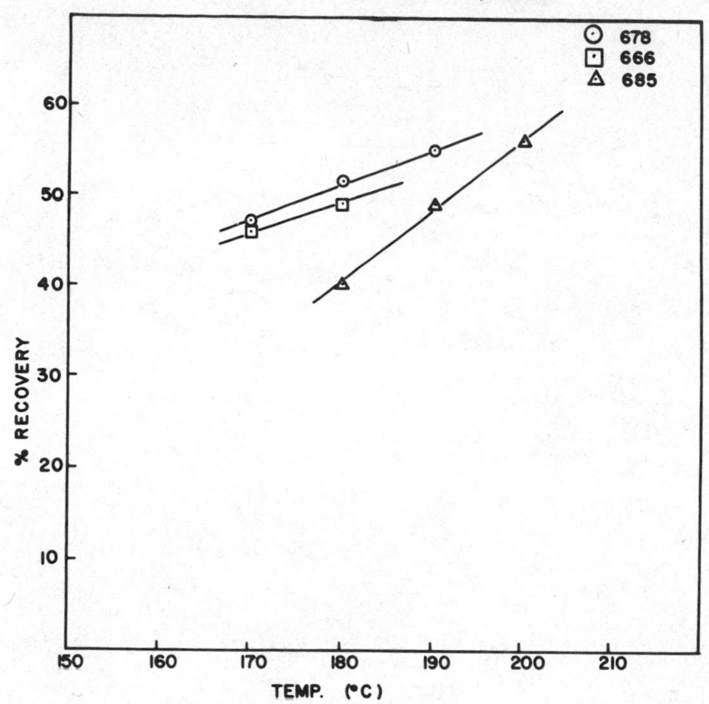

Fig. 4. Temperature dependence of stress recovery for PS samples.

Table 2. Comparison of Relaxation and Re-entanglement Times

	t_d(sec)	t_e(sec)
Styron 678	7.8	180.0
Styron 666	11.5	350.0
Styron 685	20.1	445.0
Nylon 6.6	0.5	18.5
Mineral filled Nylon	1.5	35.0

The tensile strength for bars with and without weld-lines is plotted against t_e for each resin in Fig 6. Tensile strength for bars without weld-lines increases with increasing t_e. This is to be expected as t_e is directly related to molecular weight. However, the opposite is found for bars with weld-lines, as the breaking strength decreases with increasing t_e. This may be explained by the fact that as t_e increases, less molecular entanglements are reformed across the weld interface within the time taken for the polymer to solidify in the cavity. Consequently there is a reduction in tensile strength.

The concept of re-entanglement of polymer molecules following deformation is quite different when dealing with filled systems due

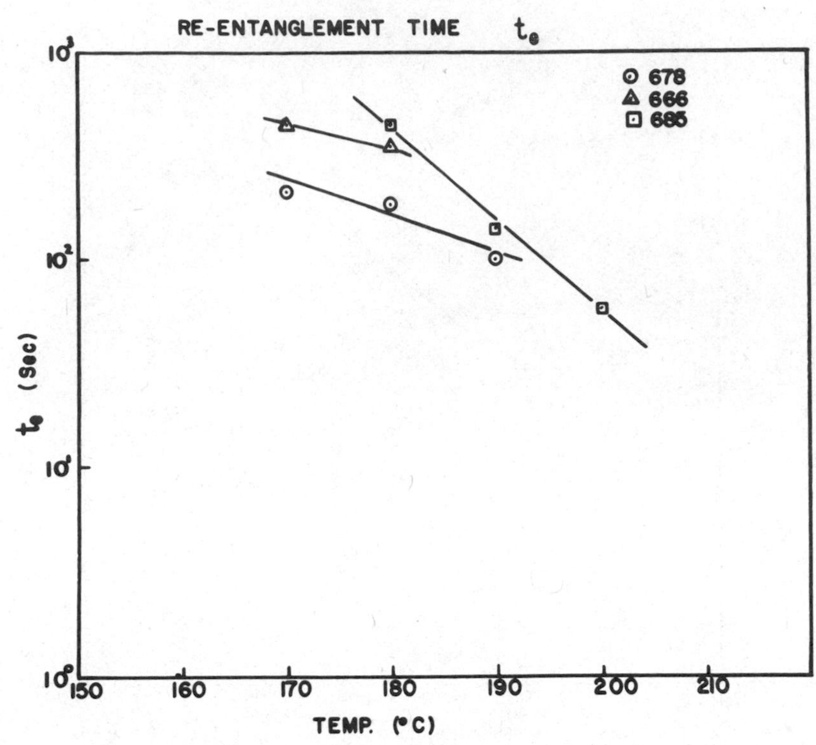

Fig. 5. Re-entanglement times obtained from the Double Step Shear test.

to the additional interaction of molecules with the filler. For example, in the case of particles of diameters less than 0.1 micron, yield stresses have been found at very low shear rates[30]. In this case, it is more appropriate to consider the reformation of structure rather than entanglements. Furthermore, at the cessation of flow, mobility of polymer chains is impeded by the filler particles. However, when filled systems are injection molded into a double-gated cavity to form a weld-line, it is expected that the weld interface is rich in nylon and hence the re-entanglement process may be significant for the following reasons:

a) Tensile strength for bars with and without weld-lines is shown in figure 7. The lack of any composistion dependence for bars with weld-lines suggests that the nylon rich weld interface controls tensile strength.

b) In glass fiber filled melts, the extensional flow generated when the two fronts meet causes fiber alignment parallel to the weld-line, which is seen in the micrographs of figure 8. This fiber orientation would tend to reduce the load bearing contribution of the fibers.

c) At high injection speeds, micrographs of fracture surfaces indicate a relatively filler depleted zone at the surface of the

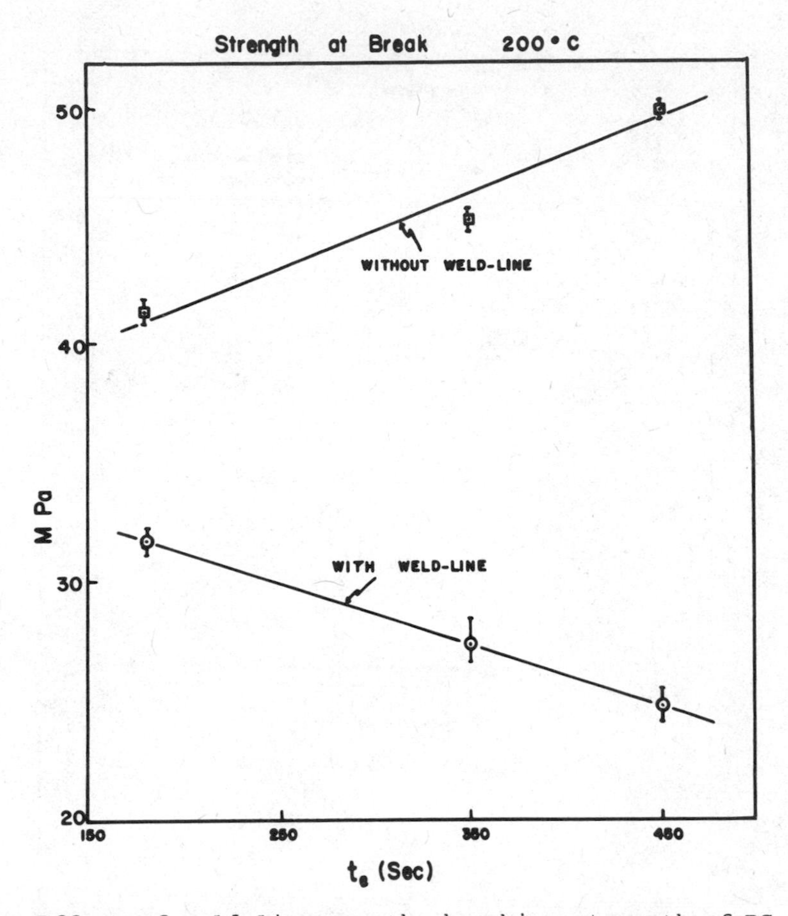

Fig. 6. Effect of weld-lines on the breaking strength of PS bars.

bar. Based on Tadmore's[14] "fountain flow" mechanism for mold filling, it is expected that the advancing front is richer in nylon than the rest of the bulk. Micrographs of the front obtained using a short shot are shown in figure 9. Fibers are seen only at aboaut 300μm behind the front.

Since the above three factors suggest that the weld interface is rich in nylon, it maybe the re-entanglement of nylon chains following deformation that would lead to an enhancement in weld strength. A plot of percent recovery against rest time for the unfilled nylon and mineral filled nylon is shown in figure 10. This yields a value for t_e of 18.2 secs. t_e for the mineral filled melt is about 35 sec., which is about the same as that of the glass filled melt. Although this suggests that the weld-line strength for the filled systems should be less than that of the unfilled nylon, they are in fact almost equal. This would result from the weld interface being richer in nylon. It

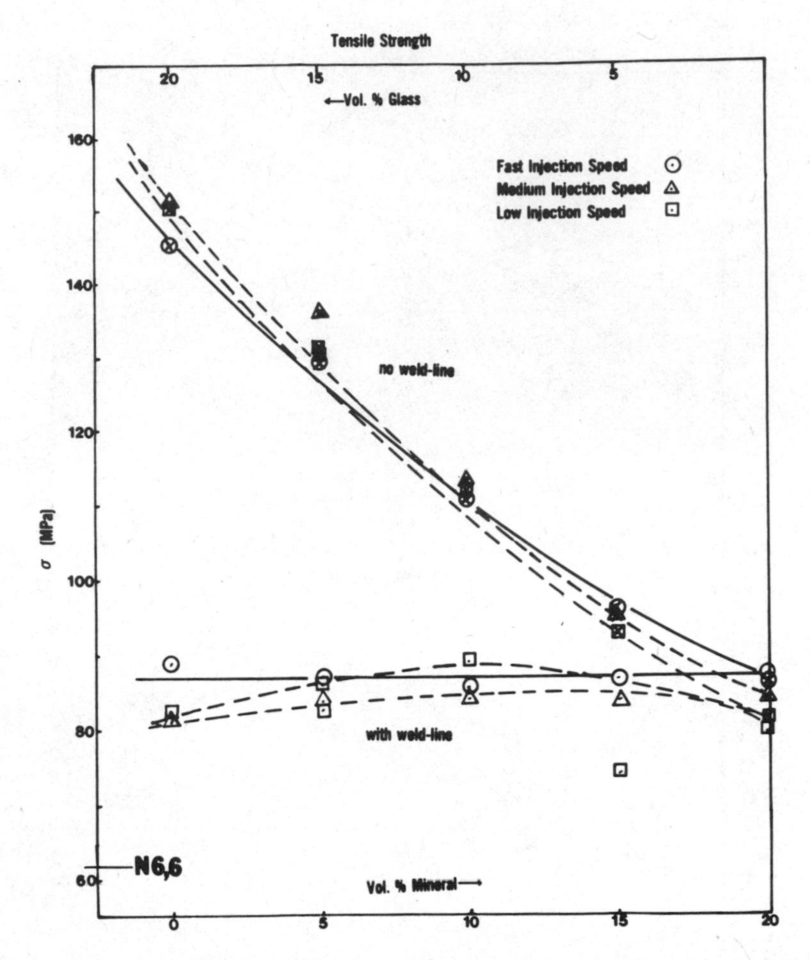

Fig. 7. Filled nylon systems - effect of weld-line on breaking strength of injection molded bars.

must be noted, however, that in the case of nylon, strain induced crystallization at the weld interface due to extensional flow can be an important factor governing weld-line strength. It is assumed here that stress relaxation and some amount of re-entanglement take place prior to crystallization.

CONCLUSION

A double-step shear experiment has been used to obtain a quantitative estimate of the re-entanglement process that takes place following a shearing deformation. The empirically defined re-entanglement time obtained from this experiment is found to correlate well with the tensile strength of bars injection molded with and without weld-lines. Factors that increase the

(a)

(b)

Fig. 8. Glass filled nylon 6,6 injection molded bars. (a) flow at some distance behind the weld-line, (b) Fiber orientation at the weld-line. Dimensioinal bar = 50μm.

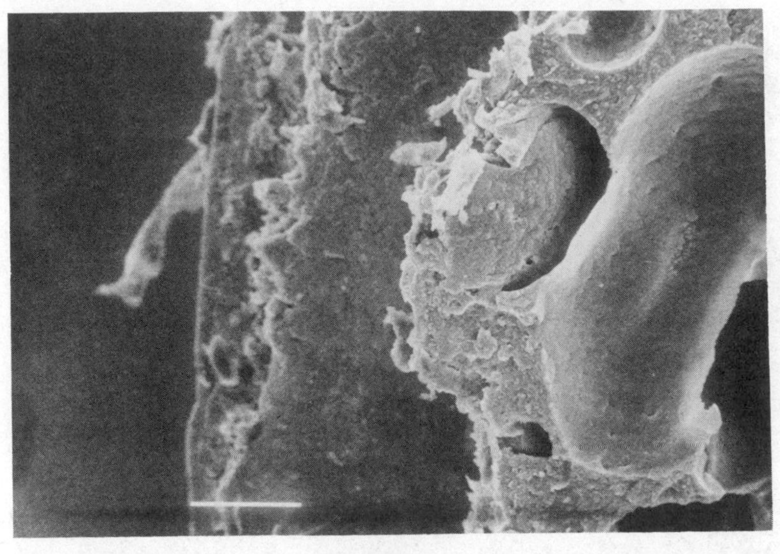

(a)

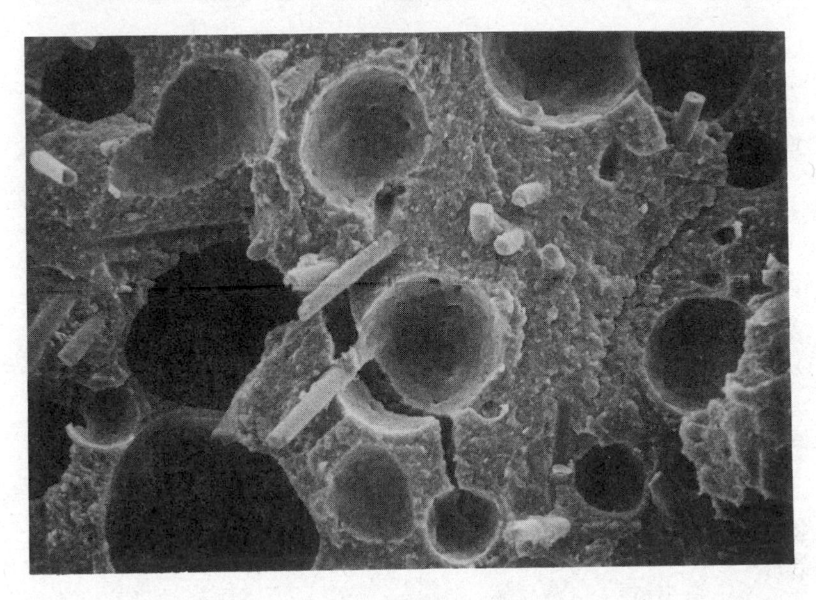

(b)

Fig. 9. Micrograph of advancing front in a glass filled nylon system. (a) The fiber free front, (b) $\sim$ 300μm behind the front. Dimension bar = 50μm.

re-entanglement time, such as molecular weight, decrease the strength of bars molded with a weld interface. To some extent, this procedure may be used to explain the reduction in tensile strength of bars of filled nylon systems with weld-lines,

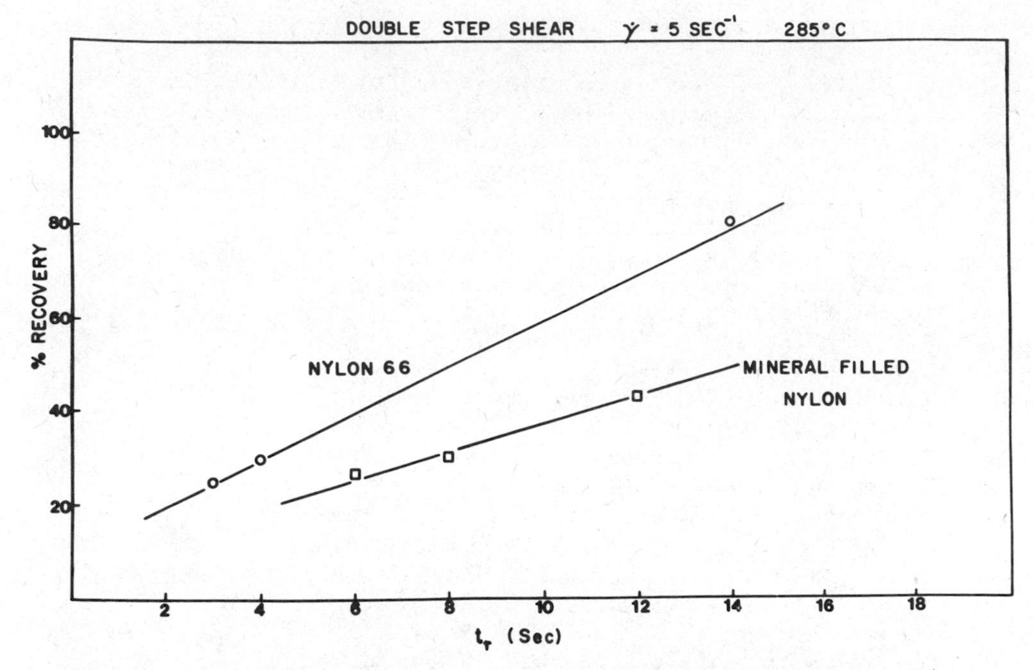

Fig. 10. Stress recovery in nylon 6,6 and mineral filled nylon as a function of rest time between deformation.

considering the phase separation that occurs at the interface. The resulting nylon rich weld interface causes the re-entanglement of nylon chains at the weld-line to control the extent of weld healing irrespective of filler used. However, with semicrystalline polymers, crystallization under processing conditions may play an important role.

REFERENCES

1. E. M. Hagerman, Plast. Eng. pg. 67, Oct. (1973).
2. P. Hubbauer, Plast. Eng., pg. 37, Aug (1973).
3. P. J. Cloud, F. McDowell and S. Gerakaris, Plast. Tech. pg. 48, Aug. (1976).
4. W. L. Krueger and Z. Tadmor, SPE 36th ANTEC, pg. 87 (1978).
5. S. Malguarnera and A. Manisali, SPE 38th ANTEC, pg. 124 (1980).
6. D. P. Isherwood, J. G. Williams and Y. T. Yap, Proc. 8th Intl. Cong. Rheol., 3:37 (1980).
7. R. A. Worth, Poly. Eng. Sci., 20:551 (1980).
8. S. C. Malguarnera and A. Manisali, SPE 39th ANTEC, pg. 775 (1981).

9. S. C. Malguaruera and A. Minisali, Poly. Eng. Sci., 21:586 (1981).
10. G. Prall, Mod. Plast., pg. 118, Nov. (1970).
11. S. Y. Hobbs, Poly. Eng. Sci., 14:621 (1974).
12. R. C. Thamm, Rubber Chem. Tech., 50:24 (1977).
13. S. C. Malguarnera and D. C. Riggs, Poly. Plast. Technol. Eng. 17:193 (1981).
14. Z. Tadmor, J. Appl. Poly. Sci., 18:1753 (1974).
15. S. Prager and M. Tirrell, J. Chem. Phys., 75:5194 (1981).
16. K. Jud, H. H. Kausch and J.G. Williams, J. Matl. Sci., 16:204 (1981).
17. R. P. Wool and K. M. O'Conner, J. Poly. Sci. Poly. Letters Ed., 20:7 (1982).
18. H. H. Kausch and K. Jud, Plast. & Rubber Proc. & Appl. 2:265 (1982).
19. H. H. Kausch, IUPAC Macromolecules, Ed. H. Benort and P. Rempp, Pergamon Oxford (1982).
20. R. P. Wool, ACS Poly. Preprints, 23:62 (1982).
21. P. G. deGennes, J. Chem. Phys. 55:572 (1971).
22. D. E. Hanson, Poly. Eng. Sci., 9:405 (1969).
23. B. Maxwell and F. Plumeri, SPE Tech. Papers 26:282 (1980).
24. E. R. Howells and J. J. Benbow, Trans. J. Plast. Inst. 30:240 (1962).
25. J. H. Prichard and K. F. Wissbrun, J. Appl. Poly. Sci., 13:233 (1969).
26. W. W. Graessley, J. Chem. Phys., 43:2696 (1965).
27. R. A. Stratton, J. Colloid & Interface Sci., 22:517 (1966).
28. R. A. Stratton and A. F. Butcher, J. Poly. Sci. Poly. Phys. Ed., 11:1747 (1973).
29. J. M. Dealy and Wm. K.-W. Tsang, J. Appl. Poly. Sci., 26:1149 (1981).
30. N. Minagawa and J. L. White, J. Appl. Poly. Sci., 20:501 (1976).

MELT STRENGTH AND EXTENSIBILITY OF HIGH-DENSITY POLYETHYLENE

F.P. La Mantia and D. Acierno

Institute of Chemical Engineering
University of Palermo
Viale delle Scienze, 90128 Palermo, Italy

The extensional flow of polymer melts has been extensively studied because of its importance in many technological processing operations and, from a more fundamental point of view, because the tensile properties of the polymer melts cannot be correlated directly with shear viscosity behavior.[1-5]

Most of the work on this subject has been carried out under isothermal conditions,[5-16] because in this manner the rheological behavior in elongational flow can readily be investigated without complications due to temperature variations and can be correlated with the molecular structure of the polymers. In practice, however, processing operations in which elongational flow is present are generally performed under nonisothermal conditions, and the melt strength and extensibility of polymers can be considered the most important parameters during their processing. In particular, a polymer can be processed more readily if it has an elevated melt strength to resist the mechanical stresses due to the process, and if it has high extensibility to achieve the desired thickness.

Only little information is available about the melt strength of polymers,[4,17-21] probably because the results found for a given class of polymers cannot be generalized to other materials, and because this property depends on a number of parameters and cannot be used to determine the elongational viscosity of polymer melts.[1,19]

In this work tensile measurements have been carried out, with the aid of a new melt tension tester, on a series of high-density polyethylenes in order to determine the influence of molecular weight and molecular weight distribution on the melt strength and on the breaking stretching ratio. Also, the effect of the temperature has been considered.

EXPERIMENTAL

The materials used in this work were five high-density polyethylenes. Their average molecular weight and molecular weight distribution are reported in Table I. A complete rheological characterization has been presented earlier.[22,23]

All the experimental runs were carried out on a new capillary rheometer, Rheoscope 1000, manufactured by CEAST (Turin, Italy). This rheometer allows the determination of both the melt viscosity as a function of the shear rate and, with the aid of the tensile module, the tensile properties of the extruded monofilament at various drawing rates. The monofilament passes through a pulley system and is then drawn by two counterrotating rolls, the rotational speed of which can be varied continuously up to 1000 rpm. The roll diameter was 120 mm. The strain gauge used as force detector is connected to the first pulley.

The measurements were performed with a capillary with D = 1 mm, L/D = 10, and inlet angle of 180°. The tensile system was situated 25 cm from the extrusion capillary. The extrusion temperature was generally 180°C; some runs were also performed at 160°C and 200°C. The stretching velocity varied with a linear acceleration of 1 rpm/sec.

The breaking stretching ratio was calculated as the ratio between the drawing speed and the extrusion velocity at the die.

TABLE I

Sample Code	Mw	β = Mw/Mn
B	113,000	5.3
C	147,000	4.7
D	158,000	7.9
E	195,000	15.6
F	395,000	22.7

RESULTS AND DISCUSSION

Figures 1 and 2 show melt strength (MS) and breaking stretching ratio (BSR) values respectively for all the samples as a function of the apparent shear rate in the capillary. In all cases MS quickly increases with $\dot{\gamma}_{app}$ in the lower range, becoming virtually insensitive to further variations of shear rate at higher values; on the other hand, BSR decreases initially, but then, like the melt strength, approaches a constant value.

These effects can be correlated with the greater molecular orientation in the extrudate achieved with increasing extrusion velocity, giving a more rigid filament.

The influence of the molecular parameters can be seen from the same figures. For the extensibility, an increase in the molecular weight, with the same molecular weight distribution (polymers B and C), gives rise to a decrease of BSR, and an increase in molecular weight distribution (polymers C and D) produces a similar effect. More quantitative determinations can be done for both the melt strength and the breaking stretching ratio.

It has already been noted that the melt strength approaches an asymptotic value, MS_∞, which depends on Mw and β. By plotting these values against the product Mwβ in a logarithmic plot, a straight line is obtained, so that

$$MS_\infty = -\ 31.57 + 2.7\ \ln(Mw\beta) \qquad (1)$$

All the dimensionless MS/MS_∞ values fall on the same curve, as shown in Fig. 3. The analytical relationship of this curve is

$$\frac{MS}{MS_\infty} = -\ 0.01 + 0.2\ \ln(\dot{\gamma}_{app}) \qquad (2)$$

Equations (1) and (2) allow one to predict the melt strength for all the high-density polyethylenes, provided, of course, the geometry of the capillary remains the same.

A vertical shift of the BSR vs. $\dot{\gamma}_{app}$ curves allows one to obtain a single curve (Fig. 4); the shift factor a_p is

$$a_p = -\ 4.53 + 0.42\ \ln(Mw\beta) \qquad (3)$$

The generalized BSR$\cdot a_p$ vs. $\dot{\gamma}_{app}$ curve, together with equation (2), allows one to obtain the breaking stretching ratio for any high-density polyethylene. The analytical relationship of the curve shown in Fig. 4 is

$$BSR \cdot a_p = 1837(\dot{\gamma}_{app})^{-0.52} \qquad (4)$$

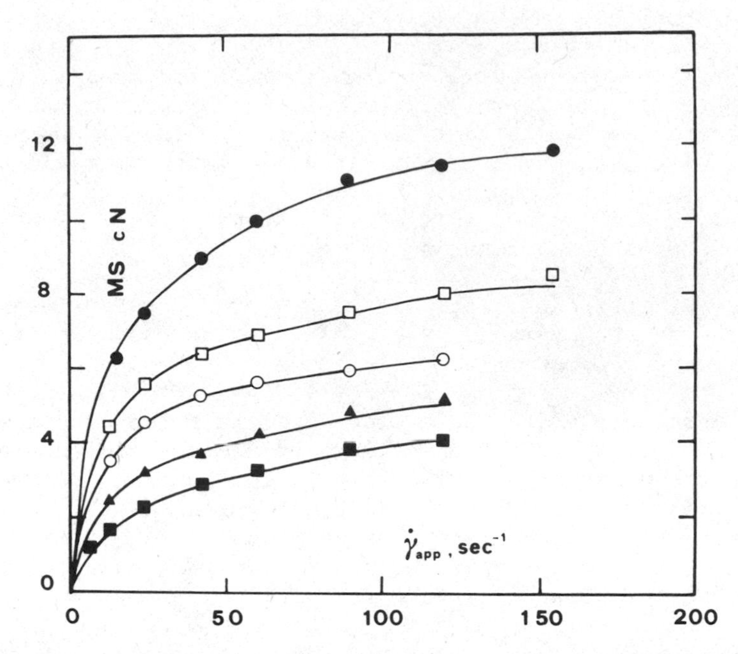

Fig. 1. Melt strength vs. apparent shear rate. Samples: ■ B, ▲ C, ○ D, □ E, ● F.

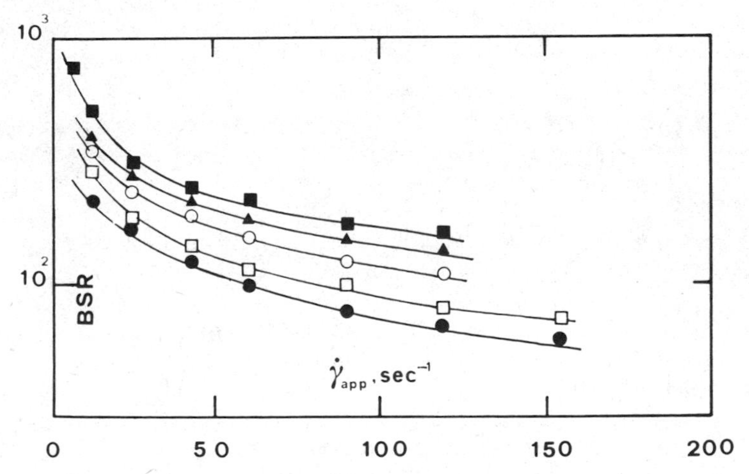

Fig. 2. Breaking stretching ratio vs. apparent shear rate. Key for symbols as in Fig. 1.

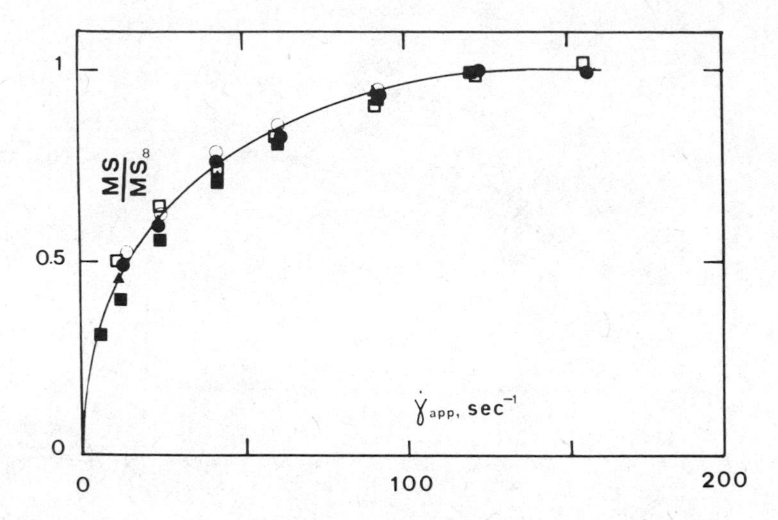

Fig. 3. Dimensionless melt strength vs. apparent shear rate. Key for symbols as in Fig. 1.

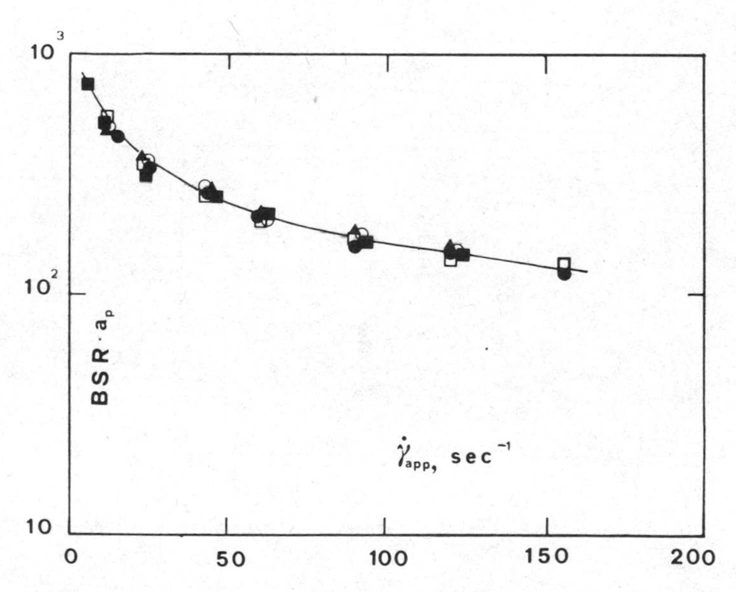

Fig. 4. Reduced breaking stretching ratio vs. apparent shear rate. Key for symbols as in Fig. 1.

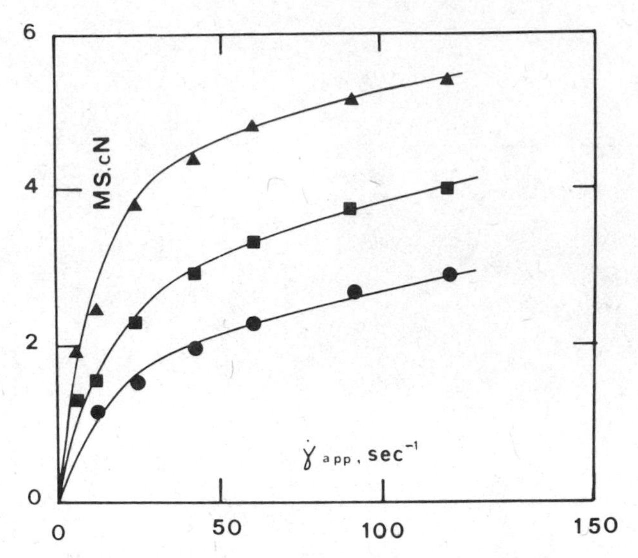

Fig. 5. Melt strength vs. apparent shear rate for sample B. ▲ T = 160°C, ■ T = 180°C, ● T = 200°C.

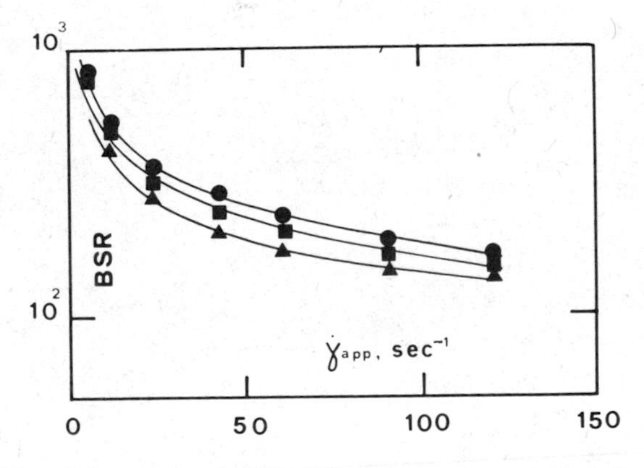

Fig. 6. Breaking stretching ratio vs. apparent shear rate for sample B. Key for symbols as in Fig. 5.

The effect of the temperature on the melt strength and on the breaking stretching ratio is shown in Figs. 5 and 6. The melt strength decreases with temperature while the breaking stretching ratio increases. For all temperatures the melt strength approaches an asymptotic value which depends on the temperature according to the relationship

$$\frac{MS_{\infty}}{MS_{\infty r}} = 1 - 0.0165(T - T_r) \qquad (5)$$

where the reference asymptotic melt strength ($MS_{\infty r}$) is the melt strength at 180°.

A good superposition can be achieved by vertical shifting also for the breaking stretching ratio. The shift factor, a_T, follows a WLF equation:

$$\log a_T = - \frac{C_1(T - T_r)}{C_2 + (T - T_r)} \qquad (6)$$

where $C_1 = -1.19$, $C_2 = -1.67$, and T_r is 180°C.

The whole body of reported relationships allows one to predict the nonisothermal behavior in elongational flow for all the high-density polyethylenes as a function of the extrusion velocity and the temperature. The quantitative features, of course, pertain to experiments performed with the same geometry (capillary and die-rolls distance) here adopted.

ACKNOWLEDGMENT

This work was carried out with the financial support of "Progetto Finalizzato del C.N.R. Chimica Fine e Secondaria." Thanks are due to CEAST for allowing us to use the tensile module of the Rheoscope 1000.

REFERENCES

1. C.J.S. Petrie, "Elongational Flow," Pitman, London (1979).
2. F.N. Cogswell, Polymer melt rheology during elongational flow, Appl. Polymer Symp. 27:1 (1975).
3. J.L. White and Y. Ide, Rheology and dynamics of fiber formation from polymer melts, Appl. Polymer Symp. 27:61 (1975).
4. J.W. Hill and J.A. Cuculo, Elongational flow behavior of polymeric fluids, J. Macromol. Sci. Rev. C14:107 (1976).

5. J. Meissner, Development of a new universal extensional rheometer for the uniaxial extension of polymer melts, Trans. Soc. Rheol. 16:405 (1972).
6. F.N. Cogswell, The rheology of polymer melts under tension, Plastics and Polymers 36:109 (1968).
7. R.L. Ballman, Extensional flow of polymer melts, Rheol. Acta 4:137 (1965).
8. J. Meissner, Dehnungsverhalten von Polyäthilen-schmelzen, Rheol. Acta 10:230 (1971).
9. H.M. Laun and H. Munstedt, Comparison of the elongational behavior of a polyethylene melt at constant stress and constant strain rate, Rheol. Acta 15:517 (1976).
10. H.M. Laun and H. Munstedt, Elongational behavior of a LDPE melt. I., Rheol. Acta 17:415 (1978).
11. H. Munstedt and H.M. Laun, Elongational behavior of a LDPE melt. II., Rheol. Acta 18:1061 (1979).
12. Y. Ide and J.L. White, Experimental study of elongational flow and failure of polymer melts, J. Appl. Polym. Sci. 22:1061 (1978).
13. H. Munstedt, New universal extensional rheometer; Measurements on a polystyrene sample, Trans. Soc. Rheol. 23:421 (1979).
14. T. Raible, A. Demarmels, and J. Meissner, Stress and recovery maxima in LDPE melt elongation, Polymer Bulletin 1:397 (1979).
15. T. Raible and J. Meissner, Uniaxial extensional experiments with large strains performed with LDPE, Polymer Bulletin 22:8 (1980).
16. H. Munstedt, Dependence of the elongational behavior of polystyrene melts on Mw and MWD, Trans. Soc. Rheol. 24:847 (1980).
17. A. Bergonzoni and A.J. Di Cresce, The phenomenon of draw resonance in polymeric melts, Polym. Eng. Sci. 6:45 (1966).
18. W.F. Busse, Mechanical structures in polymer melts. L Measurements of melt strength and elasticity, J. Polym. Sci.. A2, 5:1249 (1967).
19. K.F. Wissbrun, Interpretation of the melt strength test, Polym. Eng. Sci. 13:342 (1973).
20. G. Locati, Melt strength of polyethylene, private communication.
21. D. Romanini, A. Savadori, and G. Gianotti, Long chain branching in LDPE. 2. Rheological behavior of the polymers, Polymer 21:1092 (1980).
22. F.P. La Mantia, A. Valenza, and D. Acierno, A comprehensive experimental study of the rheological behavior of HDPE. I. Entrance effect and shear viscosity, Rheol. Acta 22:299 (1983).
23. F.P. La Mantia, A. Valenza, and D. Acierno, A comprehensive experimental study of the rheological behavior of HDPE. II: Die-swell and normal stresses, Rheol. Acta 22:308 (1983).

PART IV

PROPERTIES OF POLYMERIC MATERIALS

NEW FINDINGS ON THE MECHANICAL BEHAVIOUR OF PLASTICS

R. Knausenberger, T. Krehwinkel, G. Menges,
E. Schmachtenberg and U. Thebing

Institut fuer Kunststoffverarbeitung
Rhein.-Westf. Technischen Hochschule
Aachen

ABSTRACT

The mechanical properties of thermoplastics are formed by strong non-linear structures which are primarily determined by time, temperature and degree of multiaxial loading. For this reason the application of the classical theory of elasticity applies in very small deformation ranges.

By means of simple modifications of the idealized stress-strain correlation according to HOOKE, e.g. by introducing time and temperature dependent values, the linear elastic theory can be converted to the linear viscoelastic theory of material behaviour.

It can now be shown that, with the help of an empirically determined, non-linear (quadratic) relationship between stress and strain, an exact description and hence calculation of material behaviour is possible for uniaxial strains up to 4%. This calculation results from short-term-tests conducted at constant strain rates using time-temperature-superposition up to a standard loading time of several thousand hours.

In addition, the influence of multiaxial loading on the modulus can be shown. Pipes have been tested at constant strain rates in an axial and tangential direction under different strain ratios. The modulus decreases with increasing multiaxial loading. This can be explained by an increase in creep-behaviour under multiaxial loading.

We have developed the concept of "forced strain" which allows the multiaxial load to be regarded. Hence, it is possible, to calculate the material behaviour (strain, stress) under multiaxial loadings using material values measured in uniaxial tests.

INTRODUCTION

Mechanical loading capacity is evaluated according to relatively simple tensile tests (e.g. DIN 53455). Unfortunately however, the measured data for origin modulus, yield point and tensile strength are not suitable as characteristic data for designing building components. The deviation of both the long term load bearing capacity and the impact strength, from this data, is so great that so far, it has not been possible to generate suitable correlation functions. This unsatisfactory state necessitates expensive trials, e.g. for determining long term data. A further problem, although polymers also share this with other materials, is the uncertainty as to how multiaxial loads are best considered (1).

We have been involved with both these problems for some time, our aim being to modify a short term test in such a way as to provide data which could be used for dimensioning purposes (2). We assumed that slightly more expensive instrumentation could be used if it saved time.

UNIAXIAL TENSILE TESTS WITH CONTROLLED STRAIN RATE

The standardised tensile tests, e.g. DIN 53455, work with the specimen clamps moving at constant speeds (cf. Fig. 1). It is then generally assumed that the strain rate would also be constant. Our tests, e.g. (2), show, however, that if we monitor tests performed in this way by controlling the strain by placing a transducer at the middle of the measured length of the specimen, the strain rate contrary to expectations is not constant (Fig. 2). There are various causes for this, e.g. the positioning of the specimen clamps. Without investigating these causes, we decided to regulate the strain rates of the specimens to make them constant by taking the strain measured on the measured section of the specimen as the basis for the speed control of the machine. The strain occurring on the specimen during the tests is measured continuously and compared with a highly accurate linear set value profile plotted against time and produced by a function generator. A control circuit on the machine ensures a minimum difference between the set value and the actual value (2).

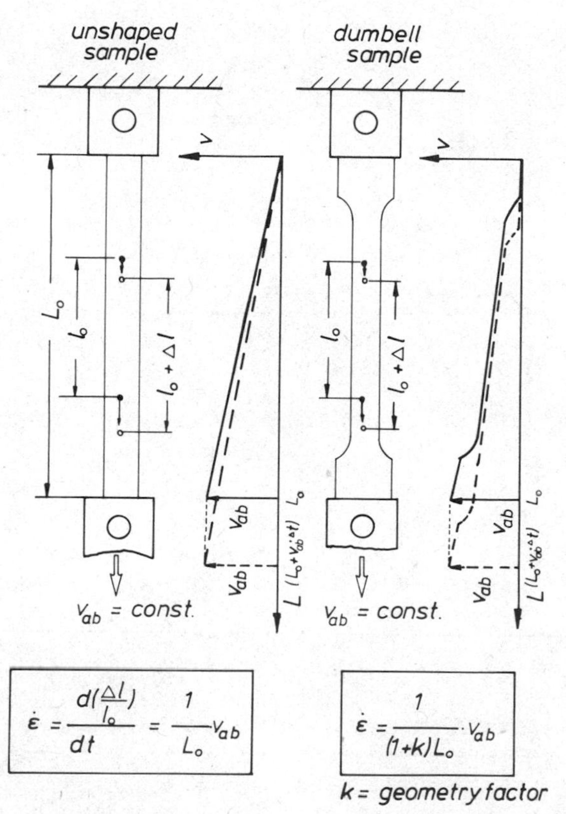

Fig. 1. Theoretical velocity distribution within sample during short-term-tensile-test.

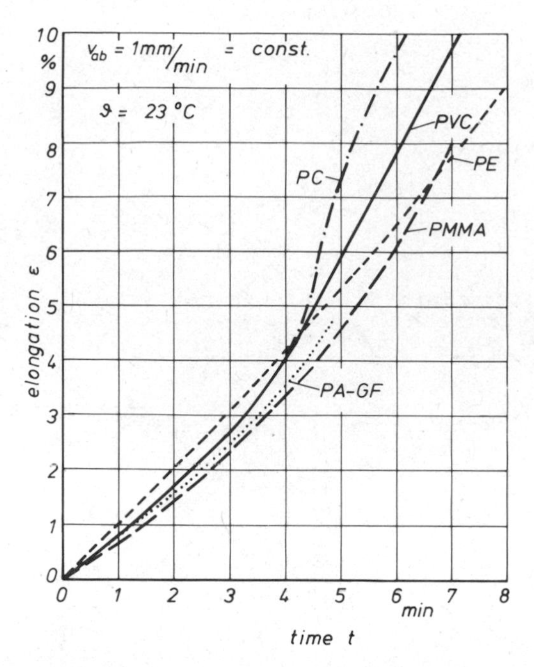

Fig. 2. Elongation behaviour of plastics during short-time testing.

Test results are normally subject to a varying degree of scatter between the measurements. This makes it necessary to perform a relatively large number of identical tests in order to ensure statistical accuracy. When testing with constant strain rates, the results show that scattering of the measurements is reduced to a remarkable extent. This is explained by the fact that it is now possible to ensure that every test specimen is subjected to the same deformation profile since possible differences in the clamping conditions or the geometry of the specimens no longer have any effect.

The stress-strain behaviour at various constant strain rates at room temperature is shown in Fig. 3 for PMMA. There is a clear relationship between the mechanical properties and the strain rate, which correspond approximately to isochronous stress-strain diagrams. The time and the strain rate have, as expected, opposite effects on the behaviour of the material. To determine the dependence on temperature of the mechanical properties at constant strain rates, we also carried out appropriate tests at elevated temperatures (Fig. 4).

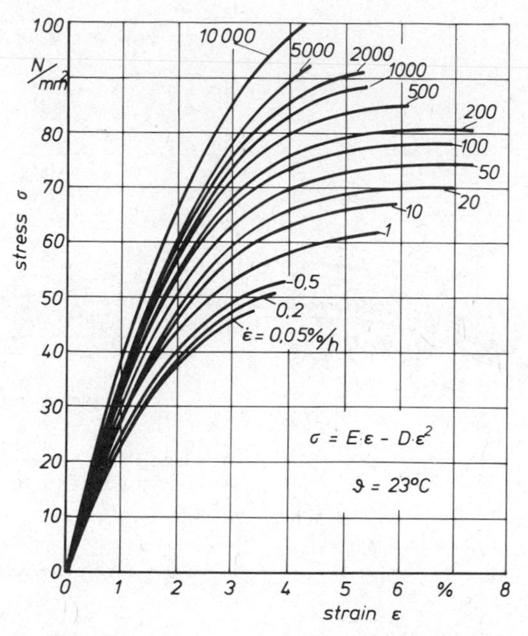

Fig. 3. Stress-strain behaviour of PMMA at constant strain rates.

Functional Description of the Mechanical Behaviour

The stress-strain profiles in Figs. 3 and 4 can be regarded as branches of a parabola, whose vertex lies at high stresses and strains and where the branches pass through the origin of coordinates. The analytic equation in Cartesian coordinates for this is:-

$$y = a_1 x - a_2 x^2. \qquad 1$$

This relationship, assumed purely on the basis of experience, signifies a non-linear relationship between stress and strain which can consequently be described by:-

$$\sigma = E\,\varepsilon - D\cdot\varepsilon^2 \qquad \text{or} \qquad 2a$$

$$\sigma = \varepsilon(E - D\varepsilon) \qquad 2b$$

Since, however, the tests were carried out at constant strain rates, we can also write for small deformations ($\varepsilon < 0{,}04$):-

$$\varepsilon = \dot{\varepsilon}\,t \qquad 3$$

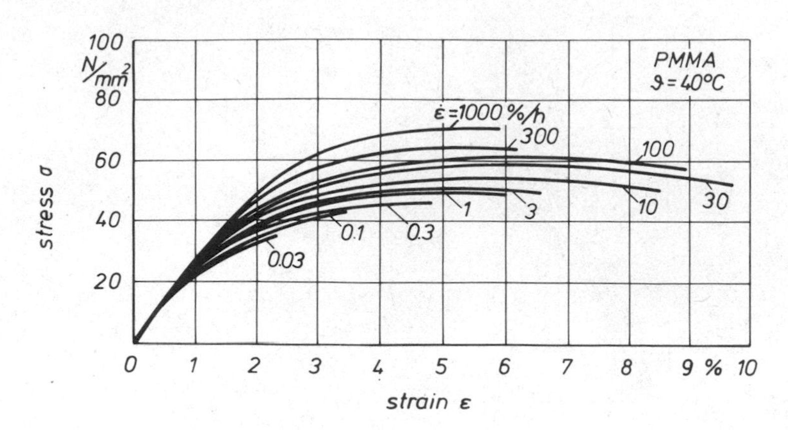

Fig. 4. Stress-strain behaviour of PMMA at constant strain rates at 40°C.

When this is inserted and account taken of the fact that different parameters are associated with every strain rate, we get:-

$$\sigma = E\,\dot{\varepsilon}\,t - D\,\dot{\varepsilon}^2 t^2 \qquad \text{or} \qquad (4a)$$

$$\sigma = \dot{\varepsilon}\,t\,(E - D\,\dot{\varepsilon}\,t) \qquad (4b)$$

This form of relationship was also selected because, for small strains, the second order quadratic term becomes small and thus extends into Hooke's law. Consequently, the factor E corresponds to a strain rate-dependent modulus of elasticity, whereas D represents a measure of the non-linearity of the material behaviour - a damping effect.

As is evident from equation 4, the stress at constant temperature depends only on the strain rate and the time (in tests at constant strain rate). Fig. 5 shows the relationship between the origin modulus and the strain rate at various temperatures.

The decrease of E with increasing temperature is in accord with experience concerning the behaviour of the elasticity modulus. Furthermore, the plot of E against the strain rate

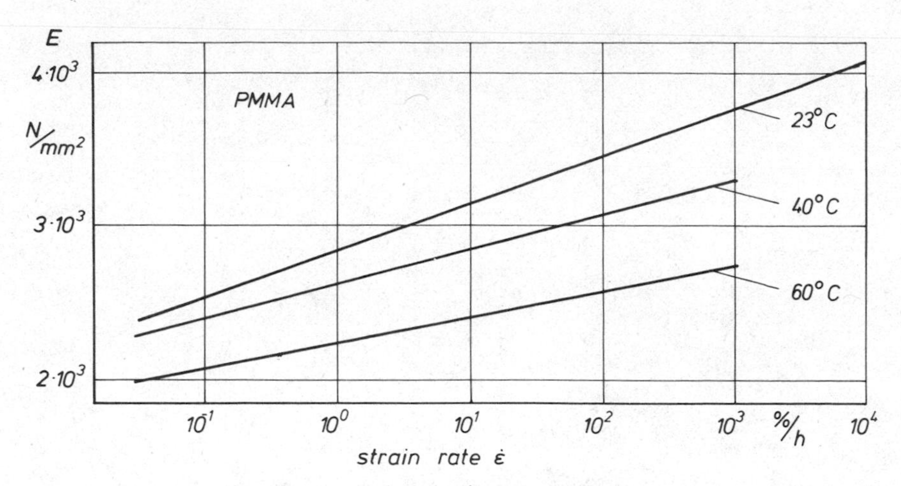

Fig. 5. Origin modulus E vs strain rate at different temperatures.

and the temperature is suitable for applying the time-temperature shift principle. If we shift the origin modulus curves with the shift factors in Fig. 6, we get, for the strain rate-dependent elasticity modulus, a single curve which, at room temperature (taken here as the reference temperature), covers the strain rate range of

$$10^{-5}\,\%/h < \dot{\varepsilon} < 10^{4}\,\%/h$$

The same shift factors applied to the damping D produce the result shown in Fig. 8. Here, too, we get a master curve from which it is evident that the non-linear influence decreases considerably for very small strain rates and thus also does so in the static creep test with very long times. It follows from the two master curves in Figs. 7 and 8 that, with tests carried out at constant strain rates, a pair of values E and D can be exactly attributed to every strain rate at constant temperature.

For the stress in equation 4, this means that it must also represent a clear and constant function of the strain rate and of the time. Mathematically, however, this means that, for the functional relationship:-

$$\sigma = f(\varepsilon, t, \vartheta) \quad = \text{const.} \qquad 5$$

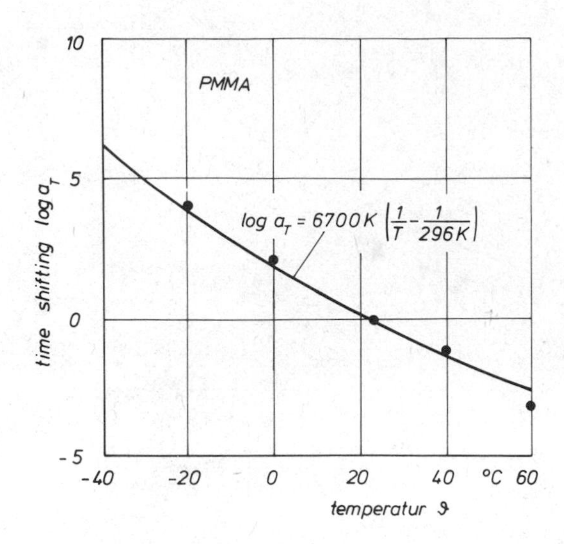

Fig. 6. Time-temperature shifting comparison:measurement-calculation.

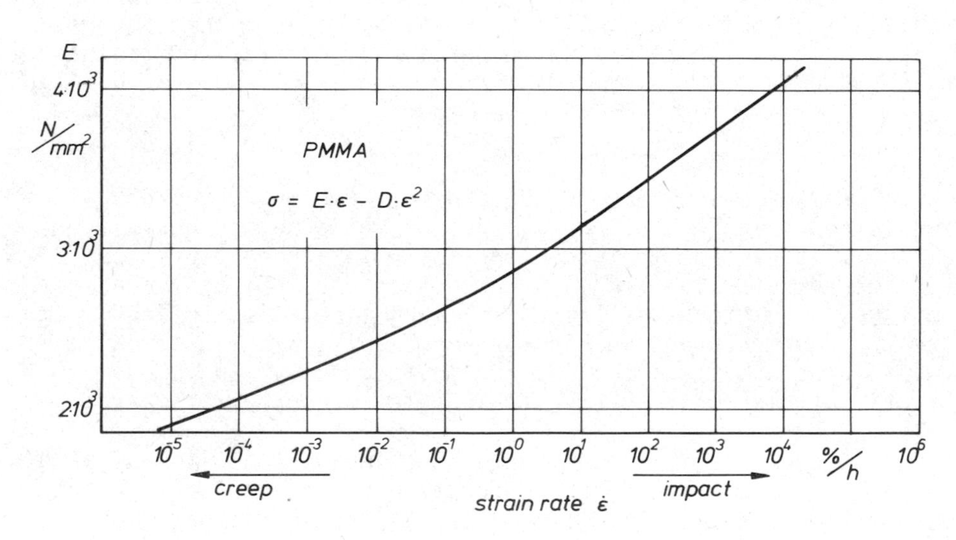

Fig. 7. "Master-curve" for the origin-modulus E at 23°C.

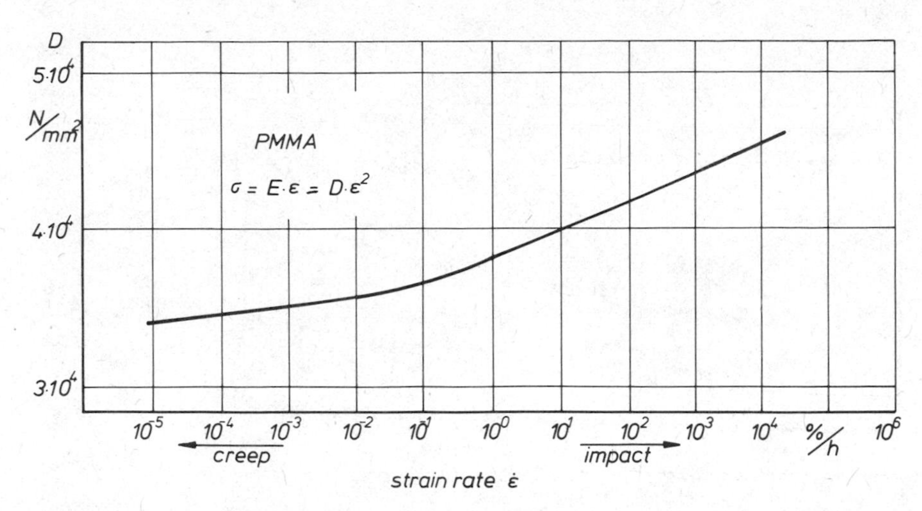

Fig. 8. "Master-curve" for the coefficient D at 23°C.

an inverse function exists of the form:-

$$\varepsilon = g(\sigma, t, \vartheta) \qquad = \text{const.} \qquad 6$$

With this inverse function, it must be possible, at given stress and time, to calculate the strain for a test with an unknown, but constant strain rate. However, the functions E ($\dot{\varepsilon}$) and D ($\dot{\varepsilon}$) contain the exponential function (sections of straight lines in semi-logarithmic representation), which is the reason why an analytical formation of the inverse function is not possible and the problem can only be solved by iteration.

Using such a method of calculation, curves were calculated to describe the strain as a function of time for a constant load. These curves were obtained by determining the strain at a given stress and time from the diagrams of the tests carried out at constant strain rate. These curves show good agreement in the deformation range $\varepsilon < 2\%$ with long-term creep tests for static loads (Fig. 9). The strains above this deformation range, which were found in the calculation process to be too small, can be explained by the first signs of damage in the material.

The similarity between the two curves over wide ranges of time testifies to the correctness of this procedure. It also

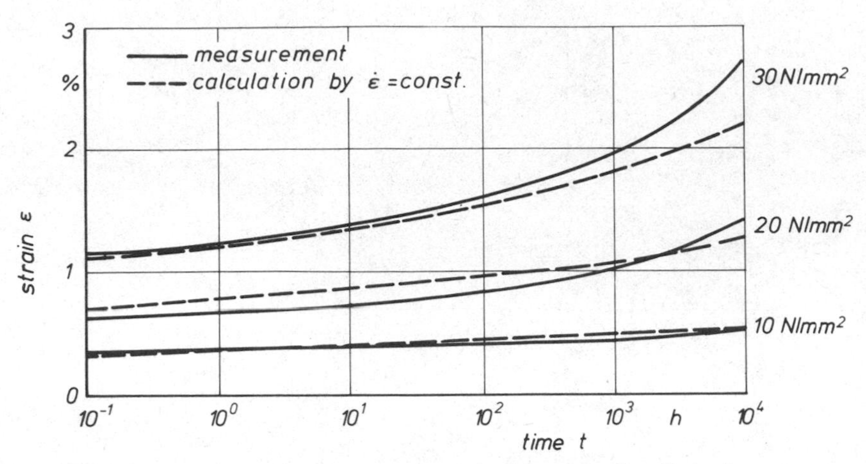

Fig. 9. Comparison between measured and calculated creep curves.

becomes clear, however, that it is not suitable for predicting the creep behaviour.

Finally, a further empirically determined relationship for the uniaxial stress state will be presented. If we apply the same shift factors to the times to fracture belonging to the tensile strengths:-

$$t_B = \varepsilon_B / \dot{\varepsilon} \qquad 7$$

from the tests carried out at constant strain rates at the various temperatures, we get the ultimate strength at room temperature shown in Fig. 10. This extends over a period up to 5×10^4 h (5.7 years) and conforms very well indeed with our experimental results. Similar agreement is obtained for the strain at ultimate load which becomes the elongation at break when the values determined at elevated temperatures are shifted to longer times using the same shift factors. This curve is shown in Fig. 11. The fact that this procedure is also successful with materials other than PMMA has been established in tests which are still continuing at the moment. In these tests, similar results have already been obtained with the semi-crystalline HDPE and other materials.

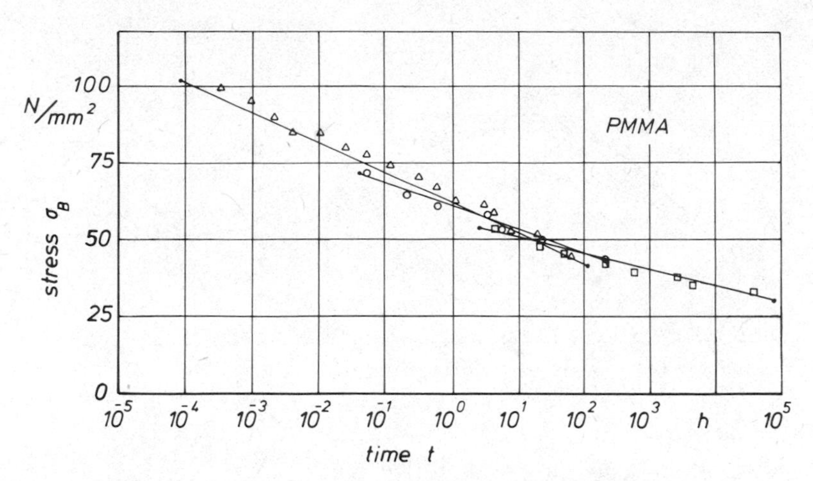

Fig. 10. "Master-curve" of ultimate strength at constant strain rates valid at room temperature.

Viewed as a whole, it can be assumed that tests carried out with constant strain rate are generally able to provide information on long-term properties.

MULTIAXIAL SHORT-TERM STRESS-STRAIN TESTS UNDER CONSTANT STRAIN RATES

Test set-up

It was assumed that the strain rate with multiaxial loads would have a similar significant influence as in uniaxial tensile tests.

For this reason, we re-designed a suitable tensile testing machine so that we could load pipe specimens with variable, regulated, axial and tangential strain rates. The two deformations could be adjusted independently of one another by altering the position of the cross-head or the internal pressure. Measurement is again made of the longitudinal strain, the circumferential strain - both of which are also fed back as measurements to the control system - as well as the thickness

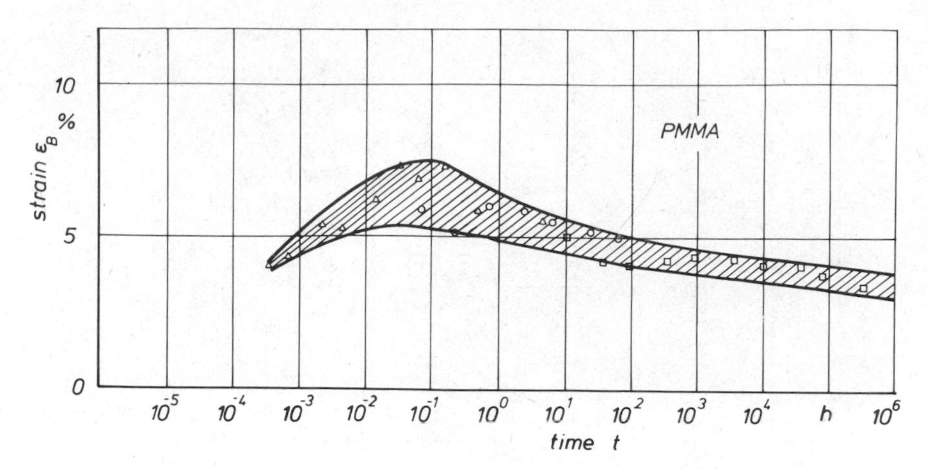

Fig. 11. "Master-curve of strain at ultimate load, calculated by time-temperature shifting principle.

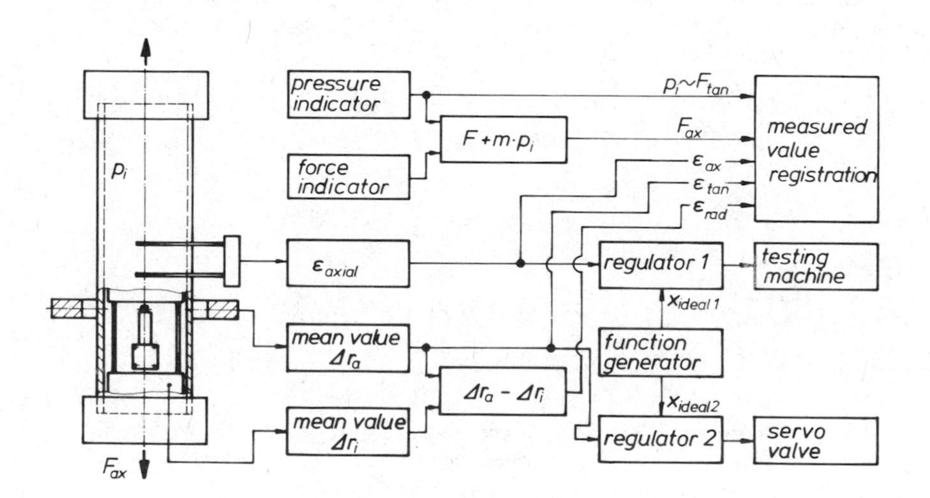

Fig. 12. Test equipment for biaxial material testing with constant strain rates.

strain. Fig. 12 shows a diagram of the set-up and Fig. 13 shows schematically the measuring sensors for the circumferential strain and for measuring the thickness of the pipe wall.

With this machine, it is thus possible to test pipe specimens under various stress states. The test parameters were either the strain rates in two directions or one strain rate and the stress in the other direction, giving certain stress and strain ratios similar to Table 1.

Table 1. Test Parameters.

Type of test	εx (%)	εy (%)	σx (N/mm^2)	σy (N/mm^2)
a	0.50	0	-	-
b	0	0.50	-	-
c	0.25	0.50	-	-
d	0.50	0.25	-	-
e	0.50	0.50	-	-
f	0.50	-	-	0
g	-	0.50	0	-

x = tangential   y = axial

The results of these tests on extruded PMMA are shown in Figs. 14 and 15.*

Results of Quasi-Uniaxial Tensile Tests on Pipe Specimens of PMMA

The tests f and g (Table 1) are the equivalent of a uniaxial tensile test. In each case, stresses occur here only in the axial or tangential direction (neglecting the low radial compressive strength in f from the internal pressure). They are thus suitable for providing information about the anisotropy of the material. Fig. 16 shows the elasticity modulus for these two directions, calculated from $E_x = \sigma_x / \varepsilon_x, E_y = \sigma_y / \varepsilon_y$. As was expected, the modulus of elasticity in the axial direction, i.e. in the extrusion direction of the pipe, is higher. This can be explained by a greater orientation in this direction. It is noticeable that the plot for E_y shows a curvature, whereas E_x is inclined linearly. This difference is also evident in the other tests. The modulus of elasticity in the thickness direction

* Extrusion material therefore of lower molecular weight than the material in 1.

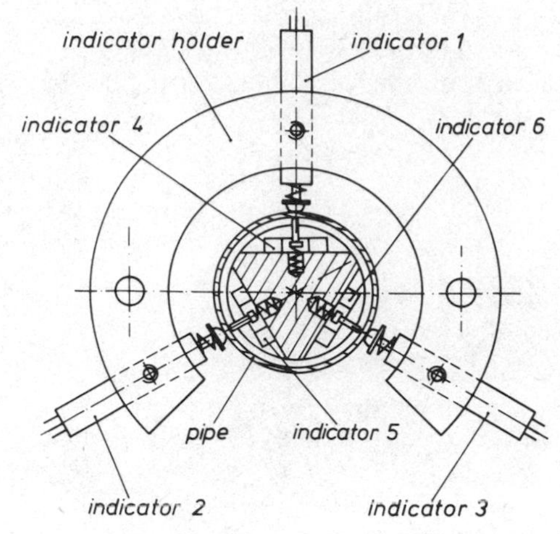

Fig. 13. Measurement of circumferential strain and wall thickness of pipes.

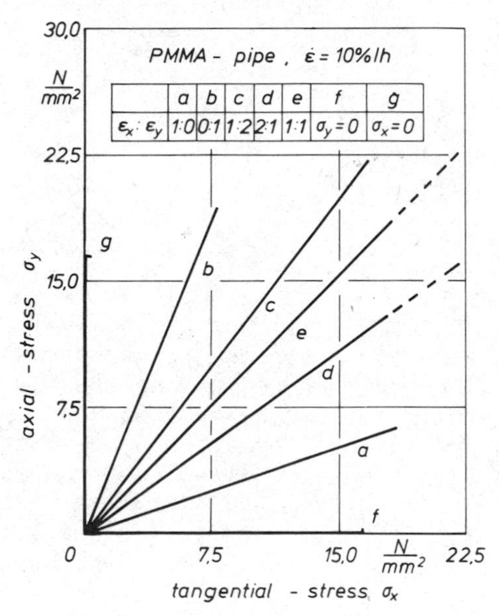

	a	b	c	d	e	f	g
$\varepsilon_x : \varepsilon_y$	1:0	0:1	1:2	2:1	1:1	$\sigma_y = 0$	$\sigma_x = 0$

Fig. 14. Biaxial stressing for different strain-ratios.

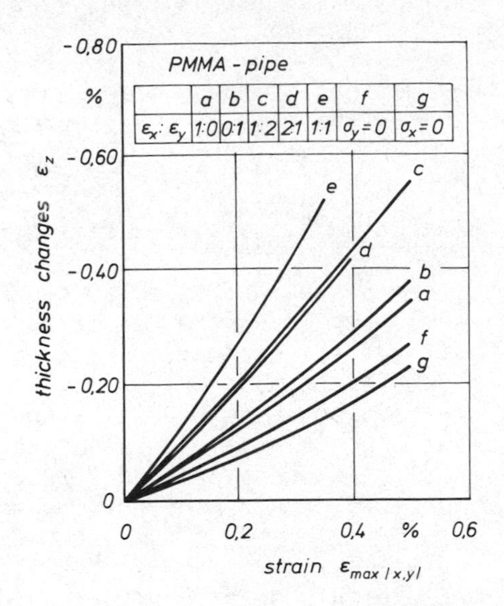

	a	b	c	d	e	f	g
$\varepsilon_x : \varepsilon_y$	1:0	0:1	1:2	2:1	1:1	$\sigma_y = 0$	$\sigma_x = 0$

Fig. 15. Thickness changes at different strains.

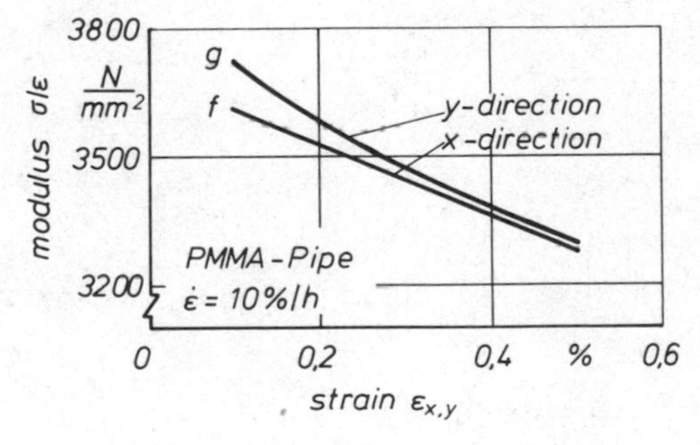

Fig. 16. Modulus in relationship to the direction in effectively uniaxialy loaded pipes.

cannot be determined directly. Because the dependence on the direction is altogether lower (maximum difference only 3.6%), we can assume a virtually isotropic material behaviour. A further point in favour of this is that the μ-values (Lame constant = shear modulus) have, for all the tests, a mean deviation of only 6.2% calculated on the smallest of these three values. This scattering already contains the measuring inaccuracy.

Determining the Material Data by a Multiaxial Pipe Test

In these tests, the stresses and strains were determined in all three spatial directions. Under the condition of isotropy, however, it is basically sufficient to measure 5 of these stress values in order to determine the material data such as elasticity modulus, transverse contraction (Poisson ratio) and the Lame constants. In the biaxial or uniaxial stress state, the number of values to be measured is reduced to 4 and 3 respectively, since the boundary conditions provide information about further stress values.

However, since measurements never provide exact values for the stress or deformation state, additional measurement of the other values can be used to provide information about the quality of the other measurements. Besides this, it has been found that it is very important which measurements are used to determine the material data. This can be explained with the example of a biaxial stress state (σ_z = 0). We then get:-

$$E = \frac{\sigma_x^2 - \sigma_y^2}{\varepsilon_x \sigma_x - \varepsilon_y \sigma_y} \qquad 8$$

or

$$\nu = \frac{\varepsilon_x \sigma_y - \varepsilon_y \sigma_x}{\varepsilon_x \sigma_x - \varepsilon_y \sigma_y} \qquad 9$$

and the characteristic data can be determined without knowing the strain in the z-direction. For stress states, = , the denominator of these expressions nevertheless approaches zero and even small measuring errors can result in major changes in the characteristic value.

For these types of tests, use therefore has to be made of the values for the strain measurement in the wall thickness direction (z-direction) for determining the material data. On observing the measuring systems and their resolution, however, we find that the accuracy of the radial strain measurement is poorer by a factor of around 15-25 than that of the longitudinal or circumferential strain measurement.

This means that, for the evaluation, it is necessary to weight these two aspects against one another in order to obtain the maximum probability for the calculated characteristic data. The procedure adopted was to calculate the data from the strains in the circumferential and axial directions. Using the characteristic data obtained in this way (points in Figs. 17 and 18), the strain state in the radial direction was determined. A comparison between measurement and calculation then provided information about the quality of the previously determined characteristic value. Where the deviations were excessive, these measurements were not taken into account for determining the characteristic data.

Therefore the following conclusions can be drawn from the characteristic data:

The Poisson Ratio

The Poisson ratio is constant for tests with constant strain rate, in other words it is dependent neither on the degree of multiaxiality nor on the strain. For the present measurements, we get $\nu = 0.37$ (Fig. 17).

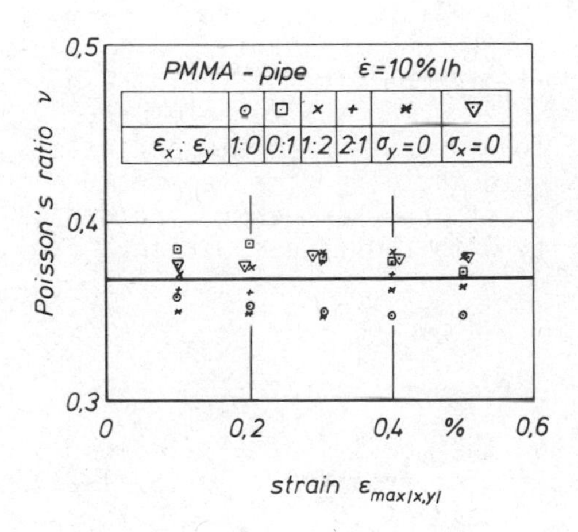

Fig. 17. Poisson's ratio for different strain-levels.

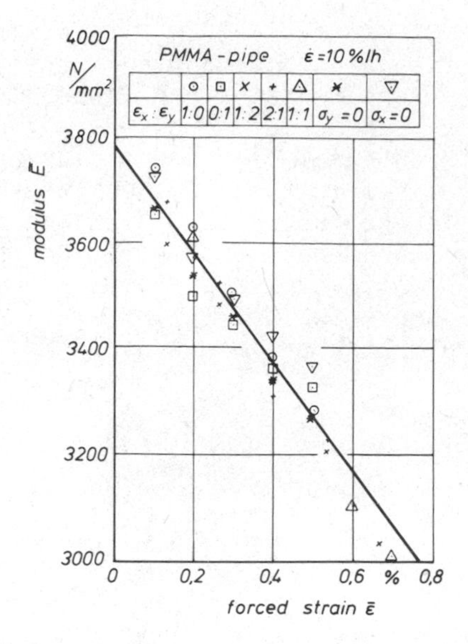

Fig. 18. Modulus vs strain for different strain-ratios.

Besides this, the Poisson ratio was determined in the uniaxial tensile test (under constant strain rate). It was also found here that it is independent of the strain and dependent only on the strain rate or on the temperature. Fig. 19 shows the Poisson ratio for various temperatures and strain rates. It is interesting to note that the time-temperature shift principle proved to be usable with the same shift factors as for the modulus. Conformity with the values obtained from the multiaxial tests is excellent.

Modulus of Elasticity

In Hooke's law, the modulus of elasticity is defined as the quotient of stress and deformation. In ideal-elastic materials, this quotient is independent of the load. In plastics, however, this quotient decreases with increasing strain. Let us discuss this once again for the uniaxial loading of PMMA. From the equation:-

$$\sigma = E \cdot \varepsilon - D \cdot \varepsilon^2 \tag{2a}$$

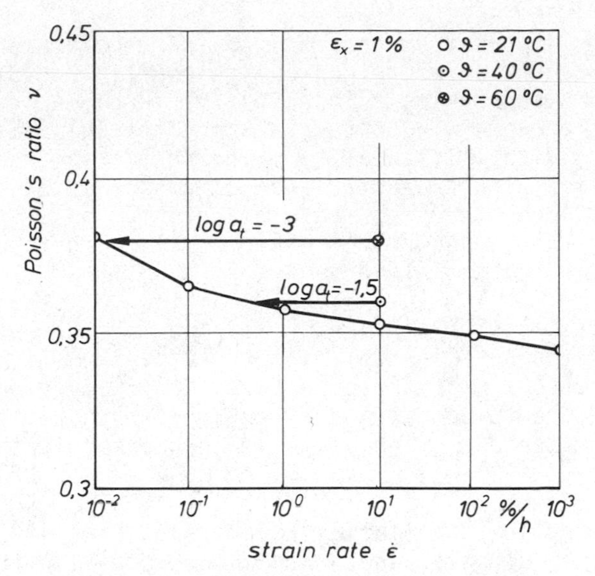

Fig. 19. Poisson's ratio in relationship to strain rate (uniaxial-tension).

the quotient of stress and deformation can be formed:-

$$\frac{\sigma}{\varepsilon} = E - D\varepsilon \tag{10}$$

For sufficiently small deformations, we get from this the origin modulus:-

$$\lim_{\varepsilon \to 0} \frac{\sigma}{\varepsilon} = E \tag{11}$$

In addition, equation 10 describes the path of the secant modulus as is determined from the stress-strain diagram when a strain is given. For the describing statement for the stress-strain behaviour in line with equation 2a, we get a linearly inclined function for the quotient $\frac{\sigma}{\varepsilon} = \bar{E}$ (described from now as the modulus).

An analysis of Fig. 18 shows that here too, a linear function is found for a multiaxial stress state. A comparison with the origin modulus measured in the uniaxial tensile test on dumbell specimens shows that the origin modulus of the pipe specimens was about 15° higher. This difference can be explained by the varying types of material (cast or extruded PMMA). Tensile specimens produced from the pipe material showed conformity in their elasticity modulus with the quasi-uniaxial tensile tests performed on the pipe specimen.

The second interesting property of the modulus function is its gradient, which is determined by the coefficient D. It was found that, with increasing multiaxiality, these curves become steeper. This property indicates that greater relaxation processes occur with multiaxiality. The multiaxial deformation states calculated with characteristic data from uniaxiality are therefore also lower than the measured deformations. An attempt was therefore made to record the higher decline in the modulus in the multiaxial stress state.

One obvious possibility is to plot the modulus against the sum of forced strains (Fig. 20), in order to take into account the higher stress, say of the pipe test, for example in the 1:1 strain state. The forced strain is defined as those strains which lie directly in the direction of the external application of force.

It was found that the curves for various biaxial stresses coincide with the uniaxial ones. In accordance with the previous considerations, the creep modulus could, in analogy to equation 10, be defined as:-

$$E_c = E_{(t,\vartheta)} - D_{(\varepsilon,t,\vartheta)} \cdot \varepsilon \qquad (12)$$

This means that the dependence on the multiaxial load is reflected only in the coefficient D for the attenuation. This is a question of practicability which will become evident on further use.

The Concept of Forced Strain

From the previous considerations, it is possible to develop a concept with which multiaxial stress-strain states can also be calculated:-

Basically, we use an extension of Hooke's law:-

$$\varepsilon_x = \frac{1}{\bar{E}} (\sigma_x - \nu(\sigma_y + \sigma_z)) \qquad (13a)$$

$$\varepsilon_y = \frac{1}{\bar{E}} (\sigma_y - \nu(\sigma_z + \sigma_x)) \qquad (13b)$$

$$\varepsilon_z = \frac{1}{\bar{E}} (\sigma_z - \nu(\sigma_x + \sigma_y)) \qquad (13c)$$

The degree of multiaxiality and the influence of time and temperature are taken into account when determining the characteristic data as follows:

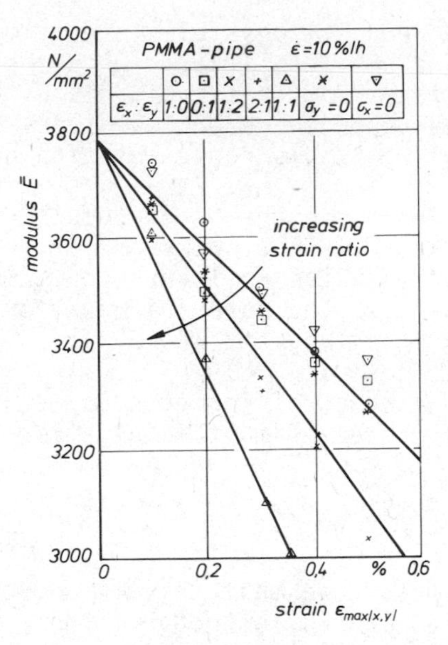

Fig. 20. Modulus for different strain-ratios vs forced strain.

For given strains, the characteristic data can be determined directly. First of all, for the deformation state, we form the characteristic strain rate from the quotient of principal strain and loading time. Fig. 18 then provides the respective Poisson ratio.

The modulus $\bar{E} = \frac{\sigma}{\varepsilon}$ can be calculated from equation 10, whereby the coefficients D and E are determined from Figs. 6 and 7 with the aid of the characteristic strain rate. For E, the sum of the forced strains is inserted in equation 10. The forced strain is defined as those strains which lie directly in the direction of the external application of force.

If, however, we proceed from the stress state, the calculation must be made in the iterative form, whereby a strain state is calculated from initially assumed characteristic data. From the calculated strain state, better characteristic data can now in turn be determined and used for a second iteration step.

With a method of calculation such as this, it is thus possible to determine multiaxial stress states from data determined in uniaxial tests.

It is, by the way, equally possible to determine the Lame constants in order to obtain simplified equation systems for calculating a stress state from the strain state. The Lame constants have the same dependences on the stress state (3). It is not intended to deal with this representation here.

It remains to be tested whether this concept can also be applied to other plastics. It has, however, at least been found that, by taking into account the influences of time, temperature and degree of multiaxiality, the material behaviour can be described using the extended Hooke's law.

Besides using the concept of forced strains, it is also possible to record the degree of multiaxial stress state with the stress vector:-

$$\bar{\sigma} = \sqrt{\sigma_x^2 + \sigma_y^2 + \sigma_z^2} \qquad (14)$$

Initial research has shown that this method covers the multiaxiality even more clearly, although it does produce additional problems in solving the equation systems for calculating the stress state. This method has therefore not been pursued any further for the time being.

REFERENCES

1. Troost, A. Einfuehrung in die allgemeine Werkstoffkunde metallischer Werkstoffe I

Wissenschaftsverlag
Bibliographisches Institut AG
Zuerich 1980

2. Knausenberger, R. Das mechanische Verhalten isotroper und anisotroper Werkstoffe mit nichtlinear-elastischen Eigenschaften

Dissertation RWTH Aachen 1982

3. Krehwinkel, T. Ermittlung der mechanischen Kennwerte von PMMA-Rohren bei mehrachsigen Beanspruchungszustaenden

Diplomarbeit IKV-RWTH Aachen 1983
Betreuer: E. Schmachtenberg

VISCOELASTICITY, ULTIMATE MECHANICAL PROPERTIES AND YIELDING MECHANISMS OF THERMALLY CRYSTALLIZED POLYETHYLENE TEREPHTHALATE

Gabriel Groeninckx

Catholic University of Leuven, Laboratory of Macromolecular and Organic Chemistry, Celestijnenlaan 200 F
B-3030 Heverlee-Leuven (Belgium)

INTRODUCTION

In recent publications, the influence of the thermal history on the morphology, melting characteristics[1-3] and modulus-temperature behavior[4-5] of polyethylene terephthalate (PET) was investigated.

In this paper the results of a thorough experimental study of the influence of different crystallization and annealing conditions upon the viscoelastic relaxation behavior and ultimate mechanical properties of PET will be presented.

EXPERIMENTAL

Isotropic partially crystalline PET-samples with a wide range of crystallinities and different microstructures can be obtained by suitable thermal treatments[1-3].

Isothermal crystallization experiments from the glassy state at different temperatures T_c ($88°C \leq T_c \leq 230°C$) and two types of annealing treatments have been carried out in a vacuum oven (Table I). The annealing treatments were performed on PET samples, isothermally crystallized at a low crystallization temperature $T_c = 100°C$ with a degree of crystallinity $X_c = 29\%$ (as calculated from density measurements). These samples were annealed at different temperatures above $T_c = 100°C$ after quick heating (series I, QH samples) or slow heating at 4°C/hr (Series II, SH samples) to the annealing temperature T_{an}; the samples were left at T_{an} until the maximum degree of crystallinity was reached (annealing time : 17 hrs).

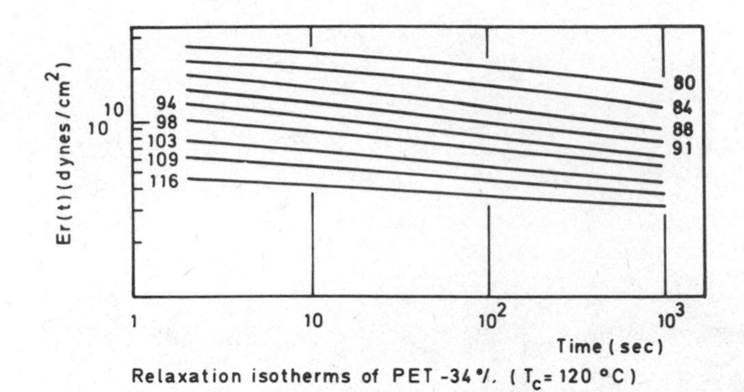

Relaxation isotherms of PET -34 %. (T_c = 120 °C)

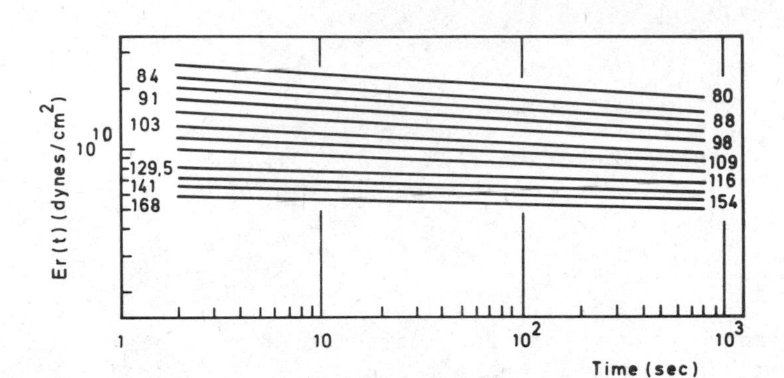

Relaxation isotherms of PET - 60 %. (T_c = 230 °C)

Fig.1. Stress relaxation isotherms of isothermally crystallized PET

Stress relaxation using a flexural mode of deformation and uniaxial stress-strain experiments were performed with an Instron testing machine.

RESULTS AND DISCUSSION

Modulus-time behavior and temperature dependence of the relaxation mechanisms.

The stress relaxation of amorphous and isothermally crystallized PET and of the two series of annealed samples was studied. The time dependence of the stress relaxation modulus $E_r(t)$ at different temperatures was investigated between 1 and 10^3 s. This behavior is illustrated in Fig.1 for isothermally crystallized PET at T_c = 120°C (X_c = 34 %) and T_c = 230° C (X_c = 60 %). The superposition of the relaxation isotherms is satisfactory and allows to conclude to the validity of the time-temperature superposition principle for all samples studied. A horizontal shift of the relaxation isotherms along the log time axis, with 94°C as the

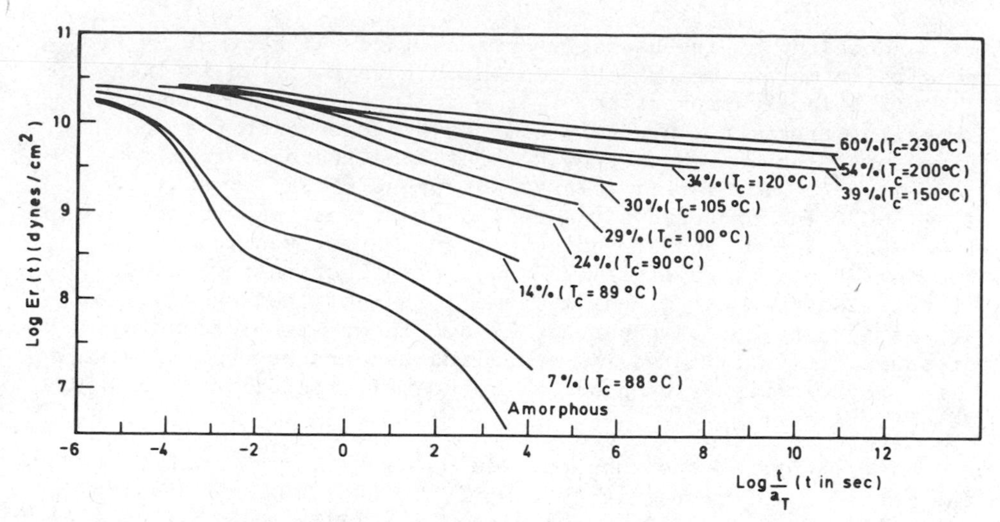

Fig.2. Modulus-time master curves of isothermally crystallized PET at 94°C.

reference temperature, results in modulus-time master curves covering many decades in time. The master curves of isothermally crystallized PET samples with different crystallinities are represented in Fig.2. Only the master curves of amorphous PET and of the sample with X_c = 7 % show a flow region at high relaxation times. The absence of flow at higher crystallinity is characteristic for highly cross-linked systems; in semicrystalline polymers, the crystalline lamellae act as physical cross-links. The value of the modulus in the rubbery region depends of both this cross-linking effect and the presence of a rigid crystalline phase. The rubbery modulus is strongly influenced by the isothermal crystallization conditions which determine the crystallinity and the morphology of the crystalline and amorphous layers. These aspects have been described in detail earlier[1,4].

The glass-rubber transition region is also strongly affected by the isothermal crystallization temperature T_c. As can be seen from Fig.2, the master curves of samples with X_c = 39, 54 and 60 % intersect the cruves of samples with X_c = 34 %. This means that the glass-transition temperature, T_i, exhibits a maximum as a function of X_c and T_c (see Table I; the values of the inflection temperature T_i have been derived from our modulus-temperature graphs reported in ref.4). Similar results have been reported in the literature from dynamic mechanical analysis[6,7].

The broadening of the master relaxation curves in the glass-transition region as a function of T_c clearly indicates a change in the nature of the interlamellar amorphous phase, composed of chain folds, tie molecules, chain ends and segregated low molecular weight molecules.

Annealing of isothermally crystallized PET above $T_c = 100°C$, results in morphological changes which have a marked influence on the viscoelastic properties. The structural changes induced by annealing treatments are very complex and depend on the heating procedure used (quick or slow heating) and the annealing temperature T_{an} (2). The master relaxation curves of PET crystallized at $T_c = 100°C$ and subsequently annealed at higher temperatures (T_{an}= 150, 200 and 230°C) after quick heating, are given in Fig.3; the degree of crystallinity X_c and the inflection temperature T_i of these samples are given in Table I (quickly heated samples, series I). As can be observed, T_i decreases with increasing crystallinity and increasing annealing temperature but is always a few degrees higher than after isothermal crystallization at the same temperature.

The change of the rubbery modulus is primarily controlled by the increase in crystallinity from 29 to 60 %. When PET samples crystallized at $T_c = 100°C$ are heated at 4°C/hr to the annealing temperature (slowly heated samples, series II), the inflection temperature T_i increases to 94°C for $T_{an} = 150°C$ and then remains nearly constant at higher annealing temperatures. This series of samples exhibit the highest modulus in the rubbery modulus.

As mentioned above, relaxation modulus master curves are constructed by sliding the experimental relaxation isotherms along the log time axis with respect to a reference temperature. The translation distances used for composing the master curves are the so-called time-temperature shift factors a_T.
The temperature dependence of the experimental shift factors has been investigated for all samples considered in this work. This temperature dependence was first examined in the temperature domain corresponding to the glass-transition region of the samples ($-10°C \leqslant T-T_i \leqslant 30°C$).

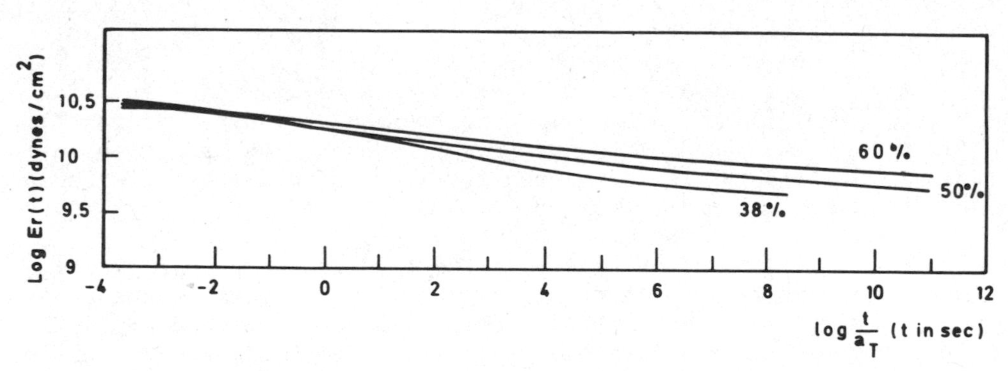

Fig.3. Modulus-time master curves of annealed PET (quickly heated samples, Series I)

Table I Characteristics of isothermally crystallized and annealed PET

Crystallization temperature T_c or Annealing Temperature T_{an} (°C)	Degree of crystallinity X_c (%)	Inflection temperature T_i (°C)
	Isothermally crystallized PET samples	
Amorphous	-	75
88	7	77
89	14	79
90	24	85
100	29	88
105	30	89
120	34	92
150	39	91
200	54	85
230	60	84
	Annealed PET samples (quick heating, Series I)	
150	38	94
200	53	88
230	60	85
	Annealed PET samples (slow heating, Series II)	
150	38	94
200	51	93
230	59	93

The experimental shift factors, log a_T, have been plotted against $T-T_i$. Such plots are shown in Fig.4 for amorphous and isothermally crystallized PET. The WLF equation given by :

$$\log a_T = - \frac{C_1 (T - T_i)}{C_2 + T - T_i}$$

where $C_1 = 16{,}14$ and $C_2 = 56$ is also represented in this figure (full line). From these plots, it can be concluded that the WLF equation can be applied to all semicrystalline samples, within their glass-transition region, with the same constants C_1 and C_2 as for the amorphous polymer. This result indicates that only the amorphous phase contributes to the relaxation in the temperature domain considered; as a consequence, the crystalline morphology of the samples does not affect the relaxation processes in this temperature domain.

With samples crystallized or annealed at high temperatures (200 and 230°C), the temperature dependence of the shift factors can be investigated up to $T-T_i \simeq 100$°C. Such plots are represented in Fig.5 for PET crystallized at T_c= 200°C and 230°C, and for PET annealed at T_{an}= 200°C and 230°C after quick heating.

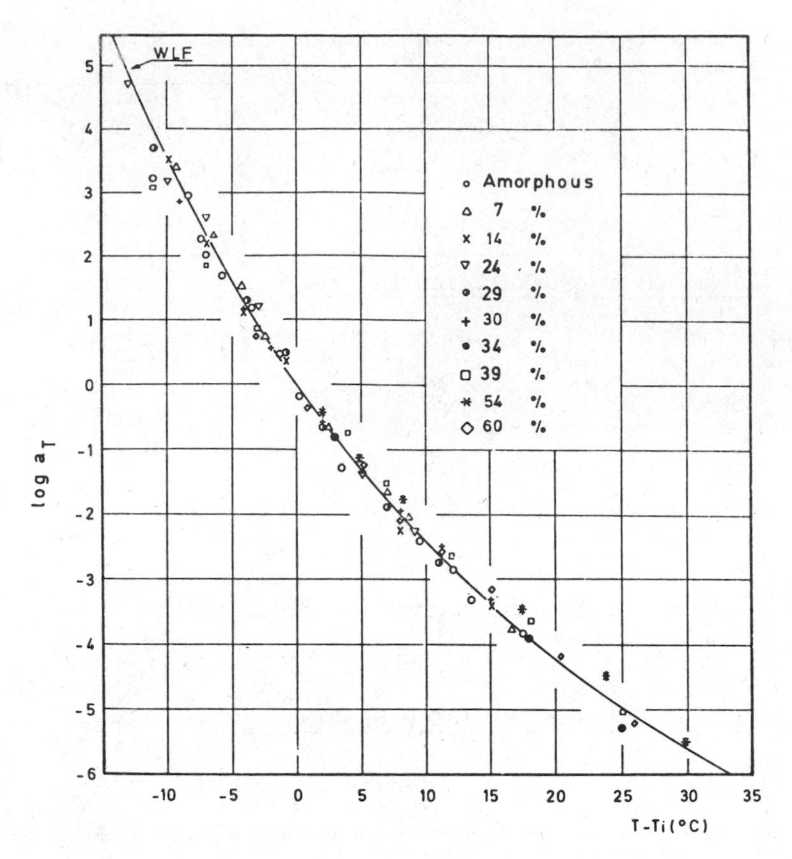

Fig.4. Experimental shift factors log a_T versus $T-T_i$ for amorphous and isothermally crystallized PET in their glass-transition region.

As can be observed for $T-T_i > 40°C$, the experimentally determined shift factors deviate noticeably from those predicted by the WLF equation, indicating the appearance of additional relaxation mechanisms. In this wide temperature range, corresponding to the rubbery state of the samples, the time-temperature superposition remains satisfactory. A similar behavior was observed by Kawai and co-workers[8] by studying the frequency dependence of the dynamic modulus at different temperatures. From Fig. 5 it is clear that the deviation from WLF behavior must be related to the crystalline morphology of the samples which depend on the crystallization and annealing conditions[1-3].

If we assume that the temperature dependence of the relaxation times of the non-crystalline phase follows the WLF equation at $T-T_i >$ 40°C, as was observed at lower temperature, it is possible to determine the contribution of the additional relaxation process for $T-T_i > 40°C$ by substracting the calculated WLF shift factors from the

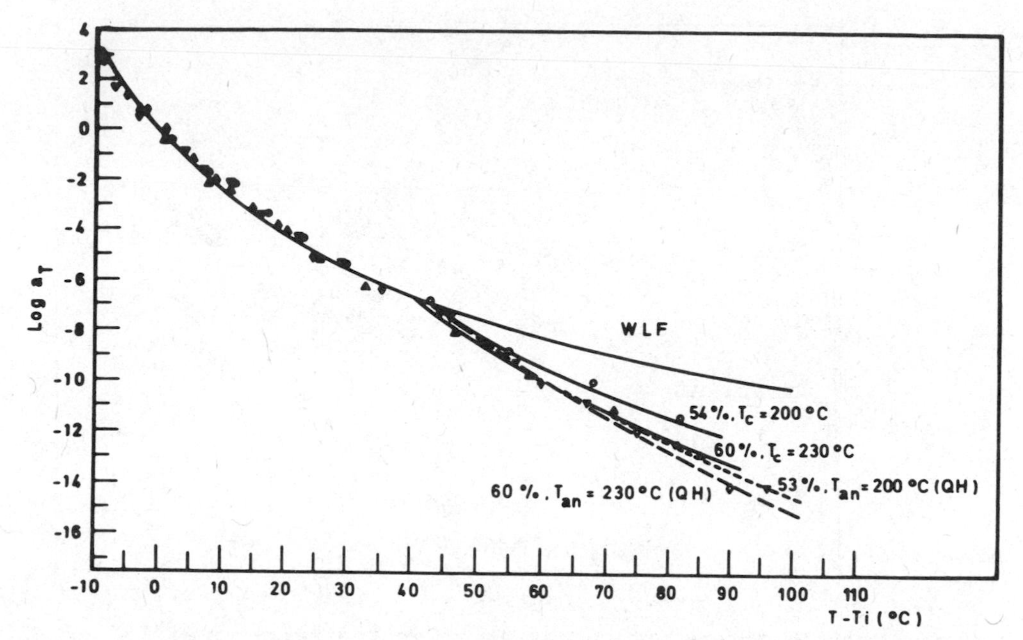

Fig.5. Experimental shift factors log a_T versus $T-T_i$ for isothermally crystallized PET and annealed PET at high temperatures.

experimentally observed shift factors. These differences have been plotted as a function of $^1/_T$ for the different samples considered. The resulting linear plots indicate that the temperature dependence of the additional relaxation mechanism is of the Arrhenius type. From the slopes of these plots, activation energies ΔE were determined which depend on the samples studied. The overall-temperature dependence of the observed shift factors can be described by an equation of the WLF form up to T_d (temperature of deviation from WLF behavior) and above T_d by an Arrhenius-type equation superimposed on the WLF equation :

$$\log a_T = -\frac{C_1\ (T-T_i)}{C_2 + T-T_i}\ h(T) + \frac{\Delta E}{2{,}303R}\left[\frac{1}{T} - \frac{1}{T_d}\right] h(T-T_d)$$

where h(T) and $h(T-T_d)$ represent Heaviside functions. Below T_d, the viscoelastic response of the semicrystalline PET specimens is due only to the non-crystalline regions, with the crystalline phase merely acting as an inert filler ; above T_d, the crystalline phase superimposes its response on the rubbery response of the amorphous regions. In the literature two types of molecular motions are reported to be responsible for these deviations in the temperature dependence of the shift factor: molecular motions in the crystal lattice(9) and reorientations of the crystalline lamellae in the amorphous matrix together with chain

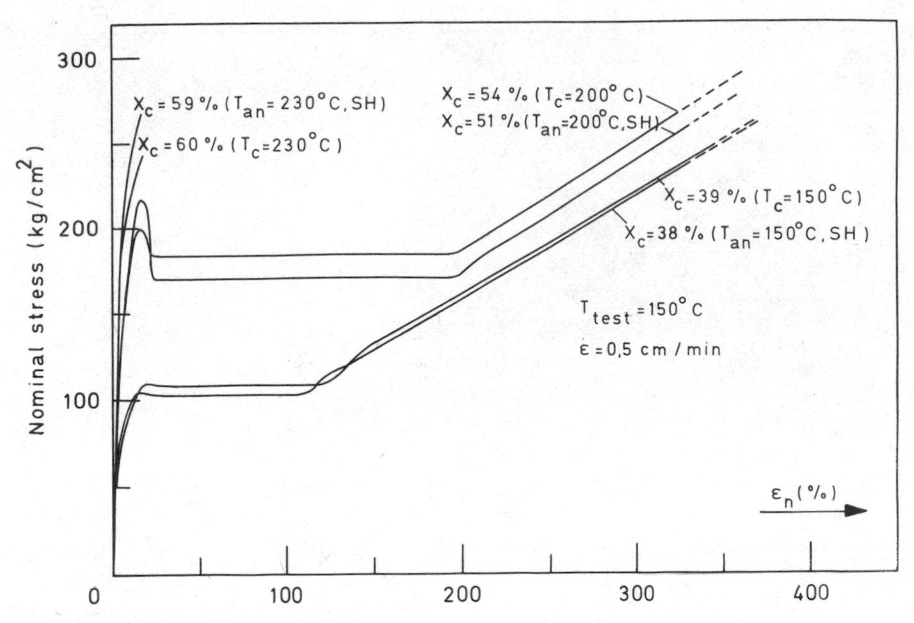

Fig.6. Tensile stress-strain curves of isothermally crystallized PET and annealed PET (SH, Series II) at 150°C

fold motions at the surface of the lamellae[10]. From our experiments, no definite conclusion can be made concerning this problem.

Ultimate mechanical properties and mechanisms of plastic deformation.

Some results regarding the influence of the degree of crystallinity and the morphology on the yield behavior will be reported in this paper. Fig.6 represents the stress-strain curves for isothermally crystallized and annealed (SH) PET samples at 150°C (strain rate : 25 %/min). Different regions can be identified corresponding to viscoelastic deformation, yielding by necking, cold drawing and strain hardening, respectively.

The change in mechanical properties is primarily determined by the increase in crystallinity; samples with completely different morphologies but with nearly the same crystallinity exhibit very similar high strain properties. The plastic deformation occurring in the necking zone of the samples transforms the original spherulitic structure into the characteristic fiber structure (microfibrils have been observed by electron microscopy). The deformation of the fiber structure in the strain hardening region of the stress-strain curves is assumed to proceed by longitudinal sliding motion of the microfibrils past each other. More details will be published elsewhere.

ACKNOWLEDGMENT

The author is indebted to the Nationaal Fonds voor Wetenschappelijk Onderzoek (NFWO-Belgium) for equipment and financial support. He also whishes to thank J.Thoré and L.Swéron (Katholieke Industriele Hogeschool Leuven) for their collaboration in the second part of this work.

REFERENCES

1. G.Groeninckx, H.Reynaers, H.Berghmans and G.Smets, J.Polym.Sci., Phys.Ed., 18, 1311 (1980)
2. G.Groeninckx and H.Reynaers, J.Polym.Sci., Phys.Ed., 18, 1325 (1980)
3. F.Fontaine, J.Ledent, G.Groeninckx and H.Reynaers, Polymer, 23, 185 (1982)
4. G.Groeninckx, H.Berghmans and G.Smets, J.Polym.Sci., Phys.Ed., 14, 591 (1976)
5. G.Groeninckx, H.Reynaers and H.Berghmans, Polymer, 15, 61 (1974)
6. K.H.Illers and H.Breuer, J.Colloid.Sci., 18, 1 (1963)
7. J.H.Dumbleton and T.Murayama , Kolloid Z.Z.Polym., 220, 41 (1967)
8. T.Tajiry, Y.Fuju, M.Aida and H.Kawai , J.Macromol.Sci., Phys., B4(1), 1(1970)
9. J.D.Hoffman, G.Williams and E.Passaglia, J.Polym.Sci., C, 14, 176 (1966)
10. Y.Ishida, J.Polym.Sci., 3, 1835 (1969)

SHEAR STRENGTH CHARACTERISTICS AND SHEAR CREEP MODELING OF LOW DENSITY RIGID POLYURETHANE FOAM

Pentti Mäkeläinen and Olavi Holmijoki

Helsinki University of Technology
Department of Civil Engineering
SF-02150 Espoo 15, Finland

INTRODUCTION

The use of rigid polyurethane plastics foam as core material in structural sandwich construction has become increasingly important in recent years. Especially light-weight sandwich panels consisting of a low density rigid foam core bonded between thin steel sheet faces have found wide application in many fields of building construction. For an efficient use of rigid polyurethane foam material in sandwich construction, it is essential to know its mechanical properties in full, particularly as regards the shear properties.

In this study, experiments were carried out for determining both the short-term shear strength and stiffness and the long-term shear creep properties of rigid polyurethane plastics foam. Tests were performed with using the standard shear test method as well as sandwich beam test method with four-point loading. Specimens for the sandwich beam tests were cut out from rigid polyurethane foam-cored sandwich wall panels with light-gauge thin steel sheet faces. Shear moduli were calculated from experimental deflections applying the theory of thin face sandwich beams. Shear creep tests were conducted at temperatures 20, 40 and 53°C for constant shear stress levels of 20, 40, 60 and 80 kPa. The duration of creep tests at 40 and 53°C was 1000 h and 9000 h at room temperature. Shear creep data were modeled with applying various linear and nonlinear creep models together with time-temperature superposition principle.

SHEAR TESTS

Shear tests for determining the shear strength and the shear modulus of low density rigid polyurethane foam material were performed with using the standard shear test method laid down in I.S.O. designation 1922-1972 (E). Shear specimens, 600 mm long x 100 mm wide x 50 mm thick, were cut out from the core centre of thin steel sheet faced sandwich wall panels of thickness 80 mm with sawing the long side of the sample to the panel lamination direction. Six different shear samples were tested at room temperature 20°C under relative humidity of the air of r.h. = 40 %. The mean foam density of the six samples was measured as 44.8 ± 0.2 kg/m^3. Recorded shear stress/strain curves for the six samples are shown in Fig. 1.(a). Values of the ultimate shear strength varied between 193-205 kPa and the corresponding ultimate shear strains between 13.5-18.3 %. In all cases the shear fracture was caused by failure of the bond between the face plate and the foam specimen. From the shear stress/strain curves of Fig. 1.(a), shear yield strength defined as 0.4 % offset yield stress was determined and its values varied from 128 to 136 kPa.

The shear modulus G of the polyurethane foam core was calculated from the shear stress/strain curves both as a secant modulus and as a tangent modulus, the latter by applying third-order approximation with finite shear strain differences of 0.2 % for the tangent values at different points of the shear stress/strain curves. In Fig. 1.(b), the calculated secant and tangent modulus values are plotted against shear stress.

Shear tests were also carried out with the aid of sandwich beams which were sawn from larger wall panels manufactured by a continuous double metal sheet lamination process. Fig. 2. shows the principle and the test apparatus of these sandwich beam tests. Beams of the span L = 1200 mm and in four-point bending were suspended from the ceiling and connected in series, as shown in Fig. 2.(a). Cross-section of the beams is shown in Fig. 2.(c).

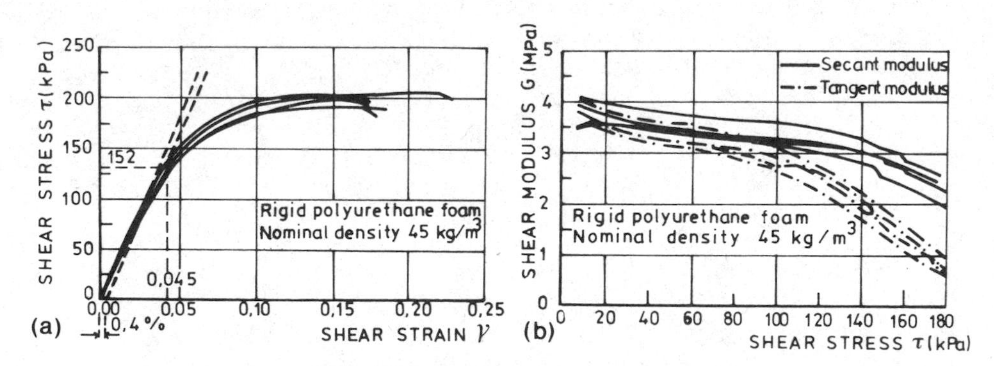

Fig. 1. Standard shear test results. (a) shear stress/strain curves; (b) secant and tangent modulus against shear stress.

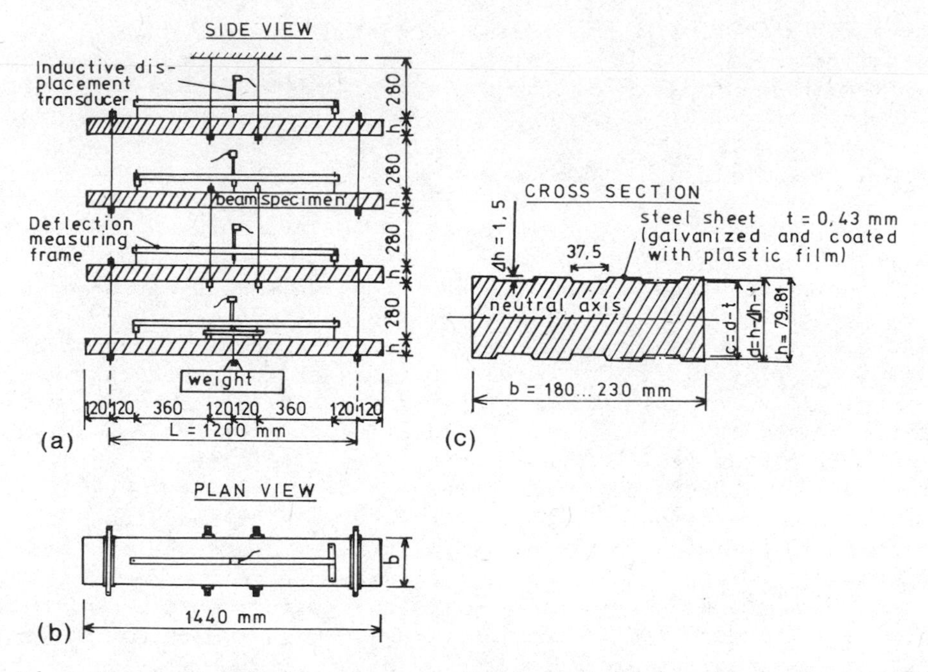

Fig. 2. Sandwich beam test apparatus. (a) side view; (b) plan view; (c) cross-section of the beam specimen.

The distance d between the centroidal axes of the lightly profiled symmetric faces was precisely measured for each beam specimen. Dimensions and loading conditions of the beams were chosen to result as great portion as possible (about 90 %) of the total deflection due to shear. The central deflection w of the beam was measured by means of inductive displacement transducer placed on a measuring frame mounted over 8L/10 of the span for eliminating the effect of local settling at the supports on measured deflection values. The deflection due to shear w_Q and due to bending w_M of the beam under two point loads 2xF/2 in the middle at distance L/5 and under the own weight gL of the beam were calculated as follows:

$$w = w_Q + w_M \tag{1}$$

$$w_Q = \frac{3}{20} FL/D_Q + \frac{2}{25} gL^2/D_Q \tag{2}$$

$$w_M = \frac{11}{800} FL^3/D_S + \frac{67}{7500} gL^4/D_S \tag{3}$$

$$D_Q = GA = Gbd(d/c) \; ; \; d = h - \Delta h - t \; , \; c = d - t \tag{4}$$

$$D_S = E_S btd^2/2 \tag{5}$$

The shear stiffness D_Q and further the shear modulus G were determined from Eqs. 1-5 by means of the measured total deflection w. The shear stress τ and the shear strain γ were then calculated by

$$\tau = (F/2 + 0.3gL)/bd \tag{6}$$

$$\gamma = \tau/G \tag{7}$$

Short-term shear tests with sandwich beams were performed at room temperature at constant rate of shear strain of 2-6 %/min according to the testing standards. Recorded shear stress/strain curves are plotted in Fig. 3.(a). Because of the presentation of a localized wrinkling instability phenomenon in the compression face of the beams, failure was already occuring at moderate low shear stress levels of 120-152 kPa. During the tests, axial strains in the tension face of the beam centre were measured with strain gages and values were compared with those calculated by sandwich beam formulas resulting a fairly close (within 8 %) agreement between the values. Shear modulus of the polyurethane foam core was determined in the short-term beam tests as secant modulus by using Eq. 4 for the measured deflection values. Secant modulus values against shear stress are presented in Fig. 3.(b).

Density variations in the polyurethane foam core of the beams were determined after the loading tests by sawing samples of size 60x60x80 mm^3 from the center beam line at distances of 300 mm along the beam axis and by cutting each sample into six pieces in the direction of foam rise. By weighing and measuring the pieces in accordance with the directions of A.S.T.M. D1622-63, the density variation in the foam rise direction was determined. A typical measured density variation is shown in Fig. 4.

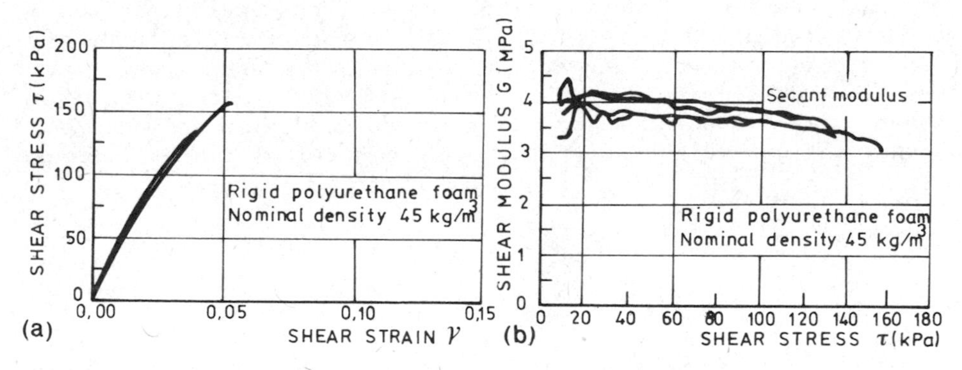

Fig. 3. Sandwich beam test results. (a) shear stress/strain curves; (b) shear modulus against shear stress.

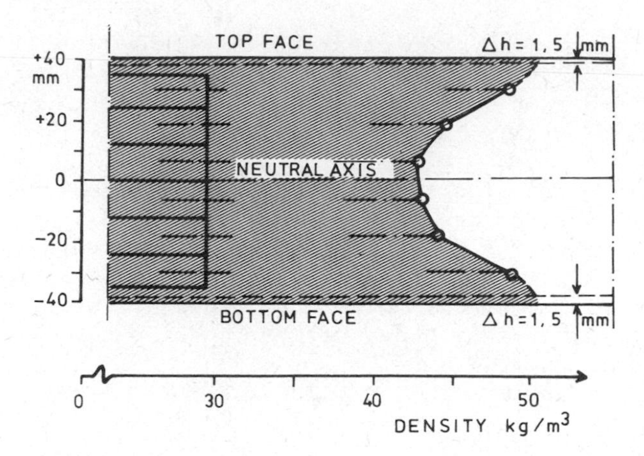

Fig. 4. Typical foam density variation over the core thickness.

SHEAR CREEP DATA AND SHEAR CREEP MODELING

Shear creep tests with the suspended sandwich beams in series were performed at three temperatures 20, 40 and 53^oC (r.h. = 40-45 %) for shear stress levels of 20, 40 and 60 kPa at 40 and 53^oC and for 20, 40, 60 and 80 kPa at room temperature. Six parallel beam specimens were used for each temperature/stress level combination. Shear creep compliance $J(t) = \gamma(t)/\tau_o$ of the polyurethane foam core was determined with time from the recorded deflection values by using Eqs. 4 and 7. Deformational behaviour in shear was characterized by the inverse of shear compliance that is the shear creep modulus $G_c(t)$:

$$G_c(t) = 1/J(t) \tag{8}$$

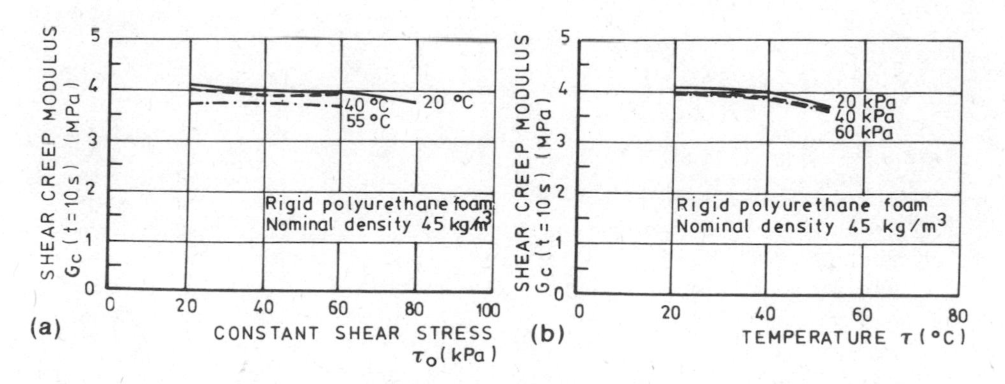

Fig. 5. Shear creep modulus values at t = 10 s. (a) modulus versus shear stress; (b) modulus versus temperature.

Average values of the shear modulus at short time period of $t = 10$ s obtained from the six tests are plotted in Fig. 5. both versus shear stress level (a) and temperature (b). The points of the curves were showing scatter within ±3...8 % of the mean.

The starting points of the modeling of shear creep data were readings $\bar{J}$ for the shear creep compliance $J(t)$ recorded at logarithmically divided intervals i.e. at equal intervals on the logarithmic time scale with 10 intervals for one decade. Parameters (α) in the creep models ($J(t,\alpha)$) were determined by the method of least squares with minimizing (Levenberg-Marquart method was used) the relative sum of squares

$$S = \sum_{i=1}^{n} \left[(\bar{J}(t_i) - J(t_i,\alpha))/J(t_i,\alpha) \right]^2 \tag{9}$$

As criterion in the evaluation of the fitness of creep models, the standard deviation

$$s = \sqrt{\frac{S}{n-1}} \tag{10}$$

was used.

Shear creep data were modeled by the following representation models:

$$J(t) = J_o + J_1 t + J_2(1-\exp(-t/\tau)) \tag{11}$$

$$J(t) = J_o + J_1(1-\exp(-t/\tau_1)) + J_2(1-\exp(-t/\tau_2)) \tag{12}$$

$$J(t) = J_o + J_1 t^n \tag{13}$$

$$J(t) = J_o + (J_1 t^n - J_2 t^{-n})/(1+J_1 t^{-n} + J_2 t^n) \tag{14}$$

$$2 \log\left(\frac{t}{t_o}\right) = \frac{M}{N}\left[1-\left(\frac{J(t)}{J_o}\right)^{-1/N}\right] - \frac{N}{M}\left[1- \frac{J(t)}{J_o}^{1/M}\right] \tag{15}$$

$$\log\left(\frac{J(t)}{J_o}\right) = A \sinh^{-1}\left(\log\left(\frac{t}{t_o}\right)\right) \tag{16}$$

Eqs. 11 and 12 are corresponding linear four and five element spring-dashpot models, Eq. 13 Findley's power law model and Eqs. 14, 15 and 16 models developed by the authors[1,2,3]. Fig. 6. shows an example of the curve-fitting by Eqs. 13, 14 and 15 with the long-term creep results at temperatures 20 and 53°C for constant shear stress level of 40 kPa. Values of the standard deviation s (%) in the curve-fitting of creep results of Fig. 6. with model Eqs. 11, 12, 13, 14 and 15 are listed in Table 1.

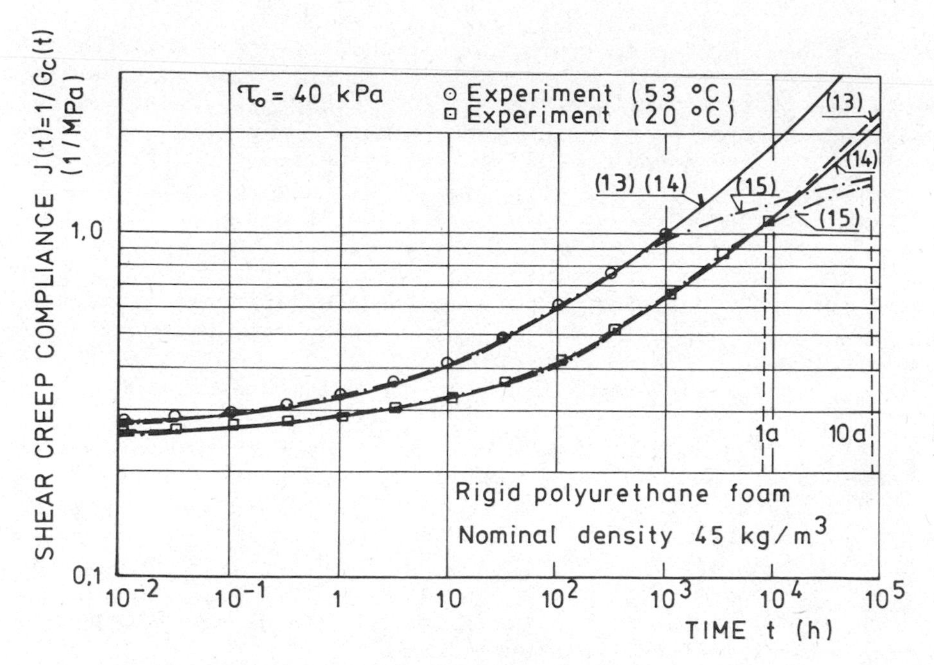

Fig. 6. Curve-fitting of the long-term shear creep results at 20 and 53°C at shear stress level of 40 kPa.

Table 1. Percentage Values of the Standard Deviation s (%) for Five Creep Models in Fitting the Creep Data of Fig. 6.

Temperature / Testing time / Number of measuring points	(11)	Creep Model Eqs. (12)	(13)	(14)	(15)
20°C / 9000 h / 61[a]	6.12	3.48	1.11	0.94	1.67
53°C / 1000 h / 51[b]	4.82	2.95	0.87	0.82	1.29

Mean values of [a]three and [b]six tests.

Time-temperature superposition principle was applied to the experimental data and it was found that horizontal shifting together with a small vertical shifting is not sufficient, some rotation in regard to the inflection points of the double-logarithmic creep curves also being necessary i.e., treatment in accordance with the McCrum-Morris superposition was applied. For instance, in superposition of the creep curves of Fig. 6. horizontal and vertical shift factors a_T and b_T were calculated with the curve-fitting programs to $a_T = 9.7422$ and $b_T = 1.0039$ after which the rotational adjustment was mady by using as inflection points of the curves those determined with Eq. 15 and as result a curve-rotation of 0.072 rad (= 4.1°) was obtained.

CONCLUSIONS

Shear test results obtained in this study give a quite well outlined illustration of both short-term and long-term shear properties of low density rigid polyurethane foam. Shear testing by suspended and in series connected sandwich beams with polyurethane foam core and thin steel sheet faces appears to be a suitable and an efficient method for determining especially the long-term shear creep properties of rigid polyurethane plastics foam. For the rather large scatter of the experimental readings in tests on low density foams, shear tests by many suspended sandwich beams in series turn out favourably in controling the scatter in the shear properties of rigid polyurethane foam.

Long-term shear creep data obtained from the creep tests can be modeled with a close curve-fitting by representation models suggested by the authors. Time-temperature superposition principle is also applicable to the experimental data when in addition to the horizontal and vertical shiftings a rotational adjustment of the double-logarithmic creep curves is made and this procedure seems to be suitable for prediction of the long-term shear creep properties of low density rigid polyurethane foam.

ACKNOWLEDGEMENTS

The authors express their appreciation to the companies MAKROTALO, KONE/Ollin Metalli, HUURRE and EKE ENGINEERING for the support of this research. Senior author thanks NESTE Oy:n SÄÄTIÖ for a grant in support of this research work.

REFERENCES

1. O. Holmijoki, "Shear strength and shear creep properties of rigid polyurethane foam evaluated by sandwich beam tests", Diploma Thesis, Helsinki University of Technology, Espoo (1983).
2. P. Mäkeläinen, "An extended reduced-time method for the modelling of long-term viscoelastic behaviour of polymeric materials", <u>Staveb</u>. <u>Čas</u>. 28(1980)2.
3. P. Mäkeläinen, "On the modelling of long-term thermoviscoelastic behaviour of glassy polymers", Research Papers 56, Helsinki University of Technology, Espoo (1976).
4. N. G. McCrum and E. L. Morris, <u>Proc</u>. <u>R</u>. <u>Soc</u>. A281(1964).
5. A. Kazmierczak, "Aus Dehnungsverhalten von PUR-Schaum", Staatsarbeit, Mitteilungen Nr. 5, IKV an der RWTH Aachen (1977).
6. J. A. Hartsock and K. P. Chong, "Analysis of sandwich panels with formed faces", <u>J</u>. of the <u>Str</u>. <u>Div</u>. Proc. ASCE, Vol. 102, No. ST4(1976).

BLENDS OF POLYSTYRENE AND THERMOTROPIC LIQUID CRYSTALS - PHASE RELATIONS AND RHEOLOGICAL PROPERTIES

R. A. Weiss[1], W. Huh[1], and L. Nicolais[2]

[1]Institute of Materials Science, University of Connecticut, Storrs, CT 06268 (USA)

[2]Dept. of Chemical Engineering, University of Naples Naples (ITALY)

INTRODUCTION

Reinforcing fibers and fillers are often incorporated into polymers in order to improve their mechanical properties and dimensional stability. Unfortunately, common reinforcements also raise the melt viscosity of the resultant composite, and in some cases may actually preclude melt processing of the material. In addition, in the case of conventional reinforcing fibers such as glass or graphite, attrition of the fiber lengths in dies and gates result in fiber aspect ratios of the order of 100 or less in the fabricated part. Since the mechanical properties are strongly dependent upon the fiber aspect ratio, this is a severe deficiency of discontinuous fiber-reinforced thermoplastics.

This investigation considers using organic molecules exhibiting liquid crystalline characteristics as molecular fibers in thermoplastic polymers. These molecules are rod-shaped and can be viewed as a fiber with effective diameters of the order of angstroms rather than microns. Therefore, high aspect ratios may be achieved utilizing relatively short chain lengths, and in addition fiber length attrition would require the breaking of covalent bonds. Because of their anisotropy, the molecules may be oriented in the shear fields characteristic of most processing operations, and as a result may lead to preferred orientations of the molecules after cooling the melt. While these oriented molecules may impart to the composite many of the same advantages characteristic of fibrous reinforcements, they may be chosen such that they are in a liquid crystalline state at the processing conditions. Being a liquid, these liquid crystals may actually plasticize the polymer melt, i.e., lower the melt viscosity.

In the work described here, two low molecular weight liquid crystals, terepthal-bis-4-n-butylaniline (TBBA) and N-(p-ethoxy benzylidine)-p-n-butylaniline (EBBA) were considered as plasticizers/reinforcements for atactic polystyrene (PS). The aspect ratio of these molecules is low, ca. 3-4, and, therefore, significant reinforcement was not expected. The concept described above is, however, demonstrated by this system and this work should be viewed as preliminary to the study of thermotropic polymeric liquid crystal-thermoplastic polymer blends. We report here the thermal and rheological characterization of these blends.

EXPERIMENTAL

Blends of TBBA/PS ($\bar{M}n$ = 18,000) and EBBA/PS ranging from 0.2 to 80 percent TBBA or EBBA were prepared with a two-roll mill at 150°C, and compression molded into 2mm thick, 2.5mm diameter disks at 140°C. Melt rheology measurements were made with a Rheometrics Mechanical Spectrometer (RMS) using eccentric rotating disks (ERD). Glass transition temperatures were determined at 20K/min with a Perkin Elmer DSC-2. Thermal Optical Analyses (TOA) were made with an optical microscope equipped with a controlled light source, cross polarizer, hot stage, and a photometer detector.

RESULTS AND DISCUSSION

Both TBBA and EBBA are crystalline solids at room temperature; their thermodynamic transitions as determined by DSC are given in Table I. TOA revealed that TBBA is miscible in PS below concentrations of 11% (wt) and EBBA is miscible in PS up to concentrations of 40% (wt). Above these concentrations, transitions are observed that are consistent with the phase transitions of the liquid crystals.

Table I. Plastic Transitions of TBBA and EBBA

TBBA transition*	temp(K)
Crystal→SmB	385
SmB→SmC	418
SmC→SmA	445
SmA→Nm	473
Nm→Isotropic Liq.	509

EBBA transition*	temp(K)
Crystal→Nm	309
Nm→Isotropic Liq.	353

*Sm = smectic phase
Nm = nematic phase

Partial phase diagrams were constructed from the DSC and TOA data and these are given in figures 1 and 2. Three regions are apparent: a single phase region, an isotropic liquid-liquid crystal region, and a glass-crystalline solid region. The boundaries were constructed from the glass transition temperatures (Tg) and the clearing points of the mixtures. Tg decreases with the addition of small amounts of either TBBA or EBBA, indicating that these compounds are effective plasticizers for PS. Above 15% TBBA (40% EBBA), Tg is independent of the additive concentration. This result is expected if phase separation occurs, a phenomenon confirmed by TOA. The clearing points of the mixtures, i.e., the transition to the single phase isotropic liquid, increases with TBBA (EBBA) concentration, which suggests a strong interaction of the PS melt on the liquid crystalline phase, i.e., some phase mixing.

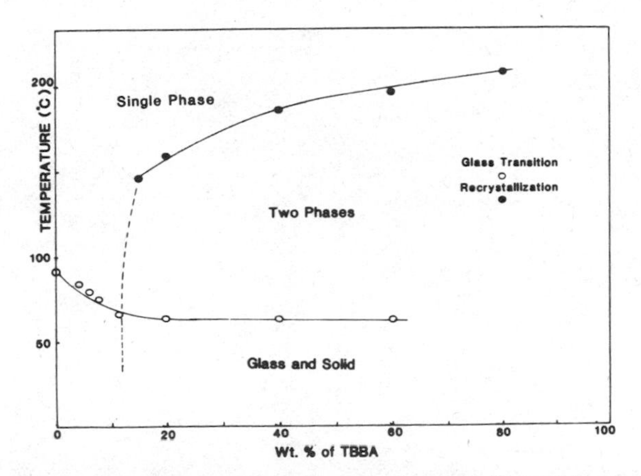

Fig. 1. Phase Diagram of TBBA/PS Blends.

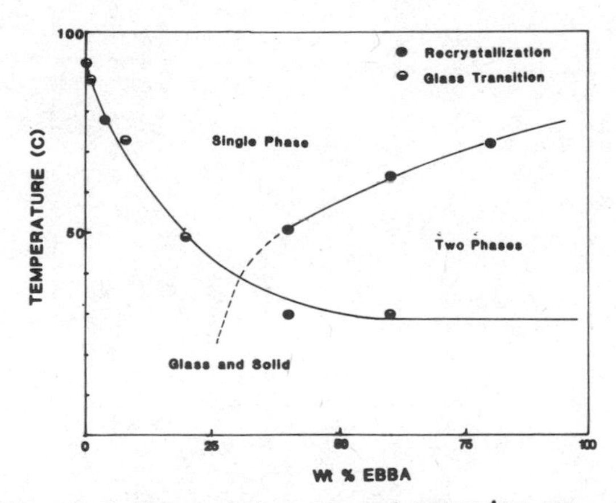

Fig. 2. Phase Diagram of EBBA/PS Blends.

The rheological properties (G' and G" vs. frequency) of the blends are given in figures 3-6. The master curves represent data at a number of temperatures shifted with respect to frequency. All compositions exhibited pseudoplastic behavior, i.e., increasing G" with increasing reduced frequency, and the dynamic modulus, G', increased with frequency. As the TBBA or EBBA concentration increased, both G' and G" decreased. This result is consistent with the Tg data and indicates that these additives are plasticizing PS in the melt.

Although the rheological data cover conditions in which a TBBA phase should be in several forms, smectic, nematic, and isotropic, the time-temperature superposition indicates that these blends are thermorheologically simple. This result was surprising and may mean that the liquid crystal state in the melt is different

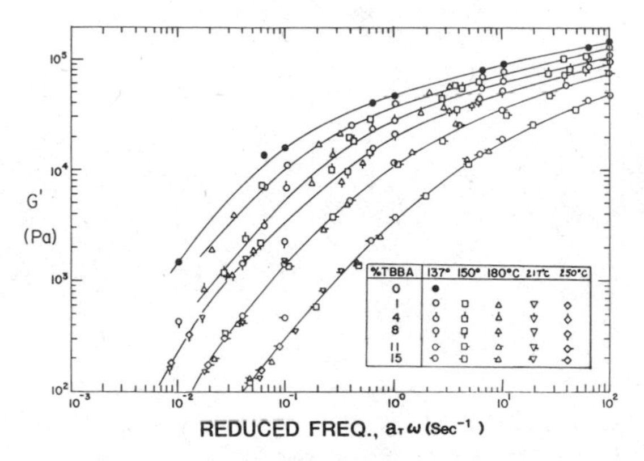

Fig. 3. Master Curves of G' vs. reduced frequency for TBBA/PS Blends as a fucntion of TBBA concentration.

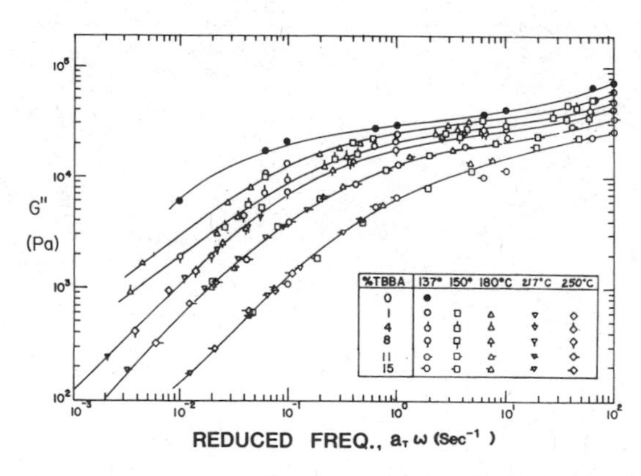

Fig. 4. Master Curves of G" vs. reduced frequency for TBBA/PS Blends as a function of TBBA concentration.

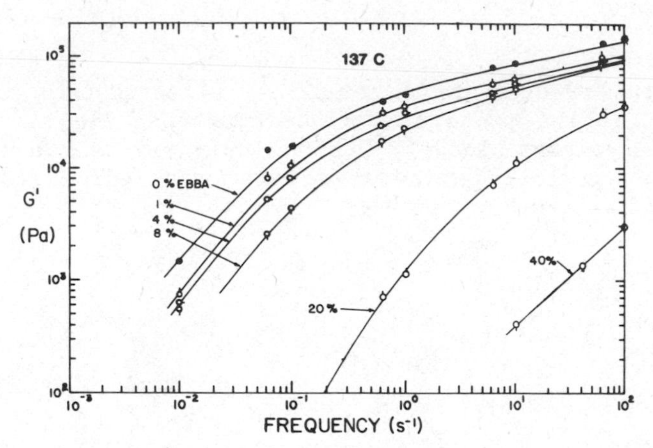

Fig. 5. G' vs. frequency for EBBA/PS Blends as a function of concentration.

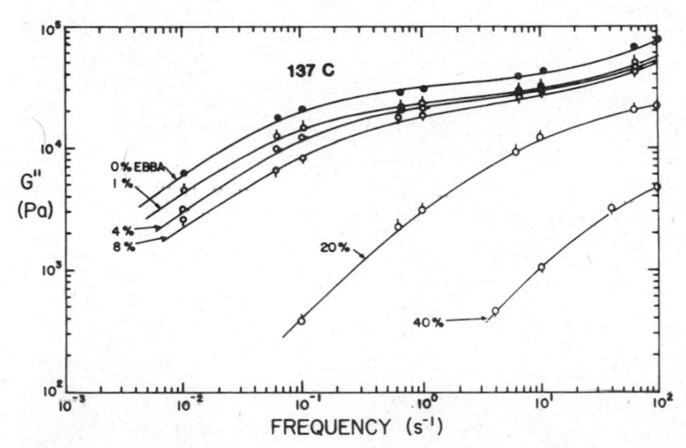

Fig. 6. G" vs. frequency for EBBA/PS Blends as a function of concentration.

then in the isolated form. Alternatively, this result may be a consequence of insignificant differences in the rheological properties of the different liquid crystal phases of TBBA. Both these possibilities are currently under investigation.

Mechanical testing of these blends indicated little if any reinforcing effect of the liquid crystal phase. This result is not unexpected, because of the small aspect ratio of these molecules and their compatibility with PS at concentrations of greatest interest to us. The results are promising, however, in that plasticization of the melt was accomplished with marginal reductions in the mechanical properties.

FUTURE WORK

This study represents an on-going investigation in our laboratories of the properties of thermotropic liquid crystal-thermoplastic polymer blends. Other liquid crystals and polymer matrices are currently being considered, including polymeric liquid crystals.

SHEAR BEHAVIOUR OF MOLTEN LINEAR LOW DENSITY POLYETHYLENE

D. Acierno, A. Brancaccio[+], D. Curto, and F.P. La Mantia

Istituto di Ingegneria Chimica, University of Palermo

Viale delle Scienze 90128 Palermo, Italy

INTRODUCTION

The rheological behaviour of molten polyethylene has been widely studied especially with respect to the different structure of the high density and low density samples[1-10]. In particular it is well known that the presence in the LDPE of long chain branching alters dramatically the whole flow curve, the activation energy, the die-swell, the instability phenomena, etc.

The birth of a new polyethylene, known as linear low density polyethylene, which presents a number of short chain branches and some interesting properties in the finished products, suggested also a rheological study. Preliminary results obtained on a commercially available sample further suggested a complete study in order to clarify the ralationship between rheological properties and molecular characteristics.

In this work six samples of linear low density polyethylene polimerized with 1-butene as comonomer and having different molecular weight but similar molecular weight distribution, have been characterized in shear flow. Generalized curves and ralationships have been obtained for some rheological properties.

+ Enoxy Chimica S.p.A., Plastics Technological Center, S.Donato Milanese, Italy

EXPERIMENTAL

The materials used in this work were six linear low density polyethylenes (LLDPE), polimerized with 1-butene as comonomer. Their main characteristics are listed in table 1.

Table 1

Sample	Mw	β	M.F.I.	ϱ, g/cm^3	$[\eta]$, dl/g	CH_3/1000C
LL 0.5	136.000	4.9	0.53	0.9216	1.83	8.7
LL 1	133.000	4.4	1.1	0.9237	1.81	12.1
LL 2	114.000	4.2	1.9	0.9215	1.44	13.3
LL 3	110.000	4.6	2.2	0.9240	1.49	10.1
LL 4	81.000	3.1	4.3	0.9383	1.30	6.6
LL 9	66.000	4.9	9.2	0.9281	1.04	9.3

Melt flow indew was determined by a capillary rheometer at constant pressure following the ASTM-D-1238/73 method, condition E.
Density measurements were performed using a gradient column filled with water/isopropylic alcohol at 23°C following the ASTM-D-1505/68 method.
Intrinsec viscosity were determined in TCB at 135 °C. The branching index g was determined through the relationship:

$$g^{0.5} = \frac{[\eta]_{br}}{[\eta]_1} \qquad 1)$$

were $[\eta]_1$ for the linear polymer is:

$$[\eta]_1 = 3.92\ 10^{-4}\ M^{0.725}$$

Short chain branching was determined by IR spectroscopy by using a Perkin Elmer spectrophotometer mod.577, following [16].
The determinations of the molecular weight and the molecular weight distribution were performed with the usual chromatographic procedure by using a Waters instrument, mod.150, and TCB as solvent.
The rheological measurements at low shear rates were carried out with the aid of a constant force viscometer, while a constant rate viscometer, Rheoscope 1000, manufactured by CEAST, Torino (Italy), was used for all the other runs. The tests were carried out at 190, 205 and 220 °C. A capillary with D = 1 mm and L/D = 40 was always used. Because of the large length to diameter ratio, the entrance corrections [11] have been neglected, while the Rabinowitsch

correction[12] has been used throughout in order to obtain the true shear rate. All the calculations have been made by using a computing program.

For the die-swell measurements samples about 5 cm long were cut to avoid changes in diameter due to their own weight.These measurements were made using a micrometer after samples solidification.

The swelling ratio, B, was evaluated as in [6]:

$$B = \frac{D'}{D} \; (\varrho V_m)^{1/3} \qquad 3)$$

In equation 3) D' is the diameter of the extruded sample, D the die diameter, ϱ the density of the polymer at room temperature, and V_m the specific volume at the extrusion temperature.

The critical shear rate has been considered as the shear rate at which the first irregularities, even only superficial,appear. This could be accurately done with the Rheoscope 1000 due to the continuous piston speed variation. Determination of $\dot{\gamma}_c$ have been performed with different capillaries (L/D = 5, 20, 40).

RESULTS AND DISCUSSION

Fig. 1 shows the flow curves for all the samples obtained at T = 220 °C. The newtonian viscosity depend strongly on the molecular weight; the flow curves, however, approach each other at high shear rates. In Fig. 2 the zero shear viscosity at 220 °C are plotted against the weight average molecular weight. η_o increases with Mw following the realtion:

$$\eta_0 = k \; Mw^a \qquad 4)$$

were a is about 4.2, larger than that generally found for linear polymers[5,13]. The value of k is 2.3 10^{-18} Pa.s at 220 °C.

By plotting the η_0 values as a function of the inverse of the temperature, the activation energy,E, of the flow can be obtained through the Arrhenius equation:

$$\eta = k_\eta \; \exp \left(- \frac{E}{RT} \right) \qquad 5)$$

All the samples have practically the same activation energy, ~7.5 Kcal/Kmole, sligthy larger than that of the high density polyethylene[8,9,14].Again,like in the HDPE[14], the activation energy remains unchanged by increasing the shear stress.

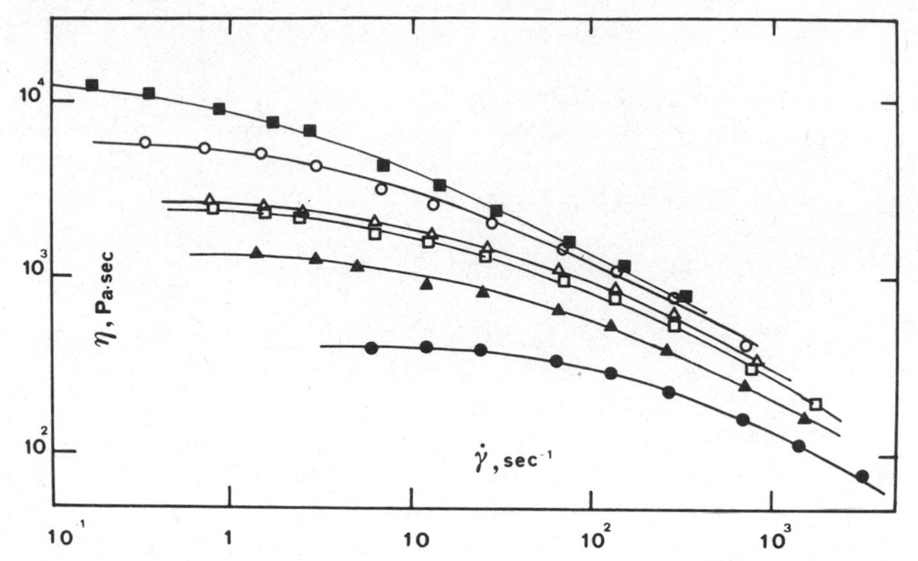

Fig. 1 Shear viscosity vs shear rate at T= 200 °C. ■ LL 0.5, ○ LL 1, △ LL 2, □ LL 3, ▲ LL 4, ● LL 9.

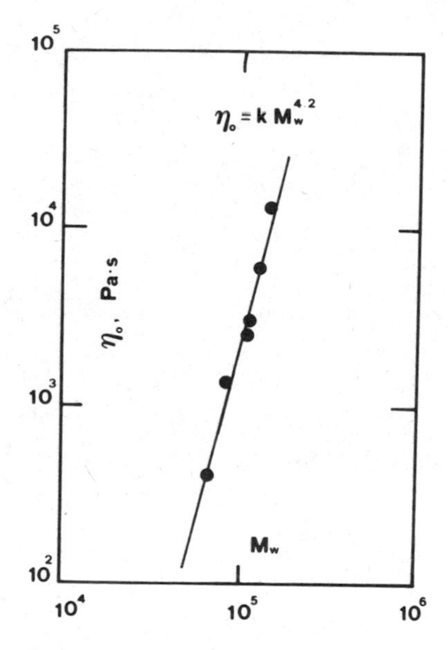

Fig.2 Zero shear viscosity vs weight average molecular weight.

Following the Graessley's analysis[15], a single generalized curve for each linear low density polyethylene, has been obtained by plotting the dimensionless viscosity η / η_o against the reduced shear rate $\dot{\gamma} \eta_o / T$ and neglecting the differences in the density due to the temperature variations. A very good superposition is obtained for all the samples, see Fig. 3.

A further superposition has been attempted with the data of Fig.3 in order to obtain a single curve. A horizontal shift is sufficient to have the curve reported in Fig.4, with a shift factor a_p, depending on molecular weight:

$$a_p = 4.8\ 10^{-6}\ Mw^{1.11} \qquad 6)$$

The swelling ratio values are reported in Fig. 5 as a function of the shear stress for five LLDPE; for the polyethylene with the lowest Mw the determination was not possible. The experimental points refer to runs at different temperatures but, in this type of plot, as also for other polymers, the data lie on the same curve.

The behaviour of the die-swell with the molecular parameters is particularly complex. And indeed no monotonic behaviour is revealed by the reported curves.The pair of samples LL 0.5-LL 1 and LL 2 -LL 3, although have a very similar molecular weight, show large differences in the die-swell. Because of the absence of the long chain branching,g values larger than 0.91 for all the samples, this behaviour can be perhaps tied with the number of short branches which is larger in the sample LL 1 and LL 2. The short branches seem to depress the swelling phenomenon. A last comment can be made concerning the absolute values of the die-swell which are larger than that obtained with similar HDPE.

The effect of the molecular structure and of the temperature on the critical shear rate is swown in Fig. 6. $\dot{\gamma}_c$ decreases with molecular weigh and with temperature. A simple generalized relationship can be found from the data reported in Fig. 6 and written as:

$$\dot{\gamma}_c = a + b\ Mw \qquad 7)$$

where a and b are depending only on the absolute temperature,T,

$$a = 1.4\ 10^{4} - 24.9\ T \qquad 8)$$

$$b = 0.078 + 1.34\ 10^{-4} T \qquad 9)$$

Finally it has to be said that experiments performed

with three different capillaries have shown that $_c$ is unaffected by the L/D ratio, differently from both the HDPE and the LDPE.

ACKNOWLEDGEMENT

This work has been carried out with the financial support of "Progetto Finalizzato del C.N.R. Chimica Fine e Secondaria".

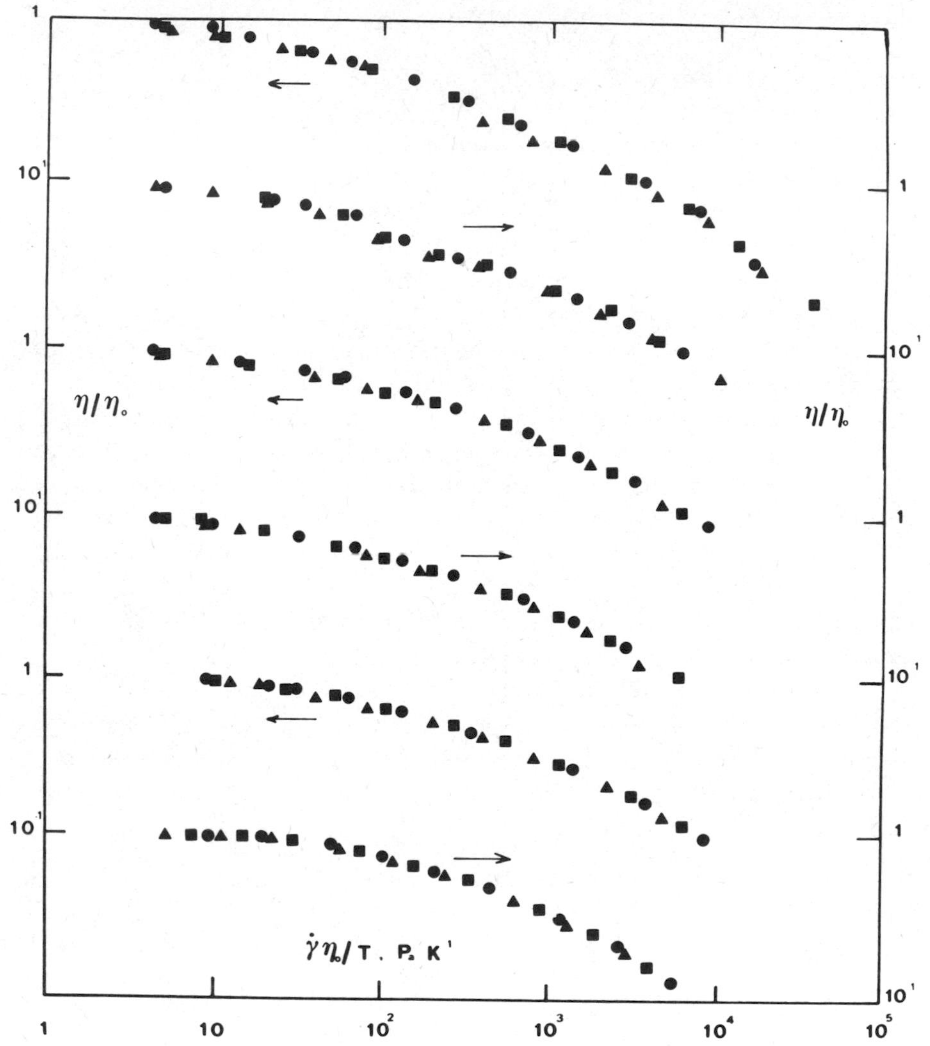

Fig. 3 Dimensionless viscosity vs a reduced shear rate.
From top to bottom LL 0.5, LL 1, LL 2, LL 3, LL 4, LL 9.

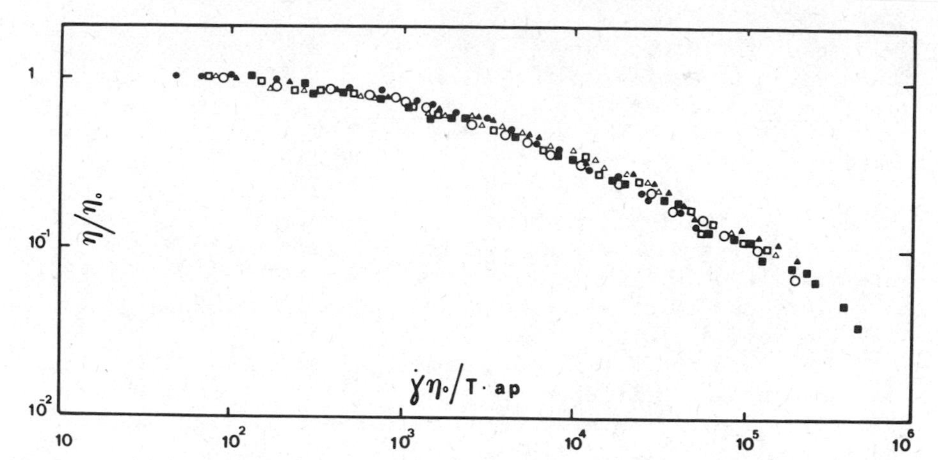

Fig. 4 Generalized flow curve for all the LLDPE samples. Key for symbols as in Fig 1.

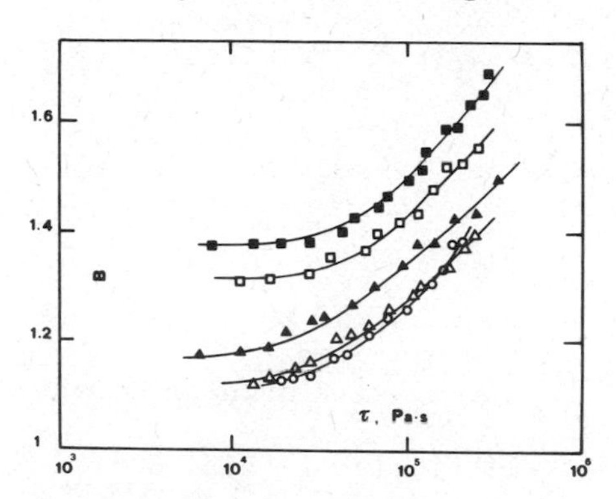

Fig.5 Swelling ratio vs shear stress Key for symbols as in Fig. 1.

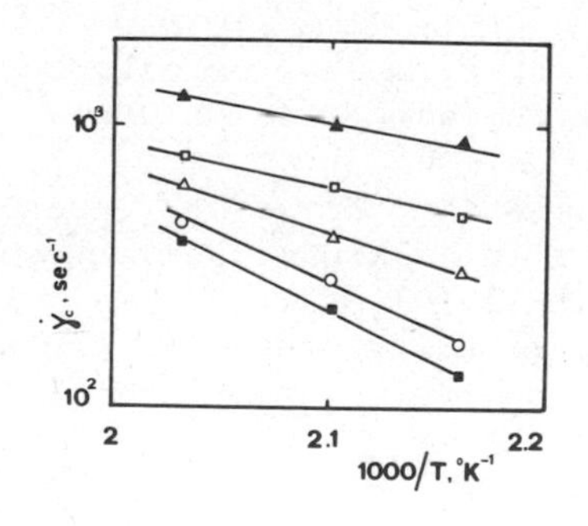

Fig. 6 Critical shear rate vs the inverse of the absolute temperature. Key for symbols as in Fig. 1.

REFERENCES

1. L.H.Tung, Melt Viscosity of Polyethylene at Zero Shear, J.Polym. Sci. 46:409 (1960)
2. W.F.Busse and R.Longworth, Effect of Molecular Weight distribution and Branching on the Viscosity of Polyethylene Melts, J.Polym.Sci. 58:49 (1962)
3. H.P.Schreiber, E.B.Bagley and D.C.West, Viscosity/Molecular Weight Relation in Bulk Polymers-1, Polymer 4:355 (1963)
4. R.L.Combs, D.F.Slonaker and H.W.Coover, Effects of Molecular Weight Distribution and Branching on Rheological Properties of Polyolefine Melts, J.Appl.Polym.Sci.13:519 (1969)
5. R.A.Mendelson, W.A.Bowles and F.L.Finger, Effect of Molecular Structure on Polyethylene Melt Rheology. I. Low-Shear Behaviour, J.Polym.Sci. A.2 8:105 (1970)
6. R.A.Mendelson and F.L Finger, Effect of Molecular Structure on Polyethylene Melt Rheology. III. Effects of Long-Chain Branching and of Temperature on Melt Elasticity in Shear, J. Appl.Polym.Sci. 17:797 (1973)
7. L.Wild, R.Ranganath and D.C.Knobelock, Influence of Long-Chain Branching on the Viscoelastic Properties of Low-Density Polyethylenes, Polym.Eng.Sci. 16:811 (1976)
8. D.Romanini, A.Savadori and G.Gianotti, Long Chain Branching in Low Density Polyethylene: 2. Rheological Behaviour of the Polymers, Polymer 21:1092 (1980)
9. F.P.La Mantia, A.Valenza and D.Acierno, A Comprehensive Experimental Study of the Rheological Behaviour of HDPE. I.Entrance Effects and Shear Viscosity Results, Rheol.Acta in press
10. F.P.La Mantia, A.Valenza and D.Acierno, A Comprehensive Experimental Study of the Rheological Behaviour of HDPE. II. Die-Swell and Normal Stresses, Rheol. Acta in press
11. E.B.Bagley, End Corrections in the Capillary Flow of Polyethylene, J.Appl.Phys. 28:624 (1957)
12. B.Rabinowitsch, The Viscosity and Elasticity of Sols, Zeit. Physik.Chem. A145:1 (1929)
13. J.D.Ferry,"Viscoelastic Properties of Polymers" J.Wiley, New York (1970)
14. S.M.Jacovic, D.Pollock and R.S.Porter, A Rheological Study of Long Branching in Polyethylene by Blending, J.Appl.Polym.Sci. 23:517 (1979).
15. R.C.Penwell, W.W.Graessley and A.Kovacs, Temperature Dependence of Viscosity Shear Rate Behaviour in Undiluted Polystyrene J. Polym.Sci. Polym.Phys.Ed. 12:1771 (1974)
16. U.Sato and T.Yashiro, The Effect of Polar Groups on the Dielectric Loss of Polyethylene, J.Appl.Polym.Sci. 22:2141 (1978)

ANALYSIS OF FAILURE BEHAVIOR OF POLYMERIC COMPOSITE MATERIALS WITH THE METHOD OF ACOUSTIC EMISSIONS (Abstract)

C. Caneva, M. Mazzola, E. Martuscelli, and M. Malinconico

Istituto di Tecnologia e Reologia dei Polimeri
Arco Felice, Napoli, Italy

Different types of composite materials have been tested with acoustic emission methods. The variables considered were the polyester foam density, the fiber content, and the fiber legth. The specimens were tested in a tensile mode until fracture; during the test the acoustic emission was monitored and registered on magnetic tape. The fracture surfaces have been analyzed by SEM. Fron the experimental data it has been observed that the efficiency of glass fibers has a maximum at a well defined value of fiber content for each polyester foam density. The fracture starts in the matrix, and after a certain amount of fiber debonding fiber fracture occurs. At high fiber contents, there is almost only matrix fracture and fiber debonding.

PLASTICIZERS MIGRATION FROM PVC: CORRELATIONS WITH STRUCTURE AND SERVICE-PERFORMANCE

Gerhard Pastuska

Department 3 "Organic Materials"
Subdepartment 3.1 "Plastics,Coatings,Elastomers"
Federal Institute For Testing Materials
1000 Berlin-Dahlem

Sheetings of plasticized PVC are often used for sealings and roof-coverings. The plasticizers are extremely important for shelf-life and service-performance of roof-coverings. A loss of plasticizer may arise by evaporation, migration, washing-out, hydrolysis or by the life history of microorganisms. This loss causes shrinkage of materials, hardening, increase of the modulus and very often damages. The loss of plasticizer depends on its chemical structure. The plasticizers in roof-coverings are very often esters of the phthalic-acid in the pure substance or in mixtures. Plasticizers analysis, after its extraction, is practicable by liquid chromatography. In practice, sheetings with plasticizers with short chains in the alcoholic components are dominant in the loss-rate. Figure 1 demonstrates a roof-covering used under a layer of gravels. About 7 years later the substances with short chains (1b) had escaped from the sheeting in comparison with the new material (1a).

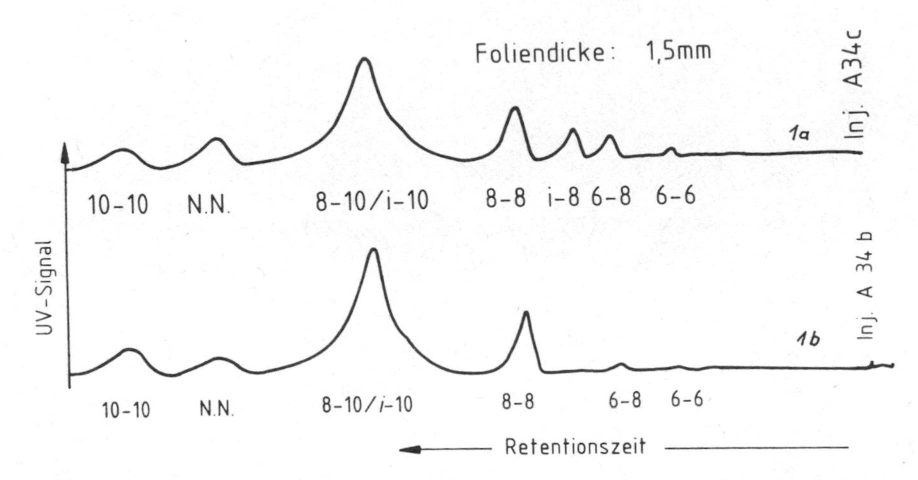

Figure 1. Loss of plasticizers with short chains

The problem is how to test in the laboratory the influence of the chemical structure of the plasticizer and the correlation to its loss. Plasticizers migration into other substances, e.g. polystyrene, is well known. We studied the migration in this way that one side of the film is covered with a plate of polystyrene and the other is open to the surrounding air, only fixed by a wire net. This specimen was stored about 600 h at 75 °C. Figure 2 demonstrates that the chain-length of the ester-alcohols and their structures are decisive for the migration of plasticizers. The migration of short chains is faster than that with long chains and those with linear structures migrate slower than those with branched structures. The thickness of the sheeting is also important for the velocity of migration and its life-time.

Figure 3 shows that the same plasticizers migrate faster out of a thin sheeting than from a thick one.

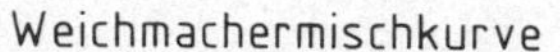

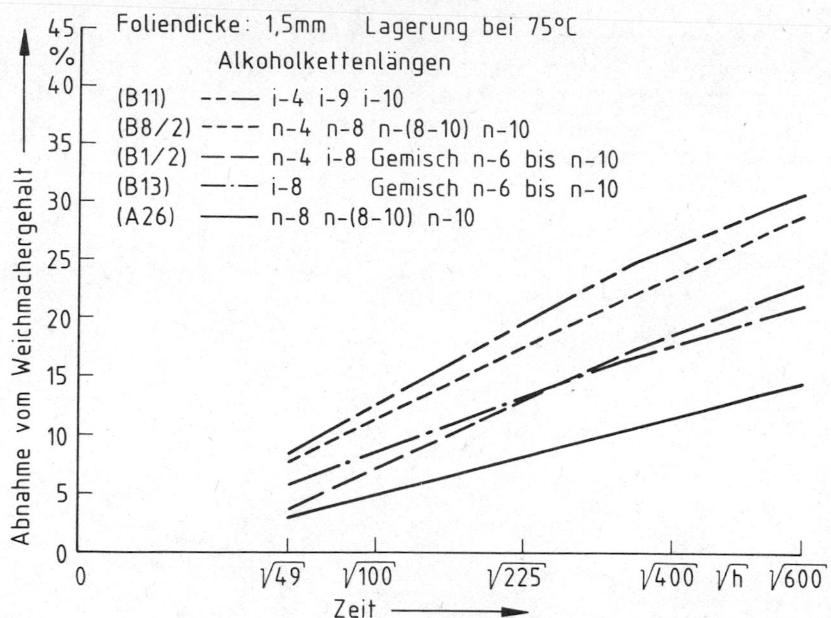

Figure 2. Migration of phthalic esters; influence of chain-length and structure

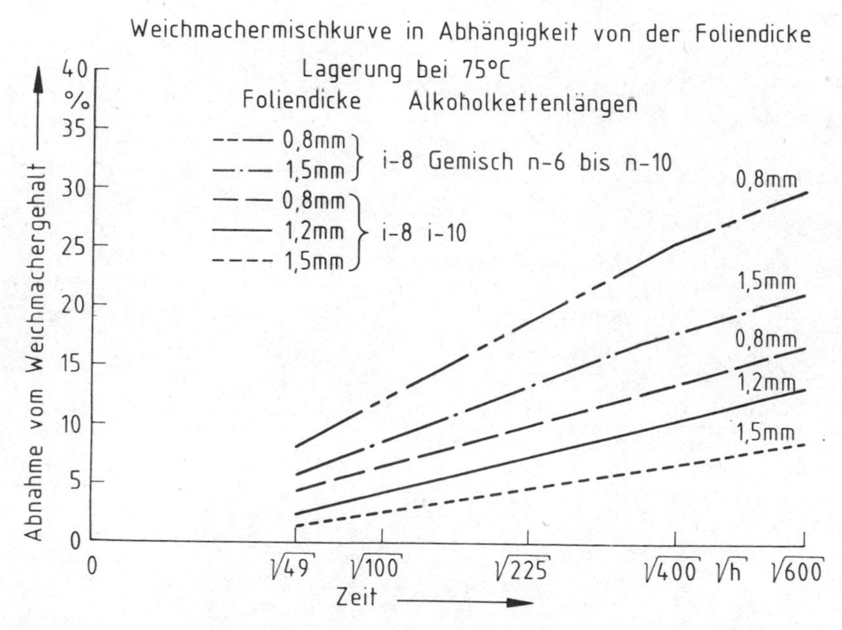

Figure 3. Migration of plasticizers in dependence to the thickness of the sheetings

The dependence between volatility and structure is different to migration. Figure 4 demonstrates the larger volatility of branched (iso-) structures than the linear (n-) structures. This implies different mechanisms of loss in practice. In Northern Europe volatility is less important to a cause of damages but it may be of importance in South Europe.

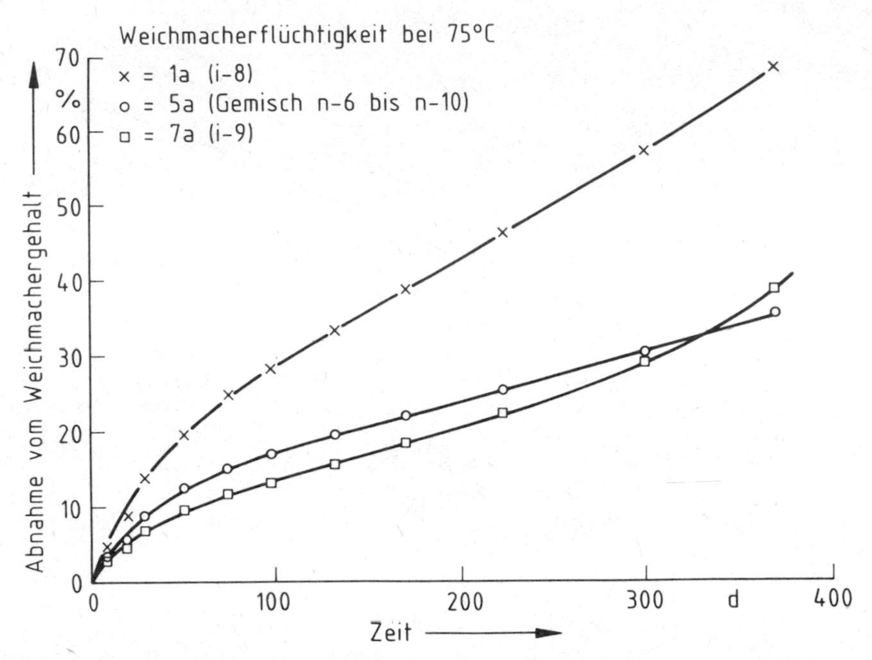

Figure 4. Volatility of plasticizers in correspondence to the structure

This loss of plasticizer by migration or evaporation is very different from the "ageing" of the PVC-material. In this work "ageing" means the decomposition of the PVC-molecules by th energy of natural UV-radiation. In dependence to the angle of incidence of sunlight in Central Europe up to 50 °N UV-radiation with a wavelength of about 300 nm reach the surface of the earth. The energy of this radiation is of such intensity to split the polymerchains in the PVC. By this molecular-mass will be diminished and the distribution becomes broader. The degree of deterioration depends on the penetration depth into the film and on the effective period of weathering. A roof-covering covered with gravels indicates no deterioration even after some years. The penetration depth of the radiation is small; in a period of ten years hardly more than 140 nm.

Figure 5 shows this deterioration.

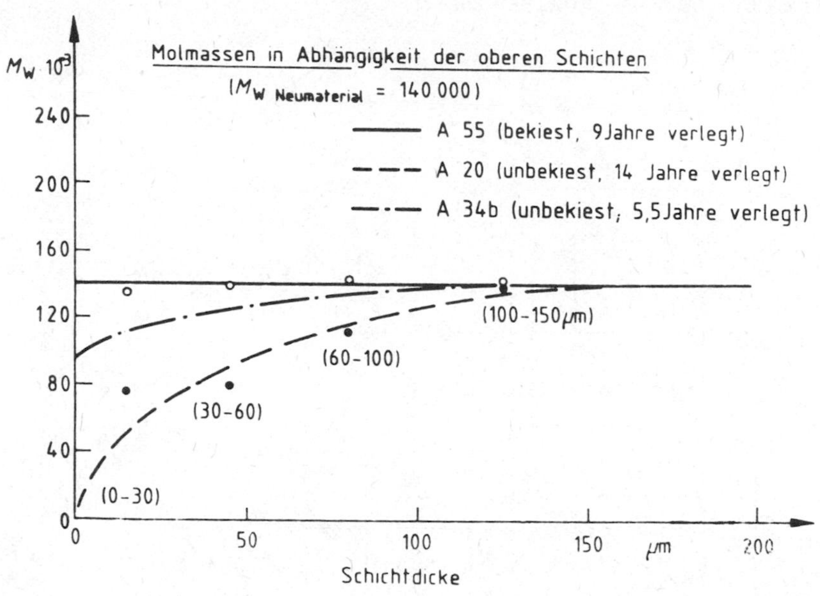

Figure 5. UV-deterioration of PVC, penetration depth

The described loss of plasticizer is different from the decomposition of the PVC-chain, but in practice often not independent. Our latest results demonstrate that at least the volatility of the plasticizers is larger from those sheetings which are covered with gravels and have no deterioration than those which are influenced in the surface by UV-radiation. The deteriorated surface acts obviously like an anticorrosive layer. The mechanism of this reaction is still unsettled.

In the beginning of this it was explained that the mechanical properties of the sheetings are changed with the loss of plasticizer. A sheeting is fixed on the edges of the roof, plasticizers loss causes shrinkage and also tensions, by cooling of the surrounding e.g. in winter-time the film shrinks in accordance with its co-efficient of expansion $\alpha \sim 20 \cdot 10^{-5}$ 1/°C; by this the tension in the sheeting will be enlarged and the attachment points will be extremely loaded. At - 10 °C this tension will be about zero for a new, flexible film. If the sheeting has after some years only 24 % plasticizer the tension at - 10 °C is enlarged to 1600 N per meter width of the sheeting. These values are worth to be known for the construction-principles. We measure these tensions biaxial with our test-equipment, shown in Figure 6. By storage in methanol it is possible to extract plasticizer out of the film and systematic experiments can be undertaken.

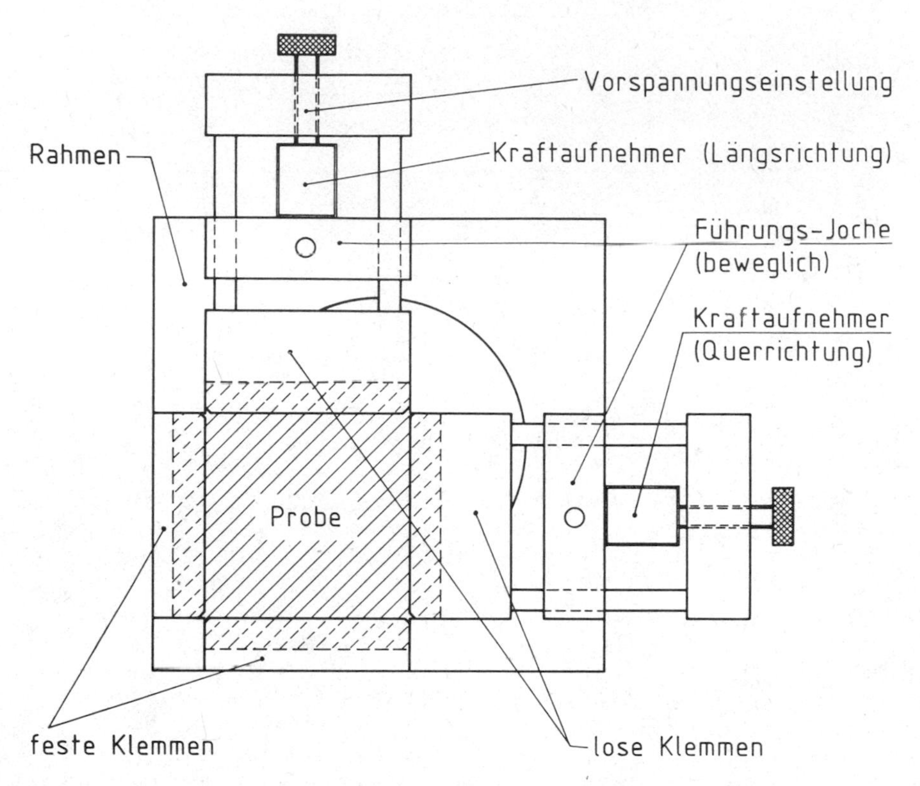

Figure 6. Test equipment for measuring biaxial tensions

TIME-DEPENDENT PROPERTIES OF STATICALLY DEFORMED GLASSY POLYMERS: DYNAMIC MODULUS AND GAS TRANSPORT

Thor L. Smith and Theonis Ricco*

IBM Research Laboratory
San Jose, California 95193, U.S.A.

INTRODUCTION

After an amorphous polymer has been cooled from above to substantially below its glass transition temperature, its volume decreases continuously with time under isothermal conditions, as has long been recognized.[1] This spontaneous volume contraction is accompanied by a progressive decrease in the mobility of short molecular segments and therefore by an increase in the relaxation times which affect the viscoelastic and other physical properties of the glassy polymer, as discussed by Struik.[2,3] This physical aging process can be reversed, at least somewhat, by applying a finite tensile strain, which gives an increase in volume. Immediately thereafter, aging begins anew at an increased rate.

In terms of the usual concepts of free volume, it is expected that a dilation of a specimen effected by a tensile strain (or stress) will initially reverse the aging process and thus increase the segmental mobility. However, it is probable that application of a finite stress or deformation in either torsion or even simple compression will also reverse the aging process initially, even though the volume of the specimen is not increased. This view has been expressed by Struik[3] who has investigated various aspects of physical aging. In particular, he has found that a finite tensile stress affects the aging process, as expected, and that essentially the same effect is produced by a stress in torsion.

---

*Permanent Address: Polytechnic Institute of Milan

In this paper, data are presented that reflect, at least qualitatively, the strain and subsequent time dependence of segmental mobility. Figures 1a and 1b show schematically the stepwise application of a finite tensile strain to a glassy polymer and the resulting relaxation of stress in the specimen. From the relaxation curve, it is impossible to determine whether the segmental mobility remains constant or changes with time.

To obtain such information, a small sinusoidal strain can be superposed periodically on the static strain to determine the storage and loss tensile moduli (E' and E''). Figure 1c shows that E' decreases when the static strain is applied, but thereafter it increases with time. The initial decrease in E' results from an increase in segmental mobility, a reflection of the reversal of the aging process. The subsequent increase in E' reflects a progressive reduction in segmental mobility.

Gas transport in a glassy polymer depends on various factors,[4] including segmental mobility. Thus, when a static tensile strain is applied to a specimen, the permeability (P) and diffusivity (D) of a gas should increase initially and thereafter decrease with time, as has in fact been found.[5,6] Such behavior is indicated by Fig. 1d, though the strain and time dependence of P and D are not the same, as might be inferred from the figure.

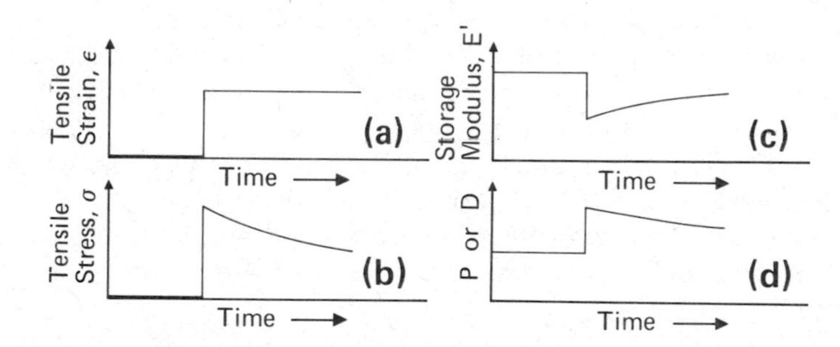

Fig. 1. Schematic representation of the strain and time dependence of stress, storage modulus, and gas-transport coefficients effected by the application of a fixed tensile strain.

MATERIALS AND EXPERIMENTAL METHODS

The materials studied were a polycarbonate film, 76 μm thick, obtained from Rodyne Industries, Mt. Vernon, NY, and a film of biaxially oriented polystyrene, 38 μm thick, termed Trycite and produced by the Dow Chemical Co.

Viscoelastic properties were measured with a Dynastat Viscoelastic Analyzer, an instrument developed by Professor S. S. Sternstein (Rensselaer Polytechnic Institute) and marketed by Imass (Accord, MA). Some characteristics of our instrument, which was the first commercial unit, and illustrative data obtained on poly(methyl methacrylate) have been discussed.[7]

In the present study, the polycarbonate film was annealed. A specimen was then subjected to a static tensile strain in 50 ms. Periodically thereafter, a sinusoidal strain whose frequency and amplitude were 10 Hz and about 0.1%, respectively, was superimposed on the specimen to determine E' and E'' and their dependence on time. Although some data were obtained at various temperatures and also on specimens subjected to a static tensile stress, only the data obtained at 50°C when the specimen was subjected to a tensile strain are discussed herein. Values of E' and E'' designated as being for the undeformed film were in fact determined on a specimen subjected to a static strain of 0.2%.

The time dependence of the effective diffusivity and steady-state permeability of several gases in the Trycite film were measured at various temperatures and static tensile strains by the gas-flow method described elsewhere.[5,8] The pressure gradient of each penetrant was 1 atm divided by the film thickness.

EXPERIMENTAL RESULTS AND DISCUSSION

Strain and Time Dependence of the Dynamic Modulus

Figure 2 shows on doubly logarithmic coordinates the time dependence of E'/E'_0, measured at 50°C and 10 Hz, for specimens of the annealed polycarbonate film at 10 static strains from 1.2 to 6.25%. (E'_0 is considered to be the storage modulus of the undeformed specimen; it actually was measured at a static strain of 0.2%, as mentioned already.) As the lowest four lines are essentially identical, they have been shifted down for clarity by the values of A given in the figure.

At each static strain (Fig. 2), $\log E'/E'_0$ increases linearly with $\log t$, where t is the time after a specimen was subjected to a static strain. Although the data in Fig. 2 were obtained only for $t \leq 3600$ s, data obtained in a separate experiment at a static strain of 3% followed the linear relation during a three-day period. Though now shown, $\log E''/E''_0$ increases substantially when a static strain is applied, but thereafter it decreases linearly with $\log t$.

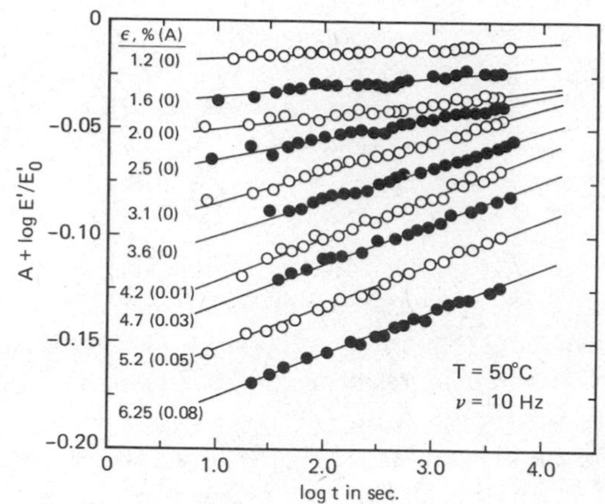

Fig. 2. Doubly logarithmic plots showing the time dependence of E'/E'_0 determined at 10 Hz on specimens of an annealed polycarbonate film at 50°C and at static tensile strains from 1.2 to 6.25%. E'_0 is the storage modulus at a static strain of 0.2%; it is sensibly time independent.

From the lines in Fig. 2 and those in a similar representation of E'', data at an elapsed time of 10 s were obtained and are plotted against the static strain in Fig. 3. As shown, log E'/E'_0 decreases and log E''/E''_0 increases until the strain becomes about 4%; thereafter these quantities are sensibly independent of strain. The line segments of zero slope represent a 20% decrease in E' and a threefold increase in E''.

When the slope of a line in Fig. 2 is multiplied by (100)(2.303), the result equals the percent increase in E' per decade of time. The left panel in Fig. 4 shows this rate plotted against the static strain. The right panel shows the rate of decrease of E'', expressed similarly. When the strain exceeds 4%, the rates become constant at +4.5% and –30% per decade of time for E' and E'', respectively.

As noted, the quantities shown in Figs. 3 and 4 become independent of the static strain when it exceeds 4%, which apparently is the yield strain. Because a 4% strain gives the same behavior as a larger strain, it appears that the strain and time dependence of the segmental mobility are unaffected by a plastic deformation.

During the determination of the data in Fig. 2, the tensile force on each specimen was recorded to obtain stress-relaxation data. At each static strain, log σ was found to decrease linearly with log t, where σ is the time-dependent stress.

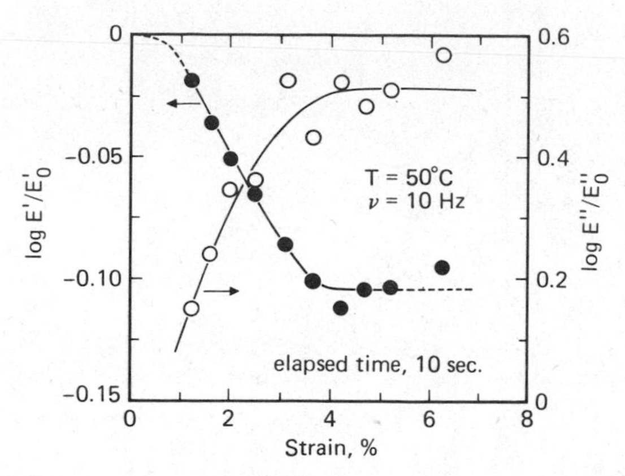

Fig. 3. Dependence of 10-second values of $\log E'/E'_0$ and $\log E''/E''_0$ on static strain.

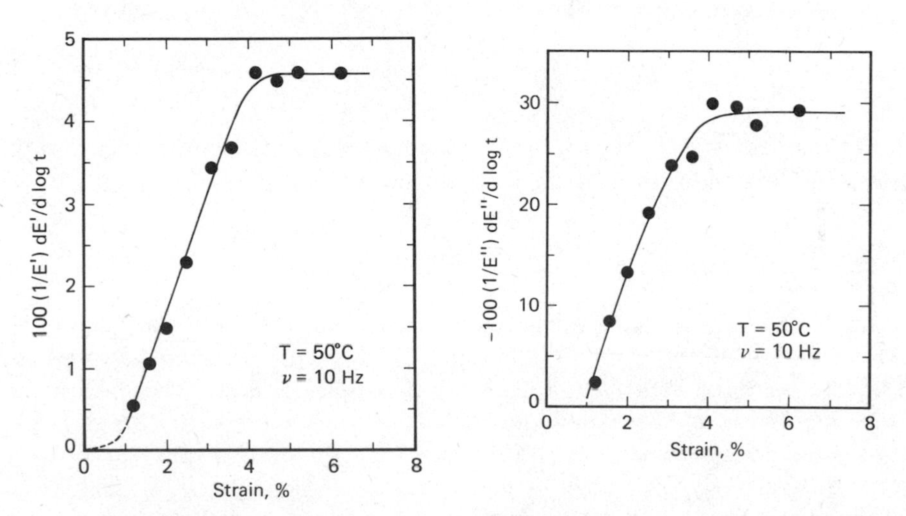

Fig. 4. Percentage increase of E' per decade of time (left panel) and the percentage decrease of E'' per decade of time (right panel) plotted against the static strain.

In Fig. 5, the stress-relaxation rate is compared with the rate of increase of E', obtained from Fig. 4. The ordinate in Fig. 5 is, in reality, $d \log E'/d \log t$ divided by $-d \log \sigma/d \log t$. As shown, this ratio attains a maximum value of 0.9 (or thereabout) at a static strain close to 4%. (Why the curve passes through a maximum is not known.) Stated otherwise, at the maximum, the stress-relaxation rate is only 10% less than the rate of increase

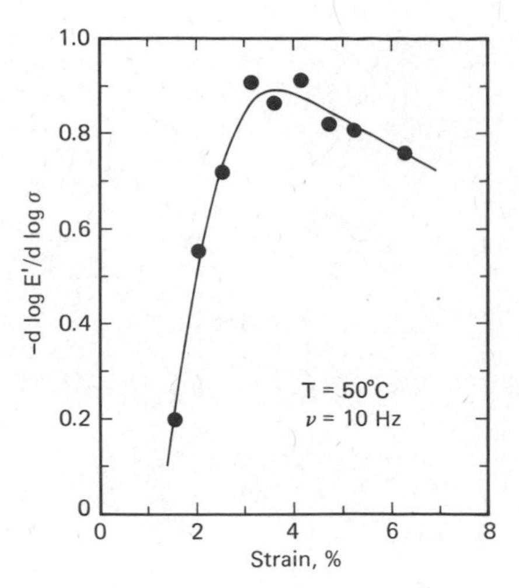

Fig. 5. Strain dependence of d log E'/d log t divided by −d log σ/d log t, where σ is the tensile stress in the specimen.

of E'. This signifies that the segmental mobility decreases significantly during the relaxation test and thus progressively reduces the stress-relaxation rate.

Strain and Time Dependence of Gas Permeability

Studies have been made of the permeability and diffusion coefficients (P and D) of carbon dioxide, argon, and xenon in biaxially oriented polystyrene, both undeformed and deformed statistically in simple tension. The data, some of which have been published,[5,6] show that the transport coefficients of a gas increase substantially when the static strain is applied, but thereafter they decrease progressively with time. These changes are considered to result from the strain and time dependence of segmental mobility as well as from certain characteristics of the gas and its interactions with the polymer. Some illustrative data are discussed below.

An empirical equation that seems appropriate to represent the time dependence of the steady-state permeability is

$$P(t) - P_0 = [P(1) - P_0]t^{-k} \tag{1}$$

where P_0 is the permeability of the undeformed film, P(1) is the permeability at unit time (here taken to be one hour), and k is a constant. Although this equation is clearly not applicable as t approaches zero, it should be valid as t becomes large, provided P(t) approaches P_0 and not a somewhat smaller value.

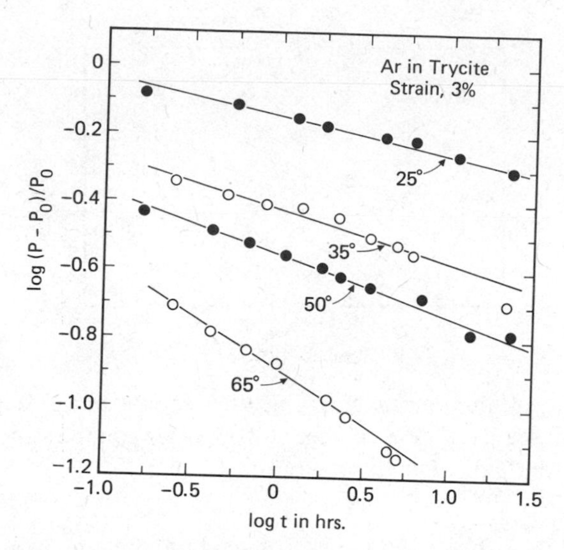

Fig. 6. Time dependence of $(P - P_0)/P_0$ at the indicated temperatures, shown on doubly logarithmic coordinates, where P is the permeability of argon in the Trycite (biaxially oriented polystyrene) film at a static strain of 3% and P_0 is the permeability in the undeformed film.

Figure 6 shows that the above equation represents P(t) for argon in the Trycite (biaxially oriented polystyrene) film at a static strain of 3% and at the indicated temperatures. The obtained values of k (i.e., the slopes of the lines) are plotted logarithmically against the temperature in Fig. 7. Also shown are data for xenon and for carbon dioxide when the static strain was 1.8%. The square symbol represents data for carbon dioxide, obtained when the strain was 2.4 and 4.2%, but only at 50°C. The lines, arbitrarily drawn parallel, fit the data reasonably well.

The permeability of a gas in a polymer above its glass transition temperature equals the product of the Fickian diffusivity of the gas and its Henry's law solubility coefficient. In the glassy state, Henry's law is not obeyed and diffusion is usually non-Fickian. Then, gas solubility and its mobility are commonly represented by the dual-sorption dual-mobility model.[4,9] While these facts were recognized, the considerable data required to evaluate the time dependence of the parameters in the dual-sorption dual-mobility model were not obtained on stretched specimens. However, it was found that the permeability of carbon dioxide at 35°C in specimens both undeformed and stretched 3% is independent of the partial pressure of the penetrant from 0.2 to 1.0 atm but that the diffusivity decreases somewhat with an increase in the pressure.

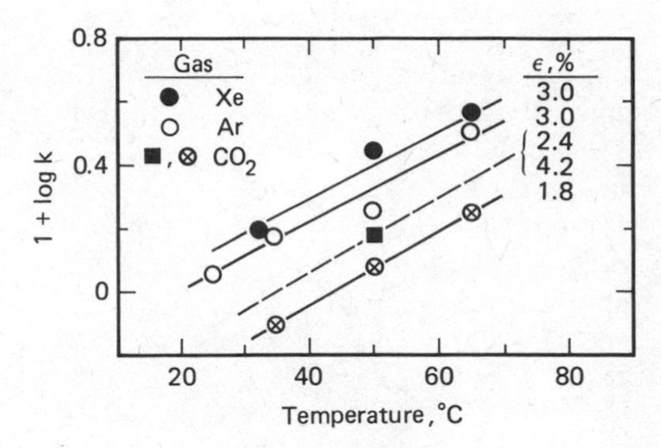

Fig. 7. Semilogarithmic plots showing the temperature dependence of k [= $-$d log ($P - P_0$)/d log t] for three penetrants in the Trycite film at the indicated static strains.

As shown in Table 1, $P(1)/P_0$ for carbon dioxide is larger than those for the other gases. Such is attributed to the distinct likelihood that the solubility of carbon dioxide increases more when the specimen is stretched than does the solubility of the other gases. It can be noted that the relative values of $P(t^*)/P_0$ for the different gases depend on the selected value of t^* because the decay rate of $[P(t) - P_0]$ is not the same for each gas but increases substantially with its diameter (Table 1).

Table 1. Data obtained at 50°C when the static strain was 3%

Quantity	Penetrant		
	CO_2[a]	Ar	Xe
diameter, nm	0.33[b]	0.34	0.41
$P(1)/P_0$	1.50	1.29	1.42
k[c]	0.15	0.21	0.25

[a]Data on strains of 2.4% and 4.2%; they should equal that at a 3% strain.
[b]Effective diameter from penetration in zeolites
[c]$k = -\,d \log [P(t) - P_0]/d \log t$

The decay rate k reflects the progressive decrease of both the solubility of the gas and its diffusivity, the latter being the dominant factor. Although (1/D)dD/d log t is probably not the best measure of the decay rate of D, it has been shown[6] that the absolute value of this quantity increases with the diameter of the gas, as is also shown by k in Table 1. Because the diffusivity of Xe depends strongly on the concentration of the larger free-volume elements, this finding indicates that the fraction of the larger free-volume elements decreases more rapidly with time than does the fraction of the smaller ones. If the progressive decrease of segmental mobility were the only relevant factor, then the decay rate of D would be the same for each gas.

ACKNOWLEDGMENTS

The data on gas transport were obtained by W. Oppermann (Technical University of Clausthal) and G. Levita (University of Pisa) during their tenure of post-doctoral fellowships sponsored by IBM Germany and IBM Italy. The data presented in this paper are derived from those in Refs. 5 and 6.

REFERENCES

1. A. J. Kovacs, Transition vitreuse dans les polymeres amorphes. Etude phenomenologique, Adv. Polym. Sci. 3:394 (1964).
2. L. C. E. Struik, Physical aging in plastics and other glassy materials, Polym. Eng. Sci. 17:165 (1977).
3. L. C. E. Struik, "Physical aging in amorphous polymers and other materials," Elsevier, New York (1978).
4. D. R. Paul, Gas sorption and transport in glassy polymers, Ber. Bunsenges. Phys. Chem. 83:294 (1979).
5. T. L. Smith and G. Levita, Effect of tensile strain, time, and temperature on gas transport in biaxially oriented polystyrene, Polym. Eng. Sci. 21:936 (1981).
6. T. L. Smith, W. Oppermann, A. H. Chan and G. Levita, Gas transport in a glassy polymer subjected to static tensile strains, Polym. Preprints 24(1):83 (1983).
7. T. L. Smith and W. Oppermann, Dynastat Viscoelastic Analyzer for studies of polymer properties, Tech. Papers of 39th Ann. Tech. Conf. Soc. Plast. Eng. 27:163 (1981).
8. T. L. Smith and R. Adam, Effect of tensile deformations on gas transport in glassy polymer films, Polymer 22:299 (1981).
9. D. R. Paul and W. J. Koros, Effect of partially immobilizing sorption on permeability and the diffusion time lag, J. Polym. Sci., Polym. Phys. Ed. 14:675 (1976).

NONLINEAR VISCOELASTIC CHARACTERISATION OF POLYMERIC MATERIALS

H.F. Brinson,* C.C. Hiel,* A.H. Cardon,☆ and W.P. De Wilde☆

*Virginia Polytechnic Institute & State University
Blacksburg, VA 24061 USA

☆Free University Brussels (VUB)
Faculty of Applied Sciences
Pleinlaan, 2
1050 Brussels, Belgium

INTRODUCTION

Polymeric materials are viscoelastic. An important problem is the prediction of their long time behaviour on the basis of results of short time tests.

The significance of short time tests is based on the validity of the accelerating procedure.

The accelerated characterisation procedure developed by Brinson et al.[1,2] for polymer based composite laminates, was started from the well known time-temperature superposition principle (TTSP). Afterwards a time-stress superposition principle (TSSP), was utilised and combined in a more general time-temperature stress superposition principle (TTSSP)[3,4].

The basic idea was that certain environmental conditions such as temperature and stress level could accelerate the viscoelastic deformation process.

In order to compute easily the long term data generated by the TSSP or TTSSP a nonlinear viscoelastic constitutive law is needed.

Dillard et al.[5,6] utilised a Findley type nonlinear creep law. This creep law is semi-empirical and his application possibilities has some severe limitations. These limitations were studied by Hiel[7].

For those reasons a thermodynamic based nonlinear viscoelastic model proposed by Schapery[8,9] was investigated by Hiel[7].

The result is a more easy computational formulation that could give more accurate long term results based on short term data than the early graphical procedures.

NONLINEAR VISCOELASTIC MODELS

A review of the different nonlinear viscoelastic representations is given by Lockett[10] and Hiel[7].

The multiple integral representations, Green-Rivlin (1957) ; Pipkin and Ropers (1968), are very difficult to work out experimentally.
Simple integral formulations were presented by Leaderman (1934), Rabotnov (1948) and Koltunov (1965).

In this study we start from the Schapery formulation of 1966[11] completely analysed by Hiel[7].

The linear thermorheological simple material law

$$\varepsilon = D_o \sigma + \int_{-\infty}^{t} \Delta D(t-u) \frac{d\sigma}{du} du \tag{1}$$

is transformed in a more general nonlinear viscoelastic law

$$\varepsilon = g_o(\sigma) D_o\sigma + g_1(\sigma) \left[\int_{-\infty}^{t} \Delta D(\psi-\psi') \frac{d(g_2\sigma)}{du} du\right] \tag{2}$$

$$\psi(t) = \int_{t_o}^{t} \frac{dt'}{a_\sigma(\sigma)} \quad ; \quad (a_\sigma > 0)$$

$$\psi' = \psi(u)$$

g_o, g_1, g_2 and a_σ are the nonlinearizing functions.
If $g_o = g_1 = g_2 = a_\sigma = 1$, equation (2) reduces to equation (1).

The Kelvin-type kernel ΔD in equation (2) is approximated by a power law, valid under restricted conditions[7], and equation (2) becomes

$$\varepsilon = g_o D_o\sigma + g_1 \int_{-\infty}^{t} C(\psi-\psi')^n \frac{d(g_2\sigma)}{du} du \tag{3}$$

For a creep test $\sigma = \sigma_o H(t)$ we obtain :

$$\varepsilon = g_o D_o \sigma_o + g_1 g_2 C\left(\frac{t}{a_\sigma}\right)^n \sigma_o \tag{4}$$

with three linear parameters (D_o,C,n) and 4 nonlinear parameters (g_o,g_1,g_2,a_σ).

A further simplification of (2) gives for low stresses, with $g_1 = 1$, $g_2 = 1/a_G$ and $a_D/a_G = a_T$

$$\varepsilon = g_o\, D_o\, \sigma + \int_{-\infty}^{t} \Delta D(\psi - \psi') \frac{d}{du} \left(\frac{\sigma}{a_G}\right) du \tag{5}$$

and $\log \psi = \log t - \log a_T$.
For $\sigma = \sigma_o\, H(t)$, equation (5) becomes

$$D_T = \frac{\varepsilon}{\sigma_o} = g_o(T)\, D_o + \frac{\Delta D(\psi)}{a_G(T)} \tag{6}$$

EXPERIMENTAL RESULTS

These experimental results were obtained on the 934 resin, (Fiberite 934). The tested 934-resin plates were fabricated at the Graftech Facilities.

In order to have a valuable homogeneous set of test specimens, each specimen was subjected to a postcure cycle, (176°C - 4 hours and slow cooldown at 5°C/hour).
A sequence of elevated temperature creep and creep-recovery tests was applied at a moderate stress level σ_o = 11 KPa.

The creep and creep-recovery curves obtained during those test sequence are very accurately modelled by the power law

$$D(T) = D_o(T) + D_1\, t^n \tag{7}$$

and the result is shown in Fig. 1.

The temperature dependence of $D_o(T)$ indicates a thermorheologically complex behaviour and shows that the simple power law breaks down above 149°C.

The master curve, obtained by horizontal and vertical shifting, is shown in Fig. 2.

The stress-strain curves reveal a brittle-ductile transition in the temperature range 93-149°C and have all initial linear path.

A test program with a sequence of creep and creep-recovery tests at a given temperature was developed, Fig. 3.

Till 160°C no significant nonlinearity was observed.

During the test program, Fig. 3, an accumulation of non-recoverable strain was observed, Fig. 4.

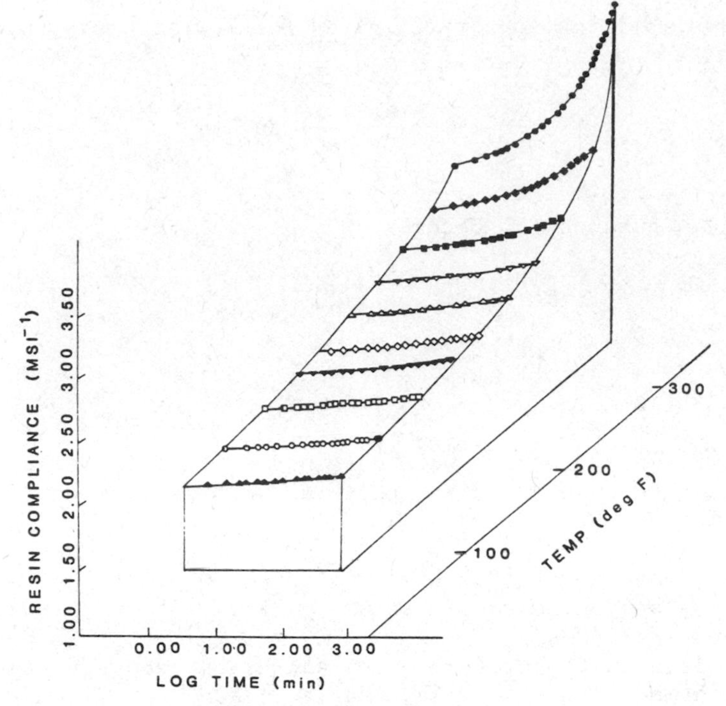

Fig. 1.

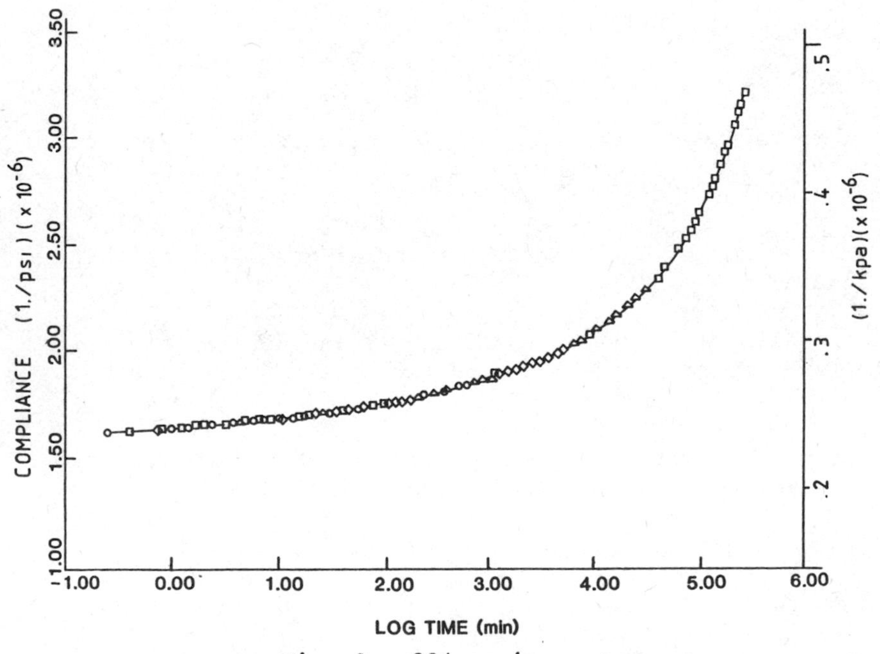

Fig. 2. 934-resin master curve

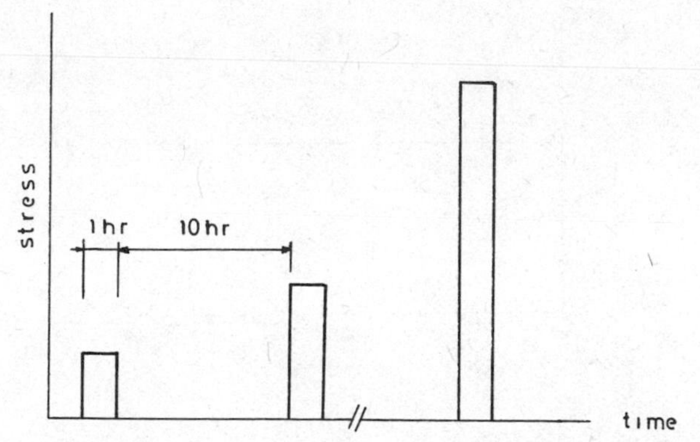

Fig. 3. Test sequence of creep and creep recovery tests at a single temperature and different stress level.

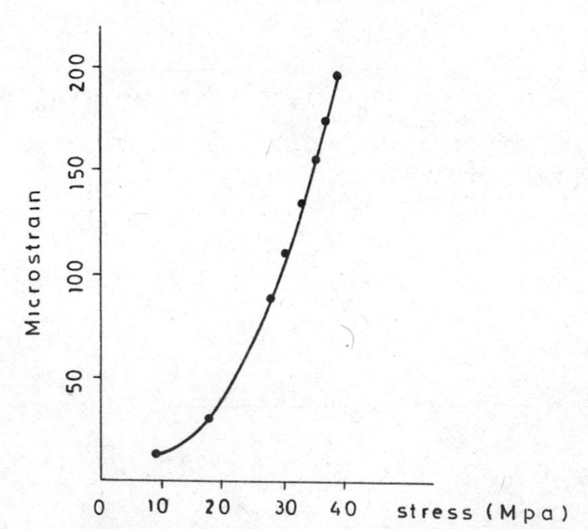

Fig. 4. Permanent strain accumulation versus stress (for 934-resin at 119°C).

The nonlinearizing parameters g_2 and a_σ were plotted with and without substraction of the accumulated permanent strain, Fig. 5 and Fig. 6.

This leads to the conclusion that this resin has a linear viscoelastic behaviour at low stress levels and a viscoelastic-plastic behaviour at higher stress levels.

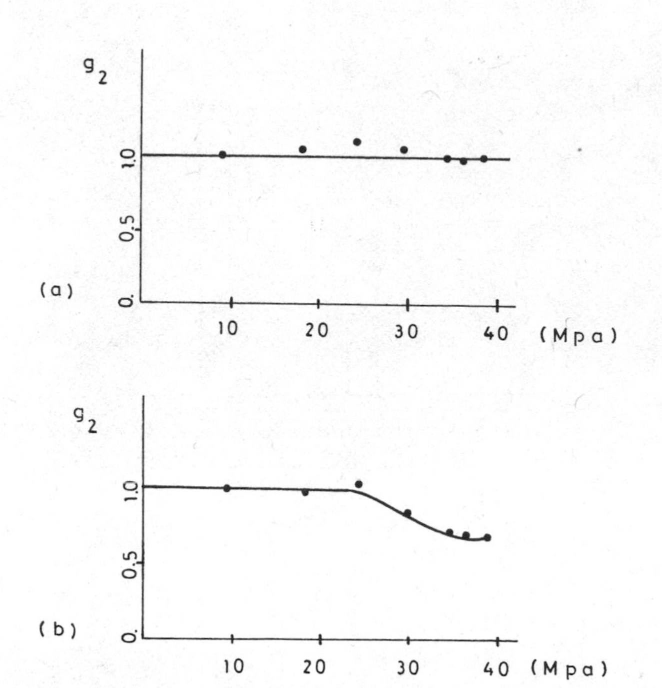

Fig. 5. For 934-resin with (a) or without (b) subtraction of accumulated permanent strain.

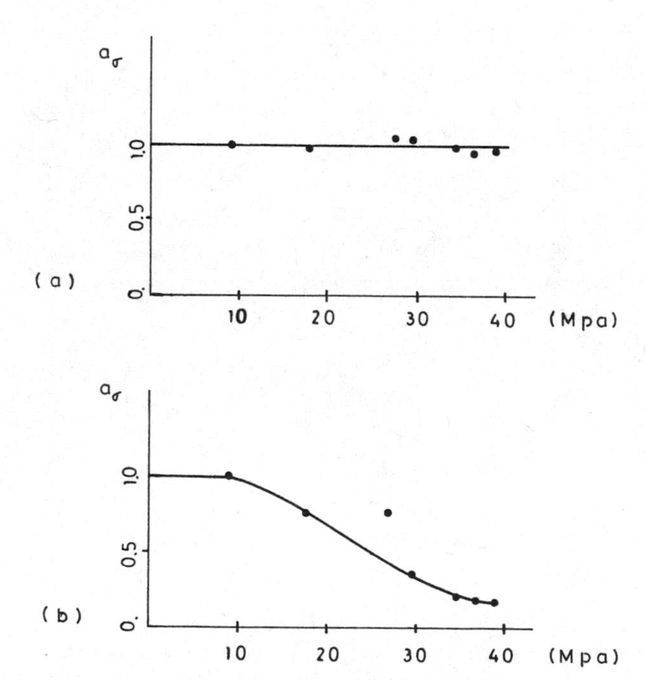

Fig. 6. For 934-resin with (a) or without (b) subtraction of accumulated permanent strain.

CONCLUSION

These results indicates that the Schapery-model and the approximative power law, even with his restrictions, are very suitable to investigate the nonlinearities in the mechanical behaviour of viscoelastic materials. Applications to composites were investigated by Brinson and Hiel.

ACKNOWLEDGEMENTS

Financial support was provided for this work by NASA Cooperative Agreement NCC-2-71. Thanks for support are also directed to Virginia Polytechnic Institute & State University, Free University Brussels (V.U.B.) and the Belgian National Science Foundation.

REFERENCES

1. H.F. BRINSON, D.H. MORRIS and Y.T. YEOW, "A New Experimental Method for the Accelerated Characterisation of Composite Materials", 6th Int. Conf. on Exp. Stress Analysis, Munich, Sept. 18-22, 1978.
2. Y.T. YEOW, "The Time-Temperature Behaviour of Graphite Epoxy Laminates", Ph.D. Dissertation, V.P.I. & S.U., Blacksburg, Virginia, 1978.
3. W.I. GRIFFITH, "The Accelerated Characterisation of Viscoelastic Composites Materials", Ph.D. Dissertation, V.P.I. & S.U., Blacksburg, Virginia, 1980.
4. H.F. BRINSON, D.H. MORRIS, W.I. GRIFFITH and D.A. DILLARD, "The Viscoelastic Response of a Graphite Epoxy Laminate", Proceedings of the 1st Int. Conf. on Comp. Structures, Paisley, Scotland, Sept. 1981.
5. D.A. DILLARD, "Creep and Creep Rupture of Laminated Graphite Epoxy Composites", Ph.D. Dissertation, V.P.I. & S.U., Blacksburg, Virginia, 1981.
6. H.F. BRINSON and D.A. DILLARD, "The Prediction of Long Term Viscoelastic Properties of Fiber Reinforced Plastics", Progress in Science and Engineering of Composites - ICCM-IV, Tokyo, Japan, Oct. 1982.
7. C.C. HIEL, "The Nonlinear Viscoelastic Response of Resin Matrix Composites", Doctoral thesis, Free University Brussels (V.U.B.), January 1983.
8. R.A. SCHAPERY, "On the Characterisation of Nonlinear Viscoelastic Materials", Polymer Eng. and Sc., vol. 9, n°4, 1969.
9. Y.C. LOU and R.A. SCHAPERY, "Viscoelastic Characterisation of a Nonlinear Fibre-Reinforced Plastics", Journal of Composite Materials, vol. 5, 1971.
10. F.J. LOCKETT, "Nonlinear Viscoelastic Solids", Academic Press, London, New York, 1972.
11. R.A. SCHAPERY, "An Engineering Theory of Nonlinear Viscoelasticity with Applications", Int. J. Solids and Structures, vol. 2, n°3, 1966.

DESCRIBING THE LONG TERM NONLINEARITY OF PLASTICS

O. S. Brüller

Institute -A- of Mechanics
Technical University of Munich
Arcisstr. 21, D - 8000 Munich 2

INTRODUCTION

A good method for the mathematical description of the response of a polymer under long term loading conditions (such as creep or stress relaxation) is provided by the use of finite exponential series. In the nonlinear viscoelastic range, the coefficients of the series are dependent upon the level of the applied load. The knowledge of this dependence enables the description of the nonlinear behavior of the material even for more complicated loading histories.

STRESS RELAXATION IN THE LINEAR VISCOELASTIC RANGE

The stress response $\sigma(t)$ of a viscoelastic material to a constant uniaxial strain applied under unchanged environmental conditions at time $t = 0$ is a function of time of the form:

$$\sigma(t) = \varepsilon_{(0)} E(t) \qquad (1)$$

where $E(t)$ is the uniaxial relaxation modulus of the considered material and $\varepsilon_{(0)}$ the applied strain.

In the linear viscoelastic range the relaxation modulus is a function of time only and can be approximated by the relation:

$$E(t) = (E_0 + \sum_{i=1}^{m} E_i e^{-\frac{t}{\tau_i}}) \qquad (2)$$

In order to obtain an accurate approximation, it is sufficient to choose the m discrete "relaxation times" τ_i one per decade in the time interval of the experiment[1,2]. The $m+1$ material parameters E_i can be computed by simple collocation[1] or by using least squares as shown here (see also[3]).

The sum of squares of the differences between the mathematical approximation (2) and the n experimentally determined values of the relaxation modulus $E^*(t_j)$ is:

$$H=\sum_{j=1}^{n}(E^*(t_j)-(E_0+\sum_{i=1}^{m}E_i\, e^{-\frac{t_j}{\tau_i}}))^2 \tag{3}$$

Its minimum is obtained if:

$$\frac{\delta H}{\delta E_k}=0 \qquad (k=0,1,2,..m) \tag{4}$$

The following linear system of $m+1$ equations results:

$$[A^T]\cdot[A]\cdot[E]=[A^T]\cdot[B] \tag{5}$$

with:

$$[A]=\begin{bmatrix} 1 & e^{-t_1/\tau_1} & e^{-t_1/\tau_2} & \cdots & e^{-t_1/\tau_m} \\ 1 & e^{-t_2/\tau_1} & e^{-t_2/\tau_2} & \cdots & e^{-t_2/\tau_m} \\ \vdots & \vdots & \vdots & & \vdots \\ 1 & e^{-t_n/\tau_1} & e^{-t_n/\tau_2} & \cdots & e^{-t_n/\tau_m} \end{bmatrix} \tag{6}$$

$$[E]=\begin{bmatrix} E_0 \\ E_1 \\ \vdots \\ E_m \end{bmatrix} \tag{7}$$

$$[B]=\begin{bmatrix} E^*(t_1) \\ E^*(t_2) \\ \vdots \\ E^*(t_n) \end{bmatrix} \tag{8}$$

and $[A]$ as transposed matrix of $[A^T]$.

Example

The relaxation curve of a styrene-acrylonitrile (SAN) is considered in a time interval between 10^{-3} and 10^{+2} h. Its relaxation modulus was approximated by taking:

$$\tau_i = 10^{i-4} \text{ h } (i = 1, 2, \ldots, 6) \qquad (9)$$

Following parameters have been obtained:

$$E_0 = 3661.34 \text{ N/mm}^2 \qquad (10a)$$

and

$$\begin{aligned} E_1 &= 19.72 \text{ N/mm}^2 \\ E_2 &= 37.34 \\ E_3 &= 77.28 \\ E_4 &= 81.67 \\ E_5 &= 134.64 \\ E_6 &= 224.39 \end{aligned} \qquad (10b)$$

Introduction of the values (9) and (10) into Eq. (2) allows the computation of the stress according to Eq. (1). As seen in Fig. 1, the agreement between experimental and computed results is within the graphical accuracy.

Here it should be pointed out that the method presented above leads also to very good fits of single relaxation curves in the nonlinear viscoelastic range. However, in such cases the determined parameters are valid only for the investigated single curve but not for curves obtained at other strain levels.

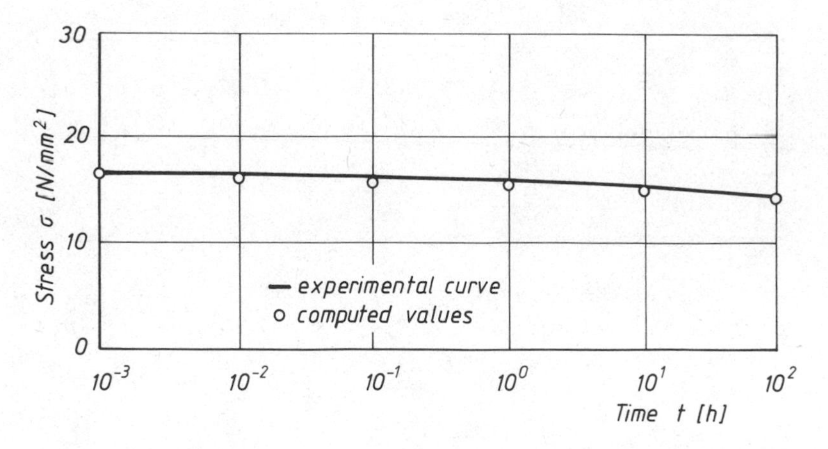

Fig. 1. Relaxation curve of SAN at $\varepsilon_{(0)} = .40$ %

STRESS RELAXATION IN THE NONLINEAR VISCOELASTIC RANGE

In order to extend the described method for computing the stress in the nonlinear viscoelastic range, a modification of Eq. (2), proposed in[2], is adopted. Accordingly, the relaxation modulus will be:

$$E(t) = h_1 E_0 + h_2 \sum_{j=1}^{m} E_j e^{-\frac{t}{a_\tau \tau_j}} \qquad (11)$$

For unfilled polymeric materials[2]:

$$a_\tau = 1 \qquad (12)$$

The values of h_1 and h_2 will be determined again by using least squares techniques at each investigated level of the strain $\varepsilon_{(0)}$. One obtains:

$$[A_h^T][A_h][h] = [A_h^T][B] \qquad (13)$$

with:

$$[h] = \begin{bmatrix} h_1 \\ h_2 \end{bmatrix} \qquad (14)$$

and:

$$[A_h] = \begin{bmatrix} E_0 & \sum_{j=1}^{m} E_j e^{-t_1/\tau_j} \\ E_0 & \sum_{j=1}^{m} E_j e^{-t_1/\tau_j} \\ \vdots & \vdots \\ E_0 & \sum_{j=1}^{m} E_j e^{-t_n/\tau_j} \end{bmatrix} \qquad (15)$$

$$[B] = \begin{bmatrix} \sigma^*(t_1)/\varepsilon_{(0)} \\ \sigma^*(t_2)/\varepsilon_{(0)} \\ \vdots \\ \sigma^*(t_n)/\varepsilon_{(0)} \end{bmatrix} \qquad (16)$$

$[A_h^T]$ is the transposed matrix of $[A_h]$.

The functions $h_1(\varepsilon_{(0)})$ and $h_2(\varepsilon_{(0)})$ should be determined, as shown, from tests conducted at different strain levels. Results obtained with different polymers show that they can be approximated by straight lines.

Example

Stress relaxation curves of SAN at various strains are depicted

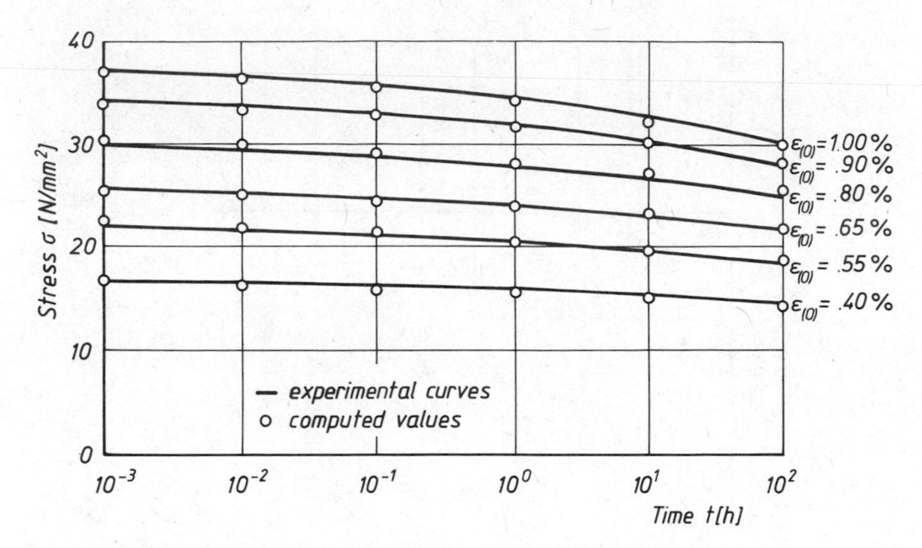

Fig. 2. Relaxation curves of SAN

in Fig. 2. For strain $\varepsilon_{(0)} \leq .4$ % the behavior of the material is linear viscoelastic.

The values of h_1 and h_2 obtained by using Eq. (13) are plotted in Fig. 3. They are approximated by the straight lines (by using linear regressions):

$$\begin{aligned} h_1 &= 1.07459 - 26.697\ \varepsilon_{(0)} \\ h_2 &= .88259 + 51.862\ \varepsilon_{(0)} \end{aligned} \qquad (17)$$

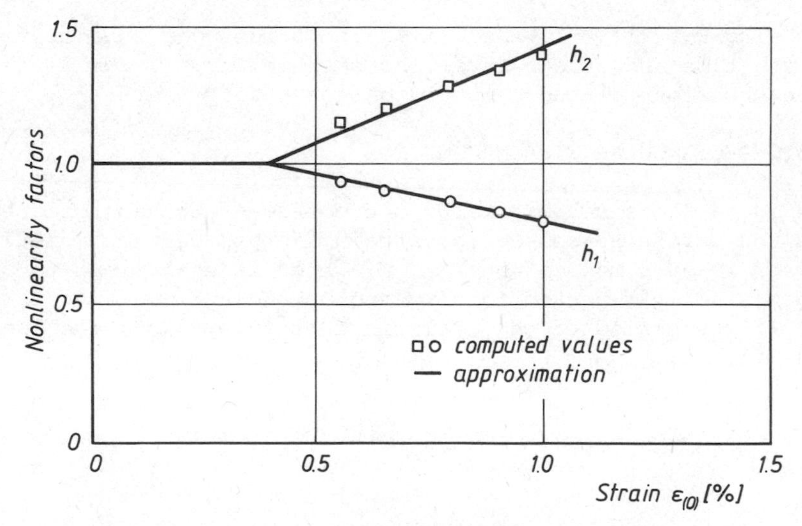

Fig. 3. The factors h_1 and h_2 of SAN

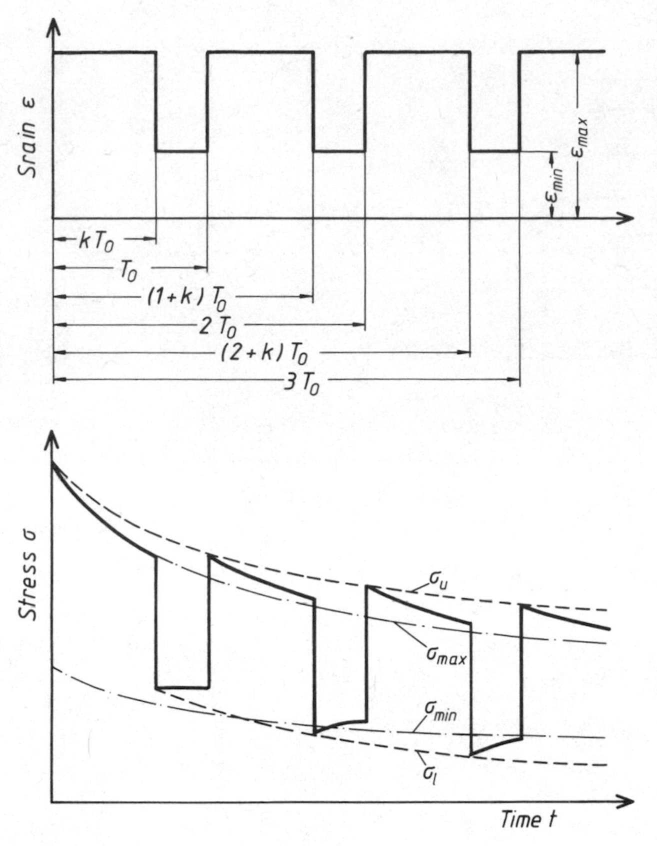

Fig. 4. Schematical representation of strain and stress under relaxation-type periodical rectangular loading and unloading

Using Eq. (1), (11) and (17), the stress can be computed for any given time and strain level. As seen in Fig. 2, the differences between experimental and computed data are very small.

PERIODIC RECTANGULAR STRAIN FUNCTION

Fig. 4 shows schematically the course of the applied strain and of the stress response. The strain "jumps" between a minimal value ε_{min} and a maximal one ε_{max}, which is equivalent to a series of equidistantly repeated loading and unloading (negative loading) steps. If the period of the cycle is denoted by T_0, the loading period will be:

$$T_L = kT_0 \tag{18}$$

where:

$$0 < k < 1 \tag{19}$$

In the schematical representation of the stress (Fig. 4) σ_{max} and σ_{min} are relaxation curves corresponding to the strains ε_{max} and ε_{min} respectively and σ_u and σ_l are curves connecting the upper and lower peaks of the stress respectively.

At time t situated in the region in which the strain has the value ε_{max}, the stress $\sigma(t)$ will be:

$$\begin{aligned}\sigma(t) = \sigma_{max}(t) &- \sigma_{max}(t-kT_0) + \sigma_{min}(t-kT_0) - \sigma_{min}(t-T_0)\\ + \sigma_{max}(t-T_0) &- \sigma_{max}(t-(1+k)T_0) + \sigma_{min}(t-(1+k)T_0) - \sigma_{min}(t-2T_0)\\ + \sigma_{max}(t-2T_0) &- \sigma_{max}(t-(2+k)T_0) + \sigma_{min}(t-(2+k)T_0) - \sigma_{min}(t-3T_0)\\ &\vdots\\ + \sigma_{max}(t-(n-1)T_0) &- \sigma_{max}(t-((n-1)+k)T_0) + \sigma_{min}(t-((n-1)+k)T_0) - \sigma_{min}(t-nT_0)\\ + \sigma_{max}(t-nT_0)&\end{aligned} \tag{20}$$

or:

$$\begin{aligned}\sigma(t) = &\sum_{j=0}^{n}\sigma_{max}(t-jT_0) - \sum_{j=0}^{n-1}\sigma_{max}(t-(j+k)T_0)\\ &+ \sum_{j=0}^{n-1}\sigma_{min}(t-(j+k)T_0) - \sum_{j=1}^{n}\sigma_{min}(t-jT_0)\end{aligned} \tag{21}$$

By introducing the notation:

$$\begin{aligned}\varepsilon_{max}\ h_1(\varepsilon_{max}) &= \varepsilon'_{max}\\ \varepsilon_{max}\ h_2(\varepsilon_{max}) &= \varepsilon''_{max}\\ \varepsilon_{min}\ h_1(\varepsilon_{min}) &= \varepsilon'_{min}\\ \varepsilon_{min}\ h_2(\varepsilon_{min}) &= \varepsilon''_{min}\end{aligned} \tag{22}$$

and:

$$t = (n+c)T_0 \tag{23}$$

where in the present case:

$$c \leq k \tag{24}$$

one obtains after some algebra:

$$\begin{aligned}\sigma(t) = &\varepsilon'_{max}E_0 + \varepsilon''_{max}\sum_{i=1}^{m}E_i e^{-\frac{t}{\tau_i}}\\ &-(\varepsilon''_{max} - \varepsilon''_{min})\sum_{i=1}^{m}E_i\sum_{j=1}^{n}\left(e^{-\frac{(j+c-k)T_0}{\tau_i}} - e^{-\frac{(j+c-1)T_0}{\tau_i}}\right)\end{aligned} \tag{25}$$

Similarily, at time t situated in the region in which the strain has the value ε_{min} (i.e. $k \leq c < 1$) one can obtain:

$$\begin{aligned}\sigma(t) = &\varepsilon_{min}E_0 + \varepsilon_{min}\sum_{i=1}^{m}E_i e^{-\frac{t}{\tau_i}}\\ &-(\varepsilon_{max} - \varepsilon_{min})\sum_{i=1}^{m}E\sum_{j=0}^{n}\left(e^{-\frac{(j+c-k)T_0}{\tau_i}} - e^{-\frac{(j+c)T_0}{\tau_i}}\right)\end{aligned} \tag{26}$$

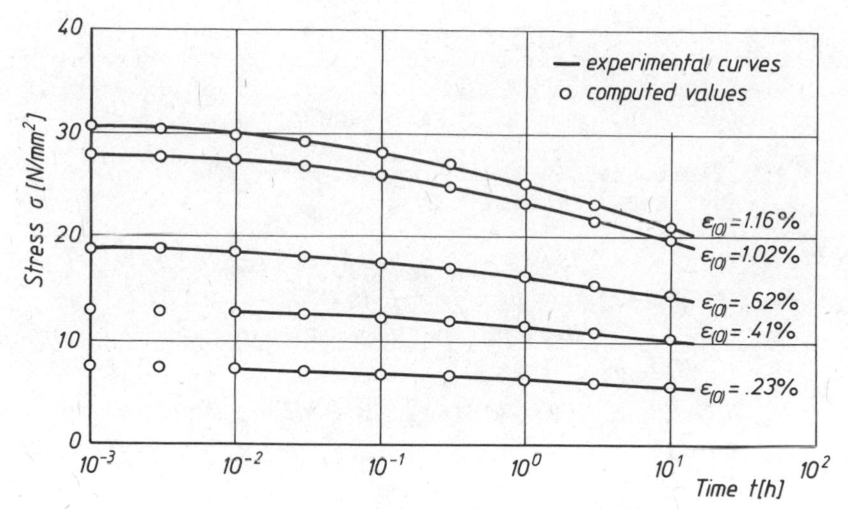

Fig. 5. Relaxation curves of POM

Examples

Relaxation curves of polyoximethylene (POM) are shown in Fig.5. The nonlinearity of the material is described by the linearly changing factors h_1 and h_2 which are depicted in Fig. 6.

Submitting now the polymer to a "jumping" stress relaxation between ε_{max} = 1.05% (10 min) and ε_{min}= .23% (30 min) the stress-response presented in Fig. 7 is obtained. In a second example, the same polymer is loaded with ε_{max}= .90% (15 min) and ε_{min}= .23% (30 min). The stress response is shown vs. log t in Fig. 8. In both cases a good agreement between experiment and computed values can be seen.

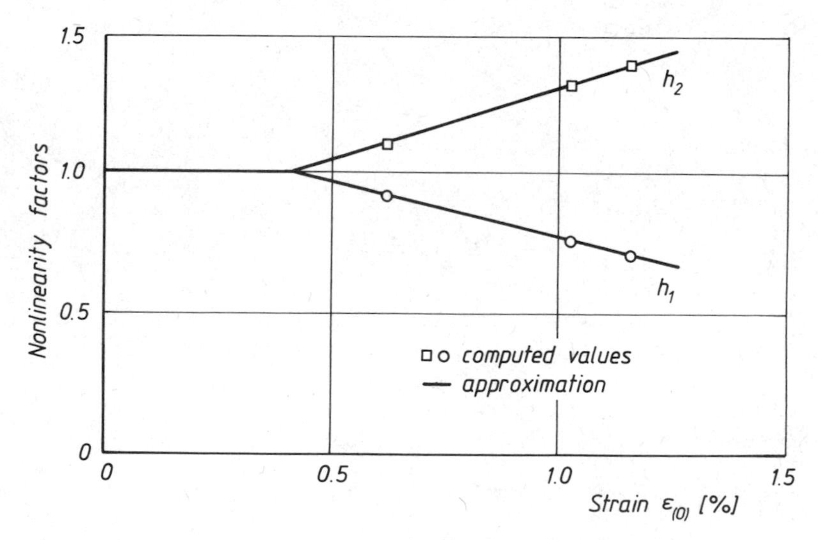

Fig. 6. The factors h_1 and h_2 of POM

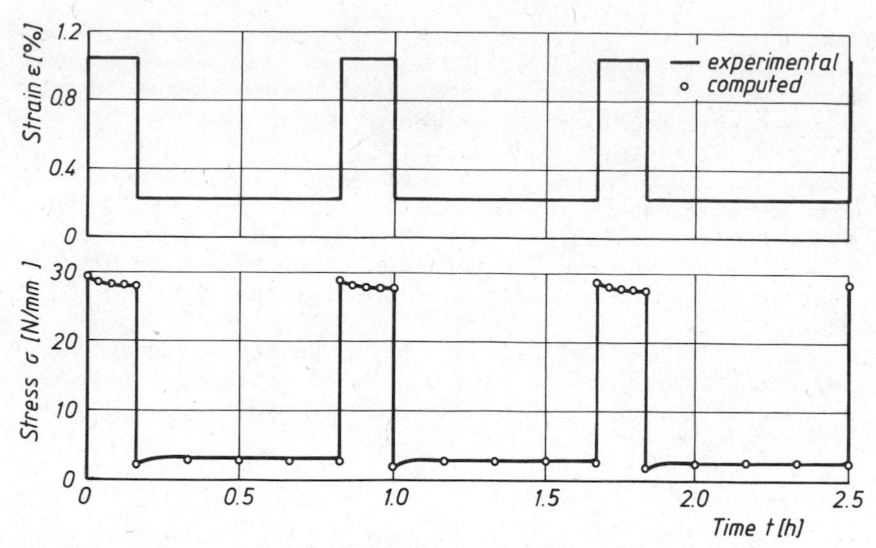

Fig. 7. Strain and stress of POM under periodic rectangular strain

CONCLUSIONS

The nonlinear long term viscoelastic behavior of polymers can be approximated by a linear variation of the material parameters. As an application of this property, the material response to a cyclical loading and unloading can be described with good accuracy. The method, exemplified on relaxation-type cyclical loading can be easily applied to creep-type cyclical loading.

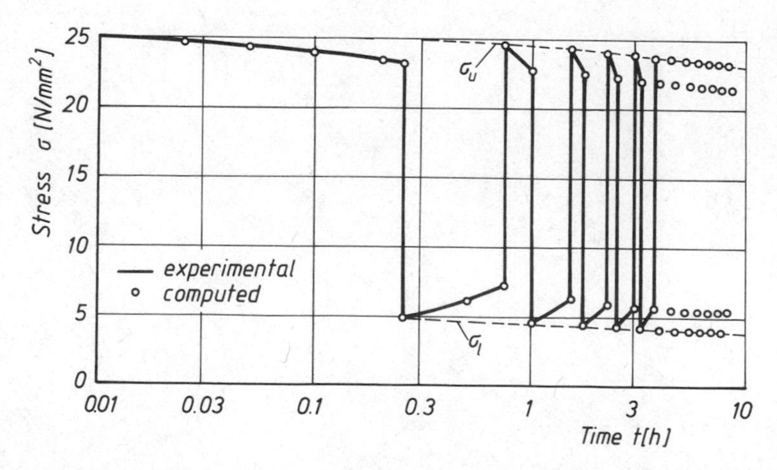

Fig. 8. Stress of POM under periodic rectangular strain

REFERENCES

1. R. A. Schapery, A Simple Collocation Method for Fitting Viscoelastic Models to Experimental Data, SM 61 - 23a, GALCIT, Calif. Inst. Technol., 1961.
2. R. A. Schapery, On the Characterization of Nonlinear Viscoelastic Materials, Polym. Eng. Sci., 32, 295, 1969.
3. O. S. Brüller and B. Reichelt, Beschreibung des nichtlinearen Verhaltens von Kunststoffen unter Belastung mit konstanter Dehngeschwindigkeit, Rheol. Acta, 20, 240, 1981.
4. O. S. Brüller, B. Reichelt and H. G. Moslé, Beschreibung des nichtlinearen Verhaltens von Kunststoffen unter Spannungsrelaxationsbelastung, Kunststoffe, 72, 796, 1982.

ORIGINS OF THE TOUGHNESS OF FIBROUS COMPOSITES

Peter W.R. Beaumont and John K. Wells*

Cambridge University Engineering Department

Trumpington Street, Cambridge CB2 1PZ, England

ABSTRACT

This paper describes the micromechanisms of fracture of fibrous composites and shows how models based on these physical processes can be used to construct fracture maps. The fracture map predicts the toughness of a fibrous composite, displays the principal mechanisms of toughening and helps in selecting a fibre and matrix system for high fracture toughness.

ANALYSIS OF FRACTURE

The interaction between fracture mechanisms of a fibrous composite are complex. The effects of certain intrinsic material parameters, like the cohesive strength of the fibre-matrix interface or ductility of the matrix on the micromechanisms of fracture and toughness are unclear. One approach is to link our understanding of the failure processes with theoretical models of fracture to predict the toughness of a composite. However, the complex interactions between various intrinsic material parameters and a set of equations would not clearly be seen. Instead, we could construct maps based on these models to display information on fracture and toughness in a useful form. The fracture map would have two axes which are labelled using any two of the intrinsic material parameters that describe the fibre composite (Table 1). The map would be divided into areas or fields, each one depicting a particular failure mechanism. The boundary between one field and another would show a change in the dominant mechanism. Contour lines of predicted toughness could be superimposed onto the map in a manner useful for design and material selection.

Table 1. Material parameters which appear in the models

FIBRE	σ_f , E_f , r_f , m , ν_f
MATRIX	σ_m , E_m , G_1 , ν_m
INTERFACE	μ , ε_o , ε_b , G_1
GEOMETRY	r_b , V_f

(f - fibre b - bundle m - matrix)

FRACTURE MAPS BASED ON MICROSCOPIC MODELLING

When a composite is loaded, failure originates at large defects (e.g. notches) or from inherent flaws (e.g. broken fibres). A matrix crack formed at such a defect propagates up to and around a strong fibre, which then bridges the crack. As the load increases, failure of the bond between fibre and matrix occurs; this process is called "debonding". Finally the fibre breaks, often out of plane of the matrix crack, producing "pull-out" of fibres or bundles of fibres. These processes absorb energy and are the origin of composite toughness.

In computing a fracture map, we require equations which predict values of the unknown terms in the fracture models, debond length and pull-out length, for instance, together with known data for material parameters, fibre strength, modulus, matrix strength, and so forth (Table 1). The actual construction of a map is done by a computer which searches incrementally over the field to find the toughness contours and field boundaries. The method is amenable to changes in the labelling of the two axes of the map, and we see the effect of altering any two material parameters simultaneously, such as fibre strength and interfacial shear strength. The fields in which the various mechanisms are dominant are shown on the map.

MICROSCOPIC MODELS OF FRACTURE

Theory of Debonding

When the matrix cracks, shear forces are set up at the fibre-matrix interface. If the shear force exceeds the adhesive bond strength then debonding occurs. Stress transfer is still possible after debonding due to friction between fibre and matrix. This arises from the differential thermal expansions of fibre and matrix which produces a residual compressive stress (or misfit strain, ε_o) during fabrication. The debond length, ℓ_d, depends on the debond stress of the fibre or fibre bundle, σ_d, and the frictional force, μ, which is calculated with allowance for the Poisson contraction of the fibre or fibre bundle under load. Fibres break when the maximum stress reaches the ultimate strength of the fibre or bundle strength, σ_f (Wells and Beaumont):

$$\sigma_f = \sigma_o - (\sigma_o - \sigma_d)\exp(-\beta\,\ell_d/2)$$

where $\sigma_o = \varepsilon_o E_f/\nu_f$, $\beta = 2\nu_f\,\mu\,E_m/E_f\,r_f\,(1+\nu_m)$.

Hence

$$\ell_d = \frac{2}{\beta}\,\ell n\left[\frac{\sigma_o - \sigma_d}{\sigma_o - \sigma_f}\right]$$

Theory of Pull-out

Knowing the stress distribution in a debonded fibre in conjunction with the distribution of fibre strengths, the average pull-out length, $\bar{\ell}_p$, can be calculated using (Wells and Beaumont):

$$\bar{\ell}_p = \frac{\sigma_f\,\eta}{2\beta(\sigma_o - \sigma_d)}$$

where η is proportional to the reciprocal of the fibre Weibull modulus, m.

Elastic Energy

The energy dissipated within the debonded region when the fibre breaks is (Wells and Beaumont):

$$G_e = \frac{V_f}{E_f}\left[\frac{\sigma_p^2\,\ell_d}{2} - \frac{(\sigma_p - \sigma_d)^2(e^{-\beta\ell_d} - 1)}{2\beta} + \frac{2\sigma_p(\sigma_p - \sigma_d)(e^{-\beta\frac{\ell_d}{2}} - 1)}{\beta}\right]$$

(σ_p is the maximum fibre stress due to friction.)

Similarly, the dissipation of stored elastic energy in a debonded fibre bundle when the bundle fractures can be calculated using the above equation, after substitution of the appropriate material properties of a bundle.

Interfacial Energy

The interfacial energy is calculated by multiplying the newly created fibre-matrix interfacial surface area and the surface energy (Wells and Beaumont):

$$E_i = \pi r_b \ell_{db} 2 \gamma_b + \pi r_f \ell_{df} 2 \gamma_1$$

where $\gamma_b = \frac{1}{a}\left[\pi r_f \gamma_1 + (a - 2 r_f) \gamma_2\right]$

γ_1, γ_2 are the surface energy terms of interface and matrix, respectively; a is fibre spacing between centres.

Pull-out Energy

The work done against friction in pulling the broken fibre from its socket is:

$$E_p = \pi r_f^2 \int_o^{\bar{\ell}_p} \sigma (x) \, dx$$

Assuming the probability distribution of the pull-out lengths is uniform, then (Wells and Beaumont):

$$E_p = V_f \sigma_p \left[\bar{\ell}_p + \left(\frac{e^{-\beta \bar{\ell}_p} - 1}{\beta}\right)\right]$$

A similar expression is used to estimate the dissipation of energy due to bundle pull-out, after distribution of the appropriate bundle properties for fibre properties.

CONSTRUCTION AND DISPLAY OF FRACTURE MAPS

First, the characteristic debond length and pull-out length are computed for given combinations of material properties. Combining these values with models of fracture enables calculation of the theoretical energy terms, which when added together, gives the theoretical toughness of the composite. A computer produces the maps by allowing any of the variables which affect the fracture process to be varied along the two axes of the map. To construct a map, the values of the two parameters are varied in sequence, all other material properties are held constant at the default values (Wells and Beaumont, 1983). Values of ℓ_d and ℓ_p are calculated at each point, and composite toughness found. A small selection of maps are presented (Figs 1 - 6).

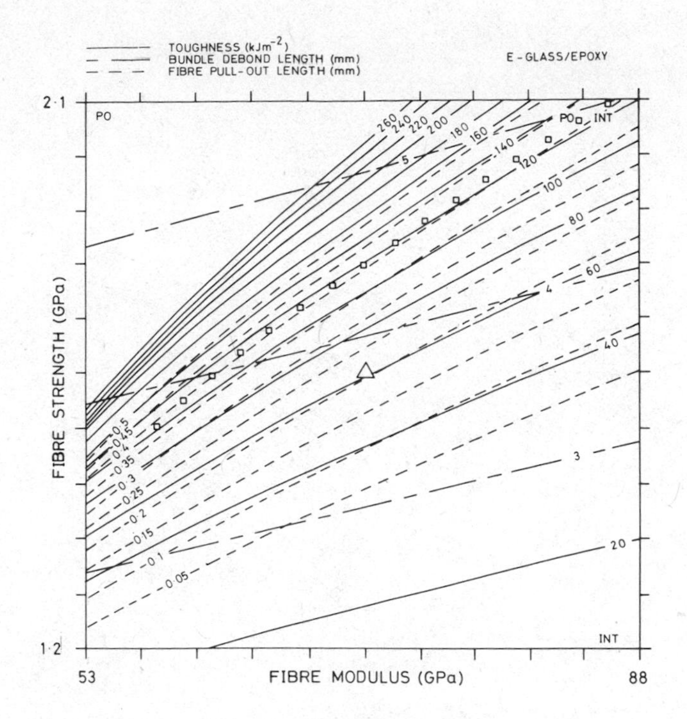
TOUGHNESS (kJm^{-2})
BUNDLE DEBOND LENGTH (mm)
FIBRE PULL-OUT LENGTH (mm)
E-GLASS/EPOXY
FIBRE STRENGTH (GPa)
FIBRE MODULUS (GPa)
2·1
1·2
53
88
PO
INT

Fig. 1.

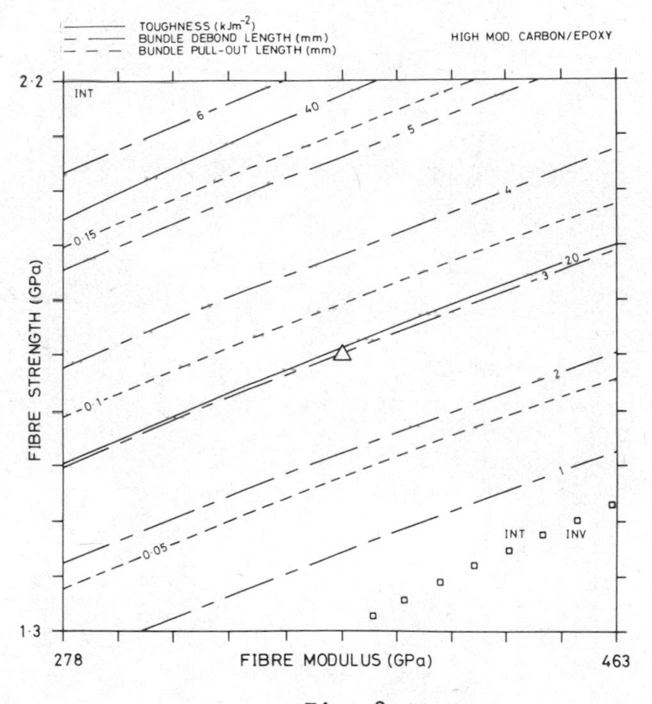
TOUGHNESS (kJm^{-2})
BUNDLE DEBOND LENGTH (mm)
BUNDLE PULL-OUT LENGTH (mm)
HIGH MOD. CARBON/EPOXY
FIBRE STRENGTH (GPa)
FIBRE MODULUS (GPa)
2·2
1·3
278
463
INT
INV

Fig. 2.

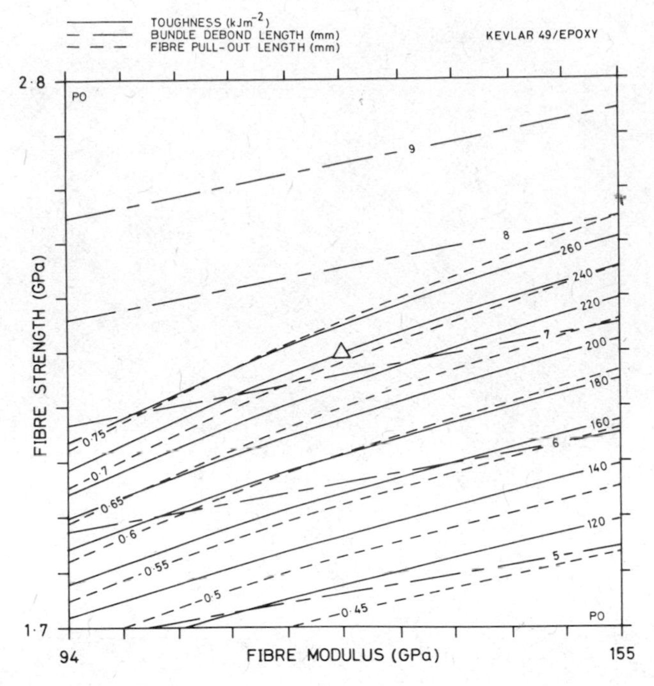

TOUGHNESS (kJm^{-2})
BUNDLE DEBOND LENGTH (mm)
FIBRE PULL-OUT LENGTH (mm)
KEVLAR 49/EPOXY
FIBRE STRENGTH (GPa)
FIBRE MODULUS (GPa)

Fig. 3.

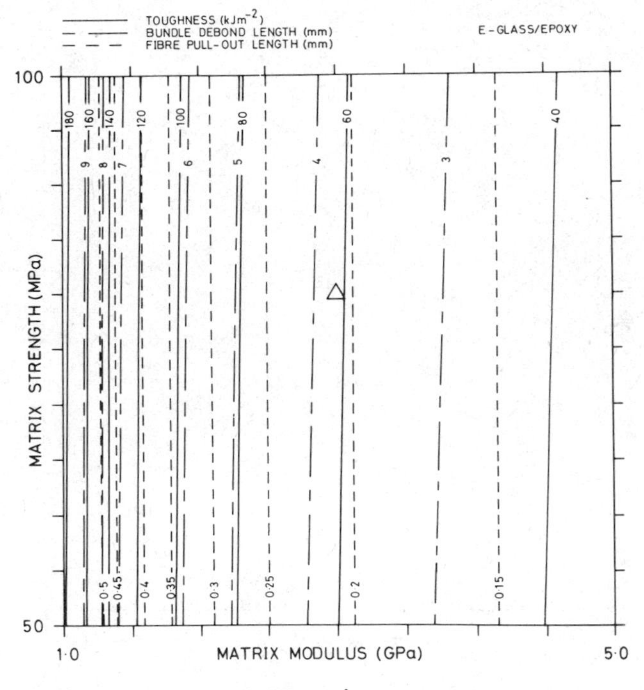

TOUGHNESS (kJm^{-2})
BUNDLE DEBOND LENGTH (mm)
FIBRE PULL-OUT LENGTH (mm)
E-GLASS/EPOXY
MATRIX STRENGTH (MPa)
MATRIX MODULUS (GPa)

Fig. 4.

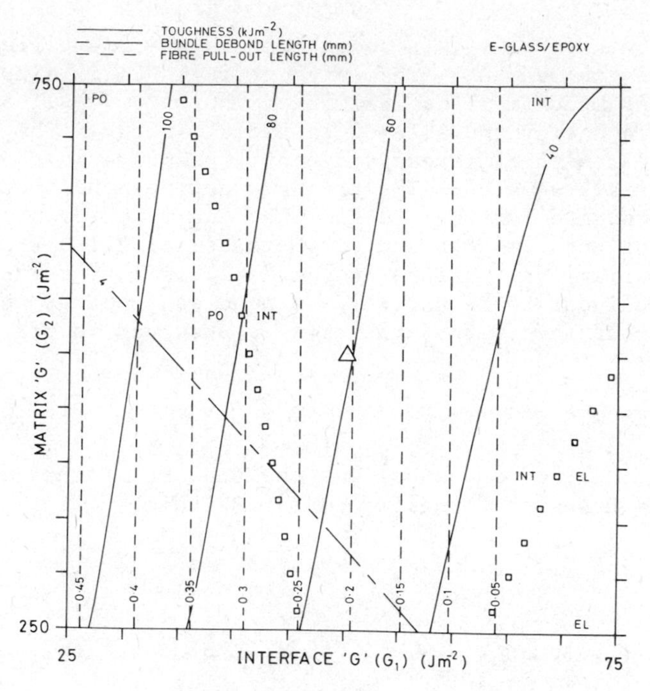

TOUGHNESS (kJm^{-2})
BUNDLE DEBOND LENGTH (mm)
FIBRE PULL-OUT LENGTH (mm)
E-GLASS/EPOXY
MATRIX 'G' (G_2) (Jm^{-2})
INTERFACE 'G' (G_1) (Jm^{-2})
750
250
25
75
PO
INT
PO INT
INT EL
EL
100
80
60
40
0·45
0·4
0·35
0·3
0·25
0·2
0·15
0·1
0·05

Fig. 5.

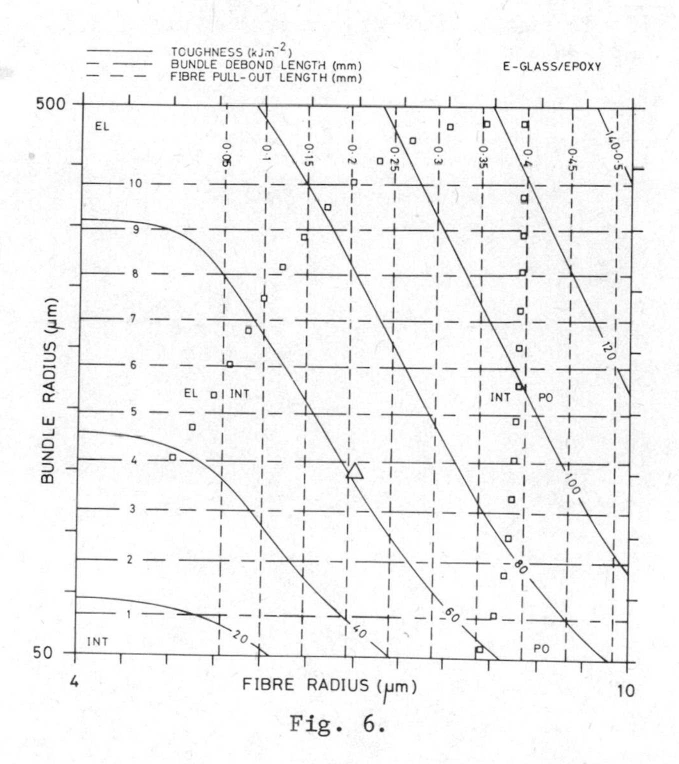

TOUGHNESS (kJm^{-2})
BUNDLE DEBOND LENGTH (mm)
FIBRE PULL-OUT LENGTH (mm)
E-GLASS/EPOXY
BUNDLE RADIUS (μm)
FIBRE RADIUS (μm)
500
50
4
10
EL
INT
EL INT
INT PO
PO
0·05
0·1
0·15
0·2
0·25
0·3
0·35
0·4
0·45
0·5
10
9
8
7
6
5
4
3
2
1
20
40
60
80
100
120
140

Fig. 6.

The contours of ℓ_d, ℓ_p and toughness are identified by the differing hatch lengths. The largest single toughening mechanism (or dominant mechanism) is identified where PO = pull-out energy, EL = elastic energy and INT = interfacial energy. The boundary between areas with different dominant mechanisms is shown by a series of small squares. The triangle defines the properties of a typical composite system. We see that the model predicts reasonable estimates of both lengths and toughnesses for all systems, despite the wide variations in fibre properties. In particular, the model predicts pull-out to be entirely of bundles in cfrp, and principally fibre pull-out in grp and krp; this is observed in practice.

ACKNOWLEDGEMENTS

We would like to acknowledge the support of the Science and Engineering Research Council of Great Britain.

REFERENCE

Wells, J.K., Beaumont, P.W.R., paper in preparation.

---

*Dr J.K. Wells is now at the BP Research Centre, Sunbury-on-Thames, Middlesex, England.

POSITIVE AND NEGATIVE SHEAR OF THE VINYL CHLORIDE

Zdeněk Sobotka

Mathematical Institute
Czechoslovak Academy of Sciences
Prague, Czechoslovakia

INTRODUCTION

The paper deals with a special mechanical behaviour of the vinyl chloride in the orientated shear. On the basis of the results of theoretical and experimental investigations, the author has made the conclusion that there are two different kinds of the opposite shear stresses which represent, in some manner, an analogy to two opposite kinds of normal stresses, i.e. to the tension and compression. He has called them the positive and negative shear. In this way, a physical meaning has been attributed to the sense and orientation of shear stresses.

The differences between both the opposite kinds of shear stresses depend considerably on the angles between the directions of shear stresses and characteristic directions of the body structure. There exists at least one neutral direction in which the differences between the positive and negative shear vanish and in which we have the neutral shear or the shear in the classical meaning.

In the positive shear, accompanied by an increase in volume, the materials have higher moduli of elasticity, yield stresses and strength than in the negative shear which is characterized by a decrease in volume.

The author has confirmed the existence of two opposite kinds of shear by the results of tests of plastics, steel, zinc, aluminium alloys, beech wood, layered soils and human bones. The present paper contains the results of tests of round specimens cut out from a board of the vinyl chloride and forming various angles with the direction of processing.

POSITIVE AND NEGATIVE SHEAR MODULI IN ELASTIC RANGE

The positive and negative shears in the elastic range are characterized by different values of shear moduli.

Let us consider a particular case of the elastic deformation of the rectangular parallelopiped acted on in the plane xy by the normal stresses $\sigma_x = -\sigma_y = \sigma$ while the stress in the direction of the z-axis is equal to zero: $\sigma_z = 0$ as shown in Fig. 1. Cutting out a square element ABCD , with the half-diagonal equal to unity, by planes parallel to the z-axis and at 45° to the x- and y-axes, it may be seen from Fig. 1, by summing up the forces along and perpendicular to AB , BC , CD and DA , that the normal stress on the sides of this element is zero and the shear stress on the sides is

$$\tau = \frac{1}{2}(\sigma_x - \sigma_y) = \sigma \ . \tag{1}$$

In this equation, the compressive stress is denoted by the positive and the tensile stress by the negative sign.

Such a condition represents the pure shear.

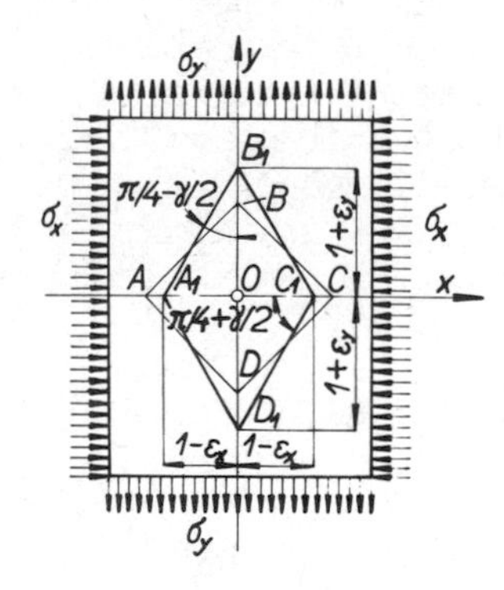

Fig. 1. Deformation due to the pure shear

Neglecting a small quality of the second order, we may conclude that the side lengths AB , BC , CD and DA of the element do not change during deformation. The right angles between the sides AB and BC as well as between CD and DA change for $\pi/2 - \gamma$ and those between the sides AB and DA as well as between BC and CD become $\pi/2 + \gamma$. The corresponding magnitude of the shear strain γ may be found from the triangles OA_1B_1 , OB_1C_1 , OC_1D_1 and OD_1A_1 so that, after deformation, we obtain

$$\tan\left(\frac{\pi}{4} - \frac{\gamma}{2}\right) = \frac{1 - \tan\frac{\gamma}{2}}{1 + \tan\frac{\gamma}{2}} = \frac{1 - \varepsilon_x}{1 + \varepsilon_y} , \tag{2}$$

from which

$$\tan\frac{\gamma}{2} = \frac{\varepsilon_x + \varepsilon_y}{2 + \varepsilon_y - \varepsilon_x} . \tag{3}$$

For small strains, we find from the preceding equation the following important relation

$$\gamma = \varepsilon_x + \varepsilon_y , \tag{4}$$

which can be used for the materials exhibiting, in the pure shear, the different magnitudes of the unit horizontal shortening ε_x and of the unit vertical elongation ε_y . This is the case of the deformation of isotropic media with different mechanical properties in tension and in compression as well as that of orthotropic media.

For the pure shear of the orthotropic elastic media with different mechanical properties in tension and compression as well as in the positive and negative shear, the Hooke law yields the following expressions for the unit horizontal shortening and the unit vertical elongation when attributed to the effects of the positive pure shear:

$$\varepsilon_{Px} = \frac{\sigma_x}{E_{Cx}} + \frac{\mu_{Cx}\sigma_y}{E_{Ty}} = \left(\frac{1}{E_{Cx}} + \frac{\mu_{Cx}}{E_{Ty}}\right)\tau_P , \tag{5}$$

$$\varepsilon_{Py} = \frac{\sigma_y}{E_{Ty}} + \frac{\mu_{Ty}\sigma_x}{E_{Cx}} = \left(\frac{1}{E_{Ty}} + \frac{\mu_{Ty}}{E_{Cx}}\right)\tau_P , \tag{6}$$

where E_{Cx} is the modulus of elasticity in compression in the direction of the x-axis, E_{Ty} is that in tension in the direction of the y-axis and μ_{Cx} , μ_{Ty} are the Poisson ratios.

In view of Eq. (4), the shear strain may be expressed by

$$\gamma_P \equiv \frac{\tau_P}{G_P} = \varepsilon_{Px} + \varepsilon_{Py} \ . \tag{7}$$

Substituting ε_{Px} and ε_{Py} from Eqs. (5) and (6) into Eq. (7), we obtain the expression for the positive shear modulus

$$G_P = \frac{E_{Cx}E_{Ty}}{E_{Cx}(1 + \mu_{Cx}) + E_{Ty}(1 + \mu_{Ty})} \ . \tag{8}$$

Let us consider the element acted on in the plane xy by the horizontal tension and by the vertical compression such that $\sigma_y = - \sigma_x$. In this case, the Hooke law yields the following expressions for the unit horizontal elongation and for the unit vertical shortening when attributed to the negative shear:

$$\varepsilon_{Nx} = \frac{\sigma_x}{E_{Tx}} + \frac{\mu_{Tx}\sigma_y}{E_{Cy}} = \left(\frac{1}{E_{Tx}} + \frac{\mu_{Tx}}{E_{Cy}}\right)\tau_N \ , \tag{9}$$

$$\varepsilon_{Ny} = \frac{\sigma_y}{E_{Cy}} + \frac{\mu_{Cy}\sigma_x}{E_{Tx}} = \left(\frac{1}{E_{Cy}} + \frac{\mu_{Cy}}{E_{Tx}}\right)\tau_N \ , \tag{10}$$

where E_{Tx} is the modulus of elasticity in tension in the direction of the x-axis, E_{Cy} is that in compression in the direction of the y-axis and μ_{Tx} , μ_{Ty} are the corresponding Poisson ratios.

The shear strain is expressed by

$$\gamma_N = \frac{\tau_N}{G_N} \ . \tag{11}$$

In view of Eq. (4), we have

$$G_N = \frac{\tau_N}{\varepsilon_{Nx} + \varepsilon_{Ny}} \; . \tag{12}$$

Substituting in the preceding expression Eqs. (9) and (10), we obtain the negative shear modulus

$$G_N = \frac{E_{Tx} E_{Cy}}{E_{Tx}(1 + \mu_{Tx}) + E_{Cy}(1 + \mu_{Cy})} \; . \tag{13}$$

If $G_P > G_N$, i.e. if

$$E_{Cx}\left[\frac{E_{Tx}}{E_{Cy}}(1 + \mu_{Tx}) + 1 + \mu_{Cy}\right] > \\ > E_{Tx}\left[\frac{E_{Cx}}{E_{Ty}}(1 + \mu_{Cx}) + 1 + \mu_{Ty}\right] , \tag{14}$$

the first shear modulus G_P really characterizes the positive shear whereas the second shear modulus G_N , having a lower value, corresponds to the negative shear.

The sufficient and necessary conditions for the existence of the positive and negative shear, characterized, in the elastic range, by the different shear moduli G_P and G_N , are:

1. anisotropy,

2. different elastic properties of the material in tension and compression.

In accordance with the inequality (14), these conditions may be expressed by

$$\frac{1 + \mu_{Tx}}{E_{Cy}} + \frac{1 + \mu_{Cy}}{E_{Tx}} \neq \frac{1 + \mu_{Cx}}{E_{Ty}} + \frac{1 + \mu_{Ty}}{E_{Cx}} \; . \tag{15}$$

If the preceding relation becomes equality, we have the neutral shear, with the unique shear modulus, or the shear in the classical meaning.

YIELD CRITERIA AND FAILURE CONDITIONS

For anisotropic materials with different mechanical

properties in tension and compression, Sobotka (1982) has formulated a yield criterion, which in terms of actual Cauchy stress components has the following form

$$R_{mn}\sigma_{mn} + R_{mnpq}\sigma_{mn}\sigma_{pq} + R_{mnpqrs}\sigma_{mn}\sigma_{pq}\sigma_{rs} = 1 \ . \quad (16)$$

Retaining on the right-hand side of this equation the first two terms only, we obtain the yield criterion for the first-order plastic flow. The tensorial coefficients R_{mn} and R_{mnpq} can then be expressed in terms of 27 particular yield stresses. Using the classical index notation for Cartesian coordinates and denoting compressive stresses by positive and tensile stresses by negative signs, we obtain the yield criterion for the plane-stress state

$$\frac{\sigma_x^2}{\sigma_{Cx}\sigma_{Tx}} + \frac{\sigma_y^2}{\sigma_{Cy}\sigma_{Ty}} + \frac{\tau_{xy}^2}{\tau_{Pxy}\tau_{Nxy}} - \Big(\frac{1}{\sigma_{Cx}\sigma_{Tx}} + \frac{1}{\sigma_{Cy}\sigma_{Ty}} -$$

$$- \frac{1}{\sigma_{CCxy}^2} - \frac{\sigma_{Cx} - \sigma_{Tx}}{\sigma_{Cx}\sigma_{Tx}\sigma_{CCxy}} - \frac{\sigma_{Cy} - \sigma_{Ty}}{\sigma_{Cy}\sigma_{Ty}\sigma_{CCxy}}\Big)\sigma_x\sigma_y -$$

$$- \Big(\frac{1}{\sigma_{Cx}\sigma_{Tx}} + \frac{1}{\tau_{Pxy}\tau_{Nxy}} - \frac{1}{\sigma_{CPxxy}^2} - \frac{\sigma_{Cx} - \sigma_{Tx}}{\sigma_{Cx}\sigma_{Tx}\sigma_{CPxxy}} -$$

$$- \frac{\tau_{Pxy} - \tau_{Nxy}}{\tau_{Pxy}\tau_{Nxy}\sigma_{CPxxy}}\Big)\sigma_x\tau_{xy} - \Big(\frac{1}{\sigma_{Cy}\sigma_{Ty}} + \frac{1}{\tau_{Pxy}\tau_{Nxy}} -$$

$$- \frac{1}{\sigma_{CPyxy}^2} - \frac{\sigma_{Cy} - \sigma_{Ty}}{\sigma_{Cy}\sigma_{Ty}\sigma_{CPyxy}} - \frac{\tau_{Pxy} - \tau_{Nxy}}{\tau_{Pxy}\tau_{Nxy}\sigma_{CPyxy}}\Big)\sigma_y\tau_{xy} -$$

$$- \frac{\sigma_{Cx} - \sigma_{Tx}}{\sigma_{Cx}\sigma_{Tx}}\sigma_x - \frac{\sigma_{Cy} - \sigma_{Ty}}{\sigma_{Cy}\sigma_{Ty}}\sigma_y -$$

$$- \frac{\tau_{Pxy} - \tau_{Nxy}}{\tau_{Pxy}\tau_{Nxy}}\tau_{xy} = 1 \ , \quad (17)$$

where σ_{Cx}, σ_{Cy} are yield stresses in simple compression, σ_{Tx}, σ_{Ty} are those in simple tension, σ_{CCxy} is the yield stress in the uniform biaxial compression, τ_{Pxy} is that in simple positive shear, τ_{Nxy} is that in simple negative shear and σ_{CPxxy} , σ_{CPyxy} are the yield stresses in the combined compression and positive shear at the same value of both the stress components. In contradistinction from yield criteria done hitherto, the criterion (17) contains a term which is linear in shear stress. Therefore this criterion depends on the sign of shear. The failure conditions have quite analogous forms as the yield criteria if the yield stresses are replaced by limiting or failure stresses.

EXPERIMENTAL VERIFICATION

In order to confirm the existence of differences between the positive and negative shear, the author has carried out tests of round specimens with the diameter 8 mm, which were cut out from a board of the vinyl chloride so that their axes formed various angles with the direction of rolling. The stress-displacement curve for the positive shear of a specimen, forming with the direction of rolling the angle 15°, is shown in Fig. 2. The curves corresponding to the negative shear have similar forms. The results of tests are listed in Table 1.

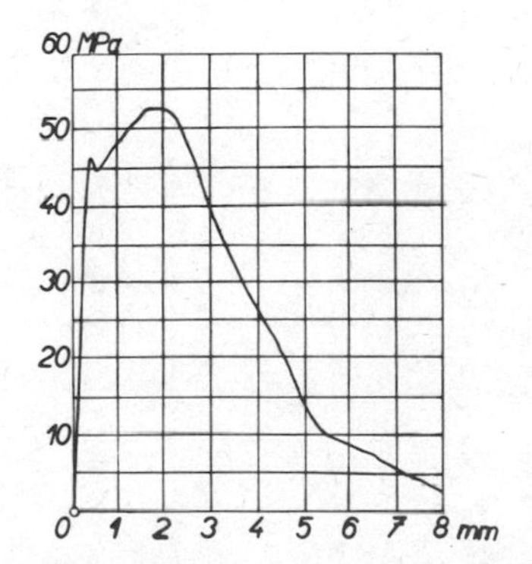

Fig. 2. Stress-displacement curve of the positive shear

REFERENCE

Sobotka, Z., 1982, Positive and Negative Shears and Body Structures, J. Tech. Phys., 23:79.

Table 1. Yield Limits and Failure Stresses of the Vinyl Chloride in Simple Shear

Angle with the Direction of Rolling	No.	Kind of Shear	Lower Yield Stress MPa	Upper Yield Stress MPa	Failure Stress MPa
0°	1	positive	45.33	47.17	48.94
		negative	43.59	46.14	47.39
	2	positive	44.83	46.80	48.40
		negative	44.07	45.96	47.08
	3	positive	44.22	45.66	47.46
		negative	44.34	44.60	46.49
15°	1	positive	44.76	46.45	52.99
		negative	42.64	43.87	48.36
	2	positive	44.58	46.10	52.36
		negative	43.22	44.16	48.31
	3	positive	43.85	45.56	48.81
		negative	43.11	45.12	47.99
30°	1	positive	45.69	46.73	52.15
		negative	45.49	45.65	50.92
	2	positive	45.52	47.98	51.79
		negative	45.43	47.70	49.58
	3	positive	45.28	47.94	51.67
		negative	44.79	47.09	48.37
45°	1	positive	45.88	47.68	55.25
		negative	44.07	46.28	51.56
	2	positive	45.38	47.87	53.94
		negative	44.43	46.45	51.36
	3	positive	45.72	47.04	53.55
		negative	44.54	46.75	51.37
60°	1	positive	46.04	47.61	54.97
		negative	44.46	46.40	53.56
	2	positive	44.34	46.35	53.52
		negative	43.70	45.46	53.39
	3	positive	43.87	46.01	52.38
		negative	42.69	45.02	50.64
90°	1	positive	44.84	47.42	54.04
		negative	44.48	46.70	52.67
	2	positive	44.70	46.85	53.36
		negative	43.20	45.45	51.86
	3	positive	44.19	46.12	52.38
		negative	43.03	44.73	51.49

PART V

STRUCTURAL MODELING OF POLYMER PROPERTIES

STRUCTURAL MODELLING OF POLYMER PROPERTIES

Albert J. de Vries

ESPCI-Laboratoire de Physico-chimie Structurale et
Macromoléculaire
75231 Paris Cédex 05
France

INTRODUCTION

It is a favourite and frequently performed exercise for scientists interested in the behaviour of real materials, to imagine and to construct hypothetical, more or less elaborate structural models whose properties may be readily described in terms of a limited number of characteristic parameters. Comparison of the observed properties of the real material with those of the structural model is supposed then to allow certain conclusions about the degree of relevancy of the chosen model in particular when the observed properties are shown to be in fair agreement with the ones calculated from the model. It should be borne in mind, however, that quite different theoretical models may exhibit similar or even identical behaviour depending on the selected parameter values, from which we may conclude that coincidence between observed and calculated properties does not necessarily imply that the essential features of the structural model bear close resemblance to those of the real material. The latter conclusion is corroborated by the well known observation that two real materials of very different structure may show similar or identical properties under appropriate conditions, which may or may not be the same for both materials.

It is hardly necessary to emphasize the technological importance of this observation which has provided the starting point for the tremendous development of the use of new synthetic materials (polymers, in particular) in substitution of other, more traditional materials of quite different structure (metals, glass, wood).

It is obvious from the above considerations that the description and explanation of properties of polymers (or any other material) in terms of structural model parameters should not be expected devoid of ambiguity, to a certain extent. In many cases, therefore, either acceptance or rejection of a particular proposed model is inevitably based on a more or less subjective judgement and the relationship of the scientist with regard to his preferred model may often be described as affectionate if not passionate. (Although the relevance of models in science and art, respectively, are not strictly comparable, it is interesting to contrast the generally admitted opinion that a true work of art is a highly subjective representation of reality, on the one hand, with the often ignored or even disregarded importance of subjectivity in scientific works, on the other).

Macromolecular materials are composed of flexible longchain molecules each containing thousands of atoms chemically bound together. Many polymer properties can be satisfactorily described in terms of structural models based on the specific long chain character of macromolecules. An outstanding example is provided by the kinetic rubber elasticity theory which presents a straightforward molecular interpretation of the observed relations between stress, strain and temperature in polymers subjected to large deformations above the glass-rubber transition temperature. One of the main premises of the rubber elasticity theory[1] states that the stress in the deformed polymer network entirely originates from the configurational entropy change in the individual macromolecular network chains whereas internal energy changes due to intra- and intermolecular interactions make negligible contributions. This statement is no longer true when deformation takes place at temperatures below the glass transition temperature (or more generally, when the mobility of molecular chain segments becomes too low with respect to the imposed deformation rate). Consequently, under these conditions, the relations between stress, strain and temperature are,in general, very different from those predicted by the theory of rubber elasticity and become closer to the ones observed in non-polymeric solids.The resistance to deformation is now mainly determined by the height of intra- and intermolecular potential energy barriers depending on the chemical constitution of the polymer chain[2].

Shaping operations during polymer processing are generally performed at temperatures above the glass transition temperature and induce large deformations of the polymer chains (extension of polymer coils and alignment of chain segments) which may be preserved in the final object to a more or less important extent, depending on the rate of cooling. The relationship between the frozen-in strains and the associated internal stresses can be described in terms of the rubber elasticity theory but the simple network model on which this theory is founded is largely inadequate to explain the general anisotropic mechanical behaviour of oriented polymeric materials, in particular if a substantial fraction of the polymer is in the crystalline state. No general models based on the specific structural features at the molecular level, have been elaborated sofar in order to describe the mechanical properties of anisotropic semi-crystalline polymers. However, quite general theoretical models which do not explicitly refer to the long chain character of polymer molecules but which take account of the several interrelated anisotropic phases, have been adapted and widely used with a certain degree of success. The larger part of the present review is devoted to a discussion of various results obtained with the aid of such more or less elaborate "aggregate models" but we will first discuss some examples concerning particular applications of the rubber elasticity theory.

INTERNAL STRESSES IN ANISOTROPIC POLYMER NETWORKS

The deformational behaviour of amorphous polymers above the glass transition temperature is similar to that of a rubber elastic network in which the chemical crosslinks have been replaced by localized physical interactions (such as entanglements, e.g.) between neighbouring molecular chains. This statement is corroborated by the following observations:

i) a large fraction of the imposed deformation is completely reversible and can be recovered after removal of the external stress.

ii) the fraction of irreversible deformation due to viscous flow may be negligibly small if the deformation rate is sufficiently high and/or the temperature sufficiently low; the fraction of irreversible deformation increases with increasing temperature and decreasing rate of deformation.

Similar behaviour is observed in the deformation of semicrystalline polymers somewhat below or above the melting temperature. Careful measurements of the respective amounts of reversible and irreversible elongation in low-density polyethylene melts, e.g. have been reported by Meissner[3] and more recently by Laun[4].

The extremely large amounts of reversible elongational strains which may be induced in a polymer melt can be readily explained by the long chain character of polymer molecules and the associated transition from a random coil to an extended chain. The inherent tendency of the molecular chains to revert to the random coil configuration is counteracted by frictional forces, exerted by the local environment of the chains, which determine the temperature- and molecular weight dependent chain relaxation times. An appreciable amount of chain extension is only obtained if the elongation rate is large with respect to the reciprocal chain relaxation times. The rubberlike network model of polymer melts is based on the hypothesis that the frictional forces which oppose themselves to irreversible flow are localized in transient network junctions (entanglements, strong secondary bonds, etc.)[5]. The finite lifetimes of the network junctions replace the chain relaxation times mentioned above.

Unfortunately, the rubberlike network theory is unable to make any predictions about the molecular weight dependence of relaxation times, in contrast with other theories such as those based on the concept of "reptation"[6,7]. In the reptation model (Fig.1)

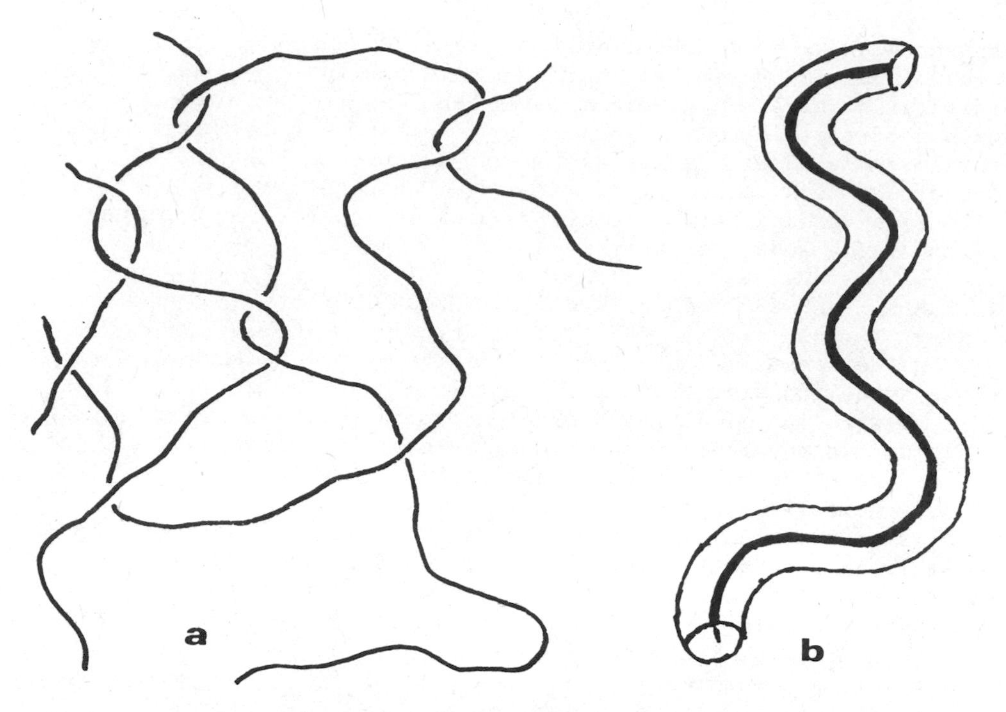

Fig. 1. Schematic picture of a) entangled network
b) chain molecule confined in flexible tube

molecular chains are supposed to move in a "wormlike" fashion as if they were confined in a contorted, flexible tube. Motions parallel to the local direction of the chain backbone are highly favoured with respect to motion in perpendicular directions, according to this model which seems to offer a physically plausible description of the (dis) entanglement process. The model predicts that the longest chain relaxation time should be proportional to the cube of the total chain contour lenght (i.e.the cube of the molecular weight) and inversely proportional to the average chain length between consecutive entanglement points. The elasticity modulus, on the other hand, is predicted to be independent of molecular weight but also proportional to the density of entanglements in accordance with the well known result of rubber elasticity theory.

It is important to point out here that the schematic picture in Fig. 1a represents one particular type of network junction viz. a physical entanglement. Different types of localized physical interaction between neighbouring chains may be assumed to exist in a real molecular network; creation and destruction of these network junctions of various nature may proceed according to mechanisms of molecular motion different from reptation.

In spite of its shortcomings and limitations,the rubberlike network model has been proved to be a useful tool for the analysis and interpretation of a whole range of physical properties among which we have chosen to emphasize those associated with the freezing-in of large reversible strains in both amorphous and semicrystalline uncrosslinked polymers.

Remarkable examples can be found in the work of Struik[8] who demonstrated that the rubber elastic entropic stresses frozen-in during cooling of a deformed polymer, make a negative contribution to the coefficient of (reversible) thermal expansion (in the case of a frozen-in tensile stress) owing to the fact that entropic stresses, in accordance with rubber elasticity theory, are proportional to absolute temperature. The frozen-in tensile stresses will tend to compress the glassy polymer and the resulting compressive force will reversibly increase on heating and decrease on cooling. Struik has published numerous experimental results in favour of his point of view, referring to both frozen-in tensile and shear stresses.

Frozen-in tensile stresses can be readily determined by measuring the retractive force exerted by a polymer sample refrained from shrinkage during heating up to a temperature above the glass transition temperature (in the case of an amorphous polymer) or into the melting range if the polymer is semi-crystalline. Depending on the processing conditions, in particular strain- and temperature history, frozen-in tensile stresses will be largely or even exclusively due to configurational entropy changes in the amorphous rubberlike network. If part of the frozen-in stress originates from configuration independent internal energy changes during straining, this part has to be eliminated from the total retractive force by means of appropriate experimental procedures[9,10] before we can check the fundamental predictions of the rubber elasticity theory:

$$\sigma_i - \sigma_j = -T\,(\lambda_i \partial S/\partial\lambda_i - \lambda_j \partial S/\partial\lambda_j)$$

$$= G\,(\lambda_i^2 - \lambda_j^2) \qquad (1)$$

where σ_i, σ_j are principal stresses ,
λ_i, λ_j principal extension ratios
T the absolute temperature , and
G the equilibrium shear modulus, which is proportional to T and to ν_e , the number of effective network chains per unit volume:

$$G = g_f\, \nu_e\, k\, T \qquad (2)$$

The front factor gf which, in general, depends on network topology and junction functionality should, according to recent theoretical estimates[11], decrease from 1 to 0.5 (for a network with tetrafunctional junctions) when the extension ratio increases, in accord with the experimental results on chemically crosslinked rubbers. If the transient network junctions are due to localized physical interactions between neighbouring molecules, the number of junctions per unit volume is expected to depend not only on rate and temperature of deformation, but also on the total amount of strain. Both the front factor and the number of effective chains per unit volume, which figure in Eq. (2), may therefore decrease with increasing strain or stress.

Early experimental results by Pinnock and Ward[12] obtained on amorphous polyethylene terephthalate fibers are well rapresented by a linear relationship between the retractive force and the recoverable strain function $(\lambda^2 - 1/\lambda)$ in agreement with Eq. (1) applied to uniaxial extension:

$$\sigma = G(\lambda^2 - 1/\lambda) \qquad (1')$$

More recent experiments in our laboratory with several vinyl polymers im which much larger amounts of recoverable strains had been induced by means of both uni-and biaxial extension[9, 10] were also found to be consistent with Eq. (1) with G either constant or slowly decreasing with increasing strain (Figures 2 and 3). For all investigated polymers the number of skeletal bonds separating two successive network junctions, estimated by means of Eq. (2), varies between a hundred and several hundreds; as expected the calculated number of bonds increases with increasing deformation temperature.

Photoelasticity of Rubber Elastic Networks

Stress-optical measurements offer further arguments in favour of the rubber elastic network model applied to the problem of frozen-in stresses and strains. The theory of photoelasticity developed by Kuhn and Grün [13] leads to the definition of a strain-independent stress-optical coefficient (SOC) given by the following expression:

$$\mathrm{SOC} = \frac{\Delta n}{\sigma} = \frac{2\pi}{45\,kT}\,\frac{(\bar{n}^2+2)^2}{\bar{n}}\,(a_1 - a_2) \qquad (3)$$

where $\bar{n}$ is the mean refractive index, Δn the birefringence a_1 the polarizability in the axial direction of a "random link" and a_2 the mean polarizability in the transverse direction.

The predicted proportionality between birefringence and frozen-in tensile stress is in excellent agreement with all experimental results, irrespective of the observed relationship between stress and strain (Figure 4).

Since the Kuhn-Grün theory is based on the model of a chain

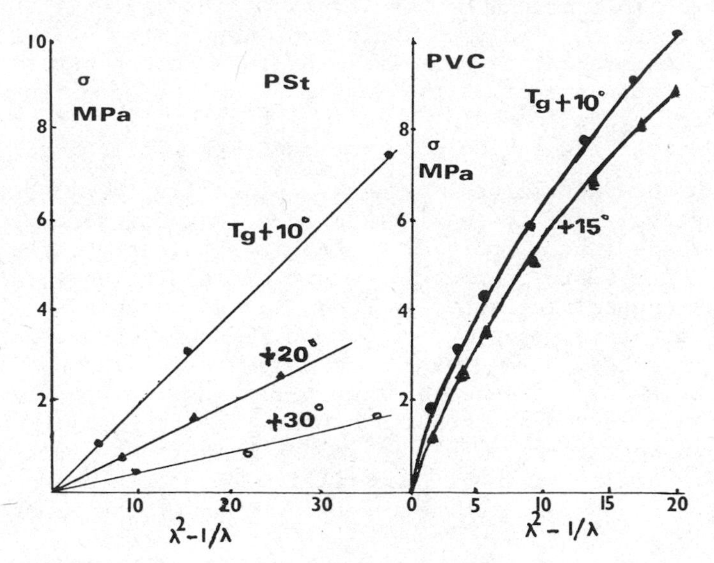

Figs. 2 and 3: Relationship between frozen-in tensile stresses and recoverable strain for polystyrene and PVC.

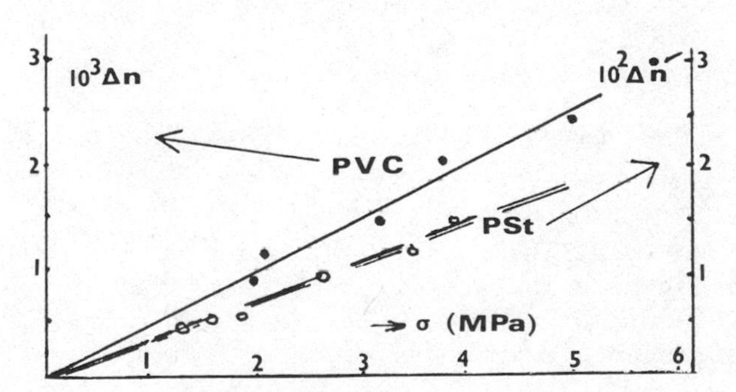

Fig.4: Birefringence versus frozen-in tensile stress for uniaxially oriented polystyrene and PVC.

consisting of freely jointed random links, the optical anisotropy $a_1 - a_2$ determined from the measured SOC by means of Eq. (3), is always larger than the optical anisotropy of a single monomer unit calculated from its known structure and conformation. The degree of steric interaction between adjacent monomer units determines the stiffness of a polymer chain and, hence, the calculated length of an equivalent random link. For this reason stress-optical investigations are of particular interest for obtaining quantitative information on chain characteristics which are essential for the understanding of the deformational behaviour of polymers.

Although the experimental methods devised for the measurement of stress-optical coefficients are founded on model concepts concerning the behaviour of polymer chains, the values obtained for these coefficients are completely determined by the chemical constitution of the polymer and may be considered independent from any particular structural model. Stress-optical coefficients for various uncrosslinked, both amorphous and semi-crystalline polymers have been determined by means of flow birefringence measurements on polymer melts[14] and the results confirm their invariance with regard to characteristics such as molecular weight. In general, however, they are found to be temperature-dependent because of conformational changes altering the anisotropy of the random link [15, 16].

Photoelasticity in Anisotropic Semi-Crystalline Polymers

In anisotropic semi-crystalline polymers the observed total birefringence n is, in general, composed of several contributions [17]; in most cases it appears justified to retain merely the two terms which represent "orientation birefringence" in the crystalline and amorphous phases, respectively:

$$\Delta n = v_X \Delta_X n + (1 - v_X) \Delta n_{am} \qquad (4)$$

where v_X is the volume fraction of crystalline polymer;

$\Delta_X n$ and Δn_{am} are related to molecular orientation in crystalline and noncrystalline phases. The contribution of the crystalline phase can be calculated from the measured crystalline orientation, determined by wide angle X-ray diffraction, if the intrinsic birefringence components of the polymer crystal are known:

$$_X n_{ij} = (n_i - n_j)_X = \frac{2}{3} \left[(n_c - n_a)(f_{ci} - f_{cj}) + (n_b - n_a)(f_{bi} - f_{bj}) \right] \qquad (5)$$

where n_c, n_b and n_a are the principal refractive indices of the polymer crystal, and f_{ci}, f_{cj} the Hermans' orientation functions of the crystal c-axis with repect to the i'th and j'th coordinate, and f_{bi} , f_{bj} Hermans' orientation functions of the crystal b-axis.

By subtraction of the calculated crystalline contribution $v_X \; {}_X n$ from the measured total birefringence, Eq.(4), we can determine the contribution of the anisotropic amorphous network and plot it as a function of the frozen-in stress. The latter is calculated from the experimentally determined retractive force by admitting this force to be active on a cross-section equal to $(1-v_X)$ times the total cross-section of the polymer sample.

Figure 5 shows the results obtained in a collaborative IUPAC-Working Party investigation of a series of uni-and biaxially oriented polypropylene films [18]. Because of the thinness of these films, the measured retractive forces were very small and difficult to determine with great accuracy. Nevertheless, in spite of the important amount of scatter in the experimental data provided by three different laboratories, the approximately linear relationship shown in Fig.5 is in fair agreement with the one expected on the basis of the independently determined stress-optical coefficient for polypropylene, derived from flow birefringence measurements [14].

It appears, therefore, that even in polymers with a high degree of crystallinity (for the polypropylene films in Fig.5, v_X was larger than 0,6) the photoelastic behaviour of the non-crystalline phase may be understood in terms of a rubberlike network model analogous with the interpretation of the behaviour of polymer melts. The results do not allow any conclusion about the details of the network structure, such as e.g. the nature of the junction points. The frozen-in tensile stresses are of comparable magnitude as those measured in completely amorphous polymers for similar amounts of frozen-in strain (Figs.2 and 3) indicating

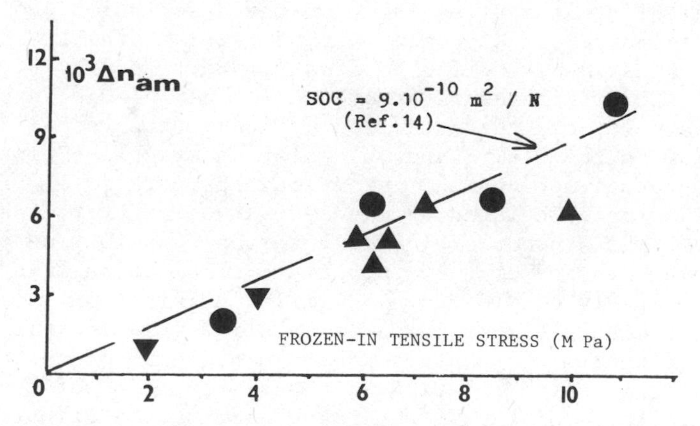

Fig.5. Noncrystalline birefringence versus frozen-in stress in polypropylene films. Adapted from Ref. 18.

that the density of network junctions is also of comparable magnitude and, apparently, unaffected by crystallinity which does not exclude the possibility that some of the network junctions may be of crystalline nature.

The numerous anisotropic noncrystalline domain are supposed to be strongly interconnected but separated by the crystalline regions whose partial fusion is required for allowing the internal stresses to be released and to be measured by means of retractive force determinations. Heating in the absence of applied external forces, on the other hand, leads to dimensional instability (shrinkage) indicative of the presence of frozen-in internal strains in the noncrystalline domains [18].

In general, the presence of numerous anisotropic domains (either crystalline or noncrystalline) does not necessarily result in macroscopic anisotropy, i.e. directional dependence of measured physical or other properties. If, indeed, the anisotropic domains are randomly distributed throughout the volume (i.e.without any preferential orientation) the material will be isotropic on a macroscopic scale.

The idea that an isotropic material may be regarded as a random assembly of anisotropic elements has first been proposed many years ago [19,20] in order to explain and to calculate the elastic constants of polycrystalline metals. In the following part of this review we will discuss the applications of "aggregate models" to the investigation of various polymer properties.

GENERAL DESCRIPTION OF AGGREGATE MODELS

Aggregate models have been adopted for the analysis of various polymer properties [21] but the majority of published works is concerned with the mechanical behaviour at small strains, starting with the well known publication by Ward [22].

McCullough et al.have presented a general formulation [23] of the mathematical treatment required for the description and prediction of anisotropic elastic behaviour of semi-crystalline polymers, which can also be applied to other properties[24].

Each of the two phases is assumed to be composed of aggregated structural elements characterized by inherent physical properties depending on the atomic and molecular interactions within and between the polymer chains. The average properties of an aggregate have to be calculated by means of a volume-averaging procedure which takes account of the orientation distribution of the structural elements and of the appropriate combination rule depending on the property considered. A similar combination rule is required for the next stage of the calculation in which the average aggregate properties of both phases are added in order to arrive at a final expression relating the bulk properties of the whole material to those of the structural elements, the orientation distribution of the latter, and the volume fraction of each phase (i.e.degree of crystallinity). In the general case the bulk material will be anisotropic and the direction dependence of the properties should be included in the final descriptive equations, which encompass as a special, limiting case, the properties of an isotropic material.

The theoretical analysis is based on several assumptions and conjectures,the first of which is concerned with the identification of the fundamental structural elements, whose properties should be completely and uniquely determined by the internal structure, irrespective of size. For the basic structural element associated with the crystalline phase, the unit cell of the polymer crystal has been proposed [24,25] but it is obvious that this can only represent a minimum size: generally, the crystalline domains with varying spatial orientations which can reasonably be identified with the constitutive entities of an aggregate (spherulite, microfibril,etc.) shall be much larger than a unit cell but their inherent properties will be those of the unit cell.

Unfortunately, the identification of the basic structural elementin a monocrystalline aggregate is much more ambiguous: Seferis [25] proposes that such an element is to be regarded as a collection of polymer chain segments not included in the crystal structure but nevertheless exhibiting some form of structural symmetry associated with anisotropic properties. The existence of locally, more or less regularly aligned chain segments in uniaxially oriented amorphous polymers is certainly consistent with the results of various experimental techniques of structural characterization, such as N.M.R., Raman spectropy, X-ray diffraction [26,27]. However, these structures are strongly orientation dependent and completely inexistent in an unoriented, isotropic amorphous polymer.

The largest structural unit to be considered as strictly invariant, in particular with regard to chain configuration, is a monomer unit which can hardly be accepted, in general, as an appropriate structural element in terms of the aggregate model theory. In spite of this fundamental conceptual difficulty the aggregate model has proved to be useful for the analysis of various properties in both amorphous and semi-crystalline polymers, which seems to signify that exact identification of model elements in terms of specific structural features at the molecular level, is not absolutely essential.

Accordingly, Ward and co-workers have adopted a somewhat different point of view and consider the aggregate model primarily as a phenomenological model likely to provide a first approximation to the actual behaviour of single-phase glassy polymers [28].In their work the intrinsic elastic properties of the hypothetical structural elements are obtained in a purely empirical way by identifying them with the (extrapolated) properties of the fully uniaxially oriented material. This procedure proved to be successful in fitting the experimental data to the equations derived from the single-phase aggregate model theory, even in the case of low-density polyethylene, a polymer whose structure and morphology are quite different from those of a homogeneous, single phase amorphous material [29].

Notwithstanding the obvious deficiencies of the single phase model, its application to the experimental data on low-density polyethylene (see below) allows a fair description of the observed behaviour in terms of a reduced set of model parameters whose relative magnitudes are expected to have some physical significance. Improved understanding of the complex relationships between structure and properties of semicrystalline polymers can only be obtained, however, if one takes full account of the heterogeneous, two-phase character of these materials.

Another fundamental limitation of the aggregate model is concerned with the combination rules required for the calculation of aggregate properties from those of its constituents, as well as for the appropriate summation of phase properties. Depending on the property under consideration, two extreme cases can easily accounted for, corresponding to a lower and an upper bound, respectively. In the classical example of the calculation of elastic constants for an isotropic polycrystalline aggregate, the assumption of uniform strain throughout the entire volume, implies a summation of stiffness constants, as first proposed by Voigt [19] and leads to a predicted upper bound for the moduli of the aggregate.

On the other hand, the assumption of uniform stress, as proposed by Reuss [20], involves summation of complicance constants and provides a lower bound for the moduli, i.e. an upper bound for the compliances. The calculated bounds may diverge more or less severely and, in extreme cases, differ by an order of magnitude. Application of improved bounding methods based on a variational treatment[30] results in a tightening of the bounds, mainly due to a significant reduction of the upper bound as demonstrated for the calculated moduli of an isotropic aggregate of orthorhombic polyethylene crystals [31].

The latter result seems to justify the suggestion that improvement in the calculated bounds for the moduli will tend to decrease the upper bounds to a greater extent than the lower bounds increase [23]. At present, improved bounding methods are only available for isotropic aggregates; McCullough et al. have proposed an empirical relationship in analogy with the Halpin-Tsai model [32] for composite materials in order to define en aggregate average intermediate between the Reuss- and Voigt averages:

$$\langle P \rangle_k = \frac{\langle P \rangle_k^R \langle P \rangle_k^V (1 + \xi_k)}{\langle P \rangle_k^V + \xi_k \langle P \rangle_k^R} \qquad (6)$$

The average value of the aggregate property in the k'th phase $\langle P \rangle_k$, as defined by Eq.(6), is intermediate between the Reuss-average $\langle P \rangle_k^R$ and the Voigt-average $\langle P \rangle_k^V$ depending on the value of the contiguity parameter ξ_k. For $\xi_k = 0$ or $= \infty$, Eq.(6) yields the Reuss- or

Voigt -average, respectively. All averages depend, of course, on the orientation distribution of the elements. The contiguity parameter is related to the size, shape and packing geometry of the aggregated unit elements, and has been associated, in particular, with the aspect ratio of the crystalline regions [23,33].

If the latter interpretation is correct, the orientation-independent contiguity parameter would contain information about morphological features of the material which are not included in the usual structural characterization at the molecular level, but whose influence on e.g. mechanical behaviour may become important in particular cases [23,25].

A number of typical examples illustrating the application of aggregate models to the analysis of polymer properties will now be examined.

Properties of Anisotropic Single Phase Polymers

Several properties of uniaxially oriented amorphous polymers have been interpreted in terms of aggregate model theory. For properties that may be represented by second rank tensors, the analysis is simple and straightforward (see Figure 6):
if the property under consideration is equal to P_3 along the molecular axis of the unit element, and equal to P_2 and P_1, respectively, in the perpendicular directions, its macroscopic value P will have the following three components:

$$\begin{aligned}
P_X &= p_1 \langle\cos^2\theta_{1X}\rangle + p_2 \langle\cos^2\theta_{2X}\rangle + p_3 \langle\cos^2\theta_{3X}\rangle \\
P_Y &= p_1 \langle\cos^2\theta_{1Y}\rangle + P_2 \langle\cos^2\theta_{2Y}\rangle + p_3 \langle\cos^2\theta_{3Y}\rangle \\
P_Z &= p_1 \langle\cos^2\theta_{1Z}\rangle + p_2 \langle\cos^2\theta_{2Z}\rangle + p_3 \langle\cos^2\theta_{3Z}\rangle \qquad (7)
\end{aligned}$$

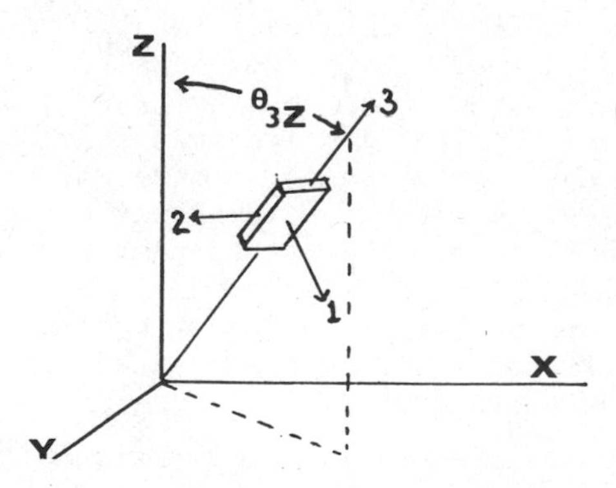

Figure 6: Spatial orientation of structural element.

The combination rule applied here, supposes simple additivity of the p-values and yields an estimated upper bound for P; if the summation is performed on the inverse p-values a lower bound is obtained.
Applications of the law of cosines: $\Sigma_j \cos^2 \theta_{ij} = 1$, leads to:

$$P_X + P_Y + P_Z = p_1 + p_2 + p_3 = 3\, p_O \qquad (8)$$

where P_O is the value of P for the isotropic material.

For an orthotropic structural element: $\Sigma_i \cos^2 \theta_{ij} = 1$, and

macroscopic anisotropy may be characterized by two independent differences between the component of P, e.g.
$P_Z - P_Y$, and $P_Z - P_X$:

$$P_Z - P_Y = (p_3 - p_1)\,(\langle \cos^2 \theta_{3Z} - \cos^2 \theta_{3Y} \rangle) + (p_2 - p_1)\,(\langle \cos^2 \theta_{2Z} - \cos^2 \theta_{2Y} \rangle) \qquad (9)$$

Equivalent expressions can be derived for the other differences. After introduction of the Hermans' orientation function:

$$f_{ij} = \frac{1}{2}\,(3 \langle \cos^2 \theta_{ij} \rangle - 1) \qquad (10)$$

Equation (9) may be written as follows:

$$P_Z - P_Y = \frac{2}{3} \left[(p_3 - p_1)\,(f_{3Z} - f_{3Y}) + (p_2 - p_1)\,(f_{2Z} - f_{2Y}) \right] \qquad (9')$$

Equations (5) for the crystalline birefringence in a biaxially oriented polymer sample is a special case of Eq.(9') in which the property considered is the refractive index N.According to a numerical analysis by Seferis and Samuels[24], for structural elements with birefringences less than or equal to 0.2 (i.e. $n_3 - n_2$ 0.2 and $n_3 - n_1$ 0.2) calculated upper and lower bounds of the refractive indices, corresponding in this case to uniform electric fields intensity or uniform electric flux density, respectively, do practically coincide which justifies the general use of Eq.(9') for the birefringence of oriented polymers. This particular situation is rather exceptional because, in general, property components along the polymer chain (which are mainly governed by covalent interatomic bonding) will be significantly different from components in directions perpendicular to the chain, resulting in a substantial degree of anisotropy in the structural elements and a relatively large difference between calculated upper and lower bounds for the aggregate properties.

In the special case of uniaxial orientation with respect to the Z-axis:

$$f_{iX} = f_{iY} = -\frac{1}{2} f_{iZ} \text{ and } f_{IZ} = f_{2Z} = -\frac{1}{2} f_{3Z} \quad (11)$$

and Eq. (9') reduces to:

$$P_Z - P_Y = P_Z - P_X = f_{3Z} \left(p_3 - \frac{p_1 + p_2}{2}\right) \quad (12)$$

$$= f_{3Z} \, \Delta^0{}_p \quad (13)$$

Equation (12) shows that even in the case of an orthotropic structural element (i.e. devoid of transverse isotropy) for which $p_1 \neq p_2$, the macroscopic state of anisotropy may be purely uniaxial.

If p represents the refractive index, the term between brackets on the right-hand side of Eq. (12) in usually designed as the intrinsic birefringence; more generally, for any property we may define $^0{}_p$ as the intrinsic anisotropy of the structural element.

The intrinsic birefringence is to be expected different, in general, for the crystalline and noncrystalline phases of the same polymer[34]; it has been conjectured, however, that the refractive index n_3 along the molecular axis in the crystal unit cell would be essentially the same as the corresponding refractive index component in the noncrystalline structural element and that, consequently, the difference in optical properties between crystalline and noncrystalline unit elements can be mainly ascribed to a difference in refractive indices perpendicular to the chain axis-direction [24].

Similar semplifications have been proposed for the analysis of other properties (stiffness, compressibility, thermal expansivity, thermal conductivity) by assuming that properties measured along the chain axis in a noncrystalline structural element will be approximately the same as for a crystal, [24,35] or in the case of polymers that do not crystallize, of the same order of magnitude as those found in compounds containing exclusively covalent interatomic bonds as e.g. inorganic glasses[21].

For the properties of noncrystalline structural elements measured perpendicularly to the chain axis direction, Hennig has estimated values assumimg transverse isotropy ($p_1 = p_2$) and further supposing that these properties are mainly determined by interchain van der Waals-forces and, hence, of the same order of magnitude as in low molecular weight organic compounds [21].

Once the intrinsic anisotropy of the structural element is known or estimated, we may use Eq. (13) in order to calculate the average degree of uniaxial orientation from the observed macroscopic anisotropy, $P_Z - P_Y$. Hennig has compared average degrees of orientation for several amorphous polymers, calculated by means of Eq. (13) from the measured anisotropic linear compressibility, thermal resistivity and thermal expansivity, respectively, and obtained reasonable agreement between the values deduced from the measurements of different properties. Henning's results are also in fair agreement with Eq. (8) which seems to indicate that for the properties considered, experimental values are close to the predicted upper bound results.

The intrinsic anisotropy for the properties considered by Hennig is, in contrast with the refractive index, very important since the values of p_1 (or p_2) are at least one order of magnitude greater than p_3. As a result upper and lower bounds are distinctly different and converge at high degress of orientation only, as illustrated in Figure 7. The upper bound values deduced from Eqs. (8) and (12) or (13), are given by:

$$P_Z = P_O + \frac{2}{3} f_{3Z} \Delta^0_p = \frac{1}{3} \quad (1 + 2f_{3Z}) P_3 + 2 (1 - f_{3Z}) p_1$$
$$= P_O \quad (1 - \frac{P_O - P_3}{P_O} f_{3Z}) \tag{14}$$

$$P_X = P_Y = P_O - \frac{1}{3} f_{3Z} \Delta^0_p = \frac{1}{3} \quad (1 - f_{3Z})p_3 + (2 + f_{3Z}) p_1$$
$$= P_O \ (1 + \frac{P_O - P_3}{2P_O} f_{3Z}) \tag{15}$$

The predicted lower bound values are obtained by replacing all the P and p in Eqs. (14) and (15) by their inverse values.

Choy[36] has compared Hennig's experimental data on thermal expansivity of PVC and PMMA with the predicted values given by Eqs. (14) and (15) and arrives at the conclusion that the experimental P_Z - values are close to the upper bound values at very low degree of orientation but decrease more rapidly with orientation than predicted by Eq. (14). Values measured in the perpendi-

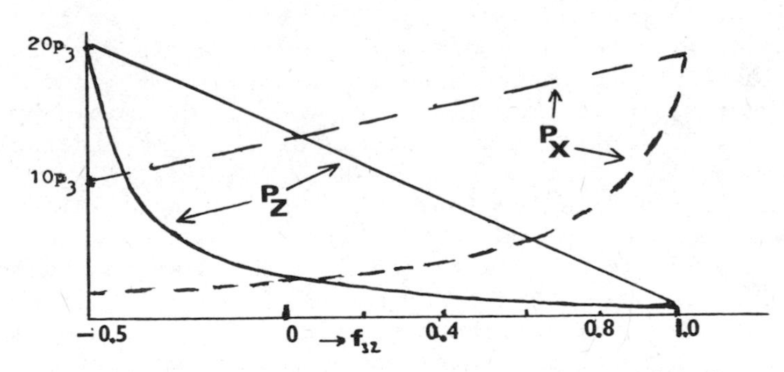

Figure 7: Upper and lower bounds of P_Z and P_X ($=P_Y$) predicted by Eqs. (14) and (15). $p_1 = p_2 = 20\ p_3$

cular direction remain, however, relatively close to those predicted by Eq. (15).

The values for the average degree of orientation, adopted by Choy are much lower than the values estimated by Hennig and it seems hazardous, therefore, to draw definite conclusions from Choy's analysis in the absence of exact data on the degree of orientation in the polymer samples investigated by Hennig. Choy suggests that the rapid decrease of thermal expansivity along orientation axis is associated with the well known existence of a small amount of crystallinity in PVC which makes it inappropriate to treat this polymer as a single-phase system.

For the properties we have discussed sofar, $p_3 \ll \frac{P_1 + P_2}{2}$ or $p_3 \ll p_0$

It then follows from Eq.(14) that:

$$P_Z = P_0 \ (1 - f_{3Z}) \qquad (16)$$

Several years ago [10] we applied Eq. (16) to the results of small strain compliance measurements on uni- and biaxially oriented PVC samples by plotting $(S_0 - S_Z)/S_0$, where S is the inverse Young's modulus, as a function of birefringence (Figure 8).

By adopting an appropriate value for the intrinsic birefringence of PVC, quantitative agreement with Eq. (16) was obtained which is a rather unexpected result: in fact, compliance is a property represented by a fourth-rank tensor whose behaviour in terms of the aggregate model cannot be treated in the simple way outlined above. Exact treatments for the case of uniaxial orientation have been published by various authors starting with Ward [22] and extended to two-phase systems by others [23,25,35].

Due to the fourth rank tensorial character of the elastic properties, the general mechanical behaviour of an orthotropic structural element needs nine independent constants for its descri-

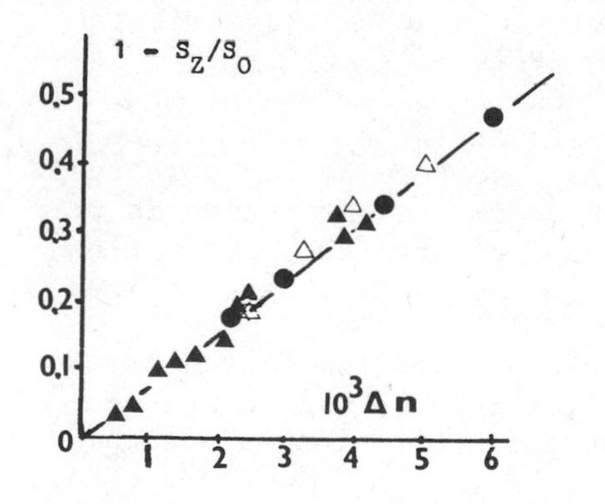

Figure 8: Relative decrease of tensile compliance versus birefringence for uni- and biaxially oriented PVC (adapted from Ref.10)

ption but in the case of uniaxial orientation this number is reduced to five.

In order to calculate the average aggregate properties not only the second but also the fourth order moment of the orientation distribuction is required. The Reuss-averaging procedure, based on uniform stress distribution in the aggregate, predicts lower bound modulus values, i.e. upper bound values for the compliances given by the compliances matrix:

$$\begin{vmatrix} s_{11} & s_{12} & s_{13} & 0 & 0 & 0 \\ s_{12} & s_{22} & s_{23} & 0 & 0 & 0 \\ s_{13} & s_{23} & s_{33} & 0 & 0 & 0 \\ 0 & 0 & 0 & s_{44} & 0 & 0 \\ 0 & 0 & 0 & 0 & s_{55} & 0 \\ 0 & 0 & 0 & 0 & 0 & s_{66} \end{vmatrix} \qquad (18)$$

Often the structural element is assumed to be transversely isotropic which means that:

$s_{11}=s_{22}$, $s_{13}=s_{23}$, $s_{44}=s_{55}$ and $s_{66}= 2(s_{22}-s_{12})$. In that case the lower bound for the Young's modulus along the stretching axis of an uniaxially oriented sample is given by:

$$1/E_A = S_{33} = s_{33}\langle\cos^4\theta\rangle + s_{11}\langle\sin^4\theta\rangle + (2\, s_{13} + s_{44})\langle\sin^2\theta\cos^2\theta\rangle \qquad (19)$$

For an orthotropic structural element, characterized by a compliances matrix with nine independent constants, the expression for the Young's modulus in uni-axial orientation is of the same form as Eq. (19), containing in particular only three independent linear combinations of the compliance constants. A general description of the elastic behaviour of uniaxially oriented specimens (including shear moduli and Poisson's ratios) requires the knowledge of five independent linear combinations of the compliance constants of an orthotropic structural element. It follows that the assumption of transverse isotropy in the structural element is of no particular interest as long as attention is restricted to uniaxial orientation; on other hand this assumption does not affect the generality of the macroscopic description, in this case [24, 25].

In order to account for the influence of the fourth order moments of the orientation distribution Seferis has introduced[25] in analogy with Herman's orientation function, a new function:

$$g\,(\langle\cos^4\theta\rangle) = \frac{1}{4}\,(5\,\langle\cos^4\theta\rangle - 1\,) \qquad (20)$$

For random orientation: $f = 0$ and $g = 0$; perfect uniaxial orientation corresponds to: $f = 1$ and $g = 1$, whereas for perpendicular orientation: $f = -1/2$ and $g = -1/4$.

The general elastic behaviour of uniaxially oriented aggregates may then be completely described by means of five independent linear combinations of the compliance constants and two orientation functions:

$$1/E_A = S_{33} = A + B\,f + C\,g$$

$$1/E_T = S_{11} = S_{22} = A - \frac{1}{2}\left(B + \frac{5}{4}C\right) f + \frac{3}{8} C\,g$$

$$1/G_A = S_{44} = S_{55} = A_G - \frac{1}{2}\left(4B_G - B - 5G\right) f - 2\,C\,g \qquad (21)$$

etc.

E_A is the Young's modulus along the axis, E_T is the modulus in the transverse direction and G_A is the axial shear modulus.

A, B, C, etc. are linear combinations of the compliance constants of the structural element, e.g. :

$$A = \frac{1}{15}\left[3\,(s_{11} + s_{22} + s_{33}) + 2\,(s_{13} + s_{23} + s_{12}) + (s_{44} + s_{55} + s_{66})\right]$$

It follows from Eq. (21) that A is the value of the extensional compliance S^0_{33} for the isotropic, unoriented material ($f=g=0$).

It can also be deduced from Eq. (21) that A_G represents the shear compliance of the unoriented material; its value is given by:

$$A_G = \frac{2}{15}\left[2\,(s_{11} + s_{22} + s_{33}) - 2(s_{13} + s_{23} + s_{12}) + 3(s_{44} + s_{55} + s_{66})\right]$$

The effect of molecular orientation on the calculated macroscopic moduli or compliances strongly depends on the various ratios of the element compliance constants as e.g. s_{33}/s_{11} and s_{44}/s_{11}; the influence of these ratios on the predicted upper and lower bounds for the Young's modulus has been examined in some detail by Kausch[37].

One the of the most interesting results, first pointed out by Ward[22], is the predicted appearance of a minimum in the curve representing the axial Young's modulus as a function of the degree of uniaxial orientation, for sufficiently large values of s_{44}/s_{11}, i.e. when the shear modulus of the structural element is much smaller than its transverse modulus. Experimentally, such a minimum has been observed for uniaxially drawn low density polyethylene, at 20°C. The transverse modulus, measured at the same temperature, increases with increasing degree of orientation and is even found to be higher than the axial Young's modulus at low degrees of orientation (see Figure 9). At temperatures below the glass transition temperature the anisotropic behaviour becomes more "conventional": the axial Young's modulus increases and the

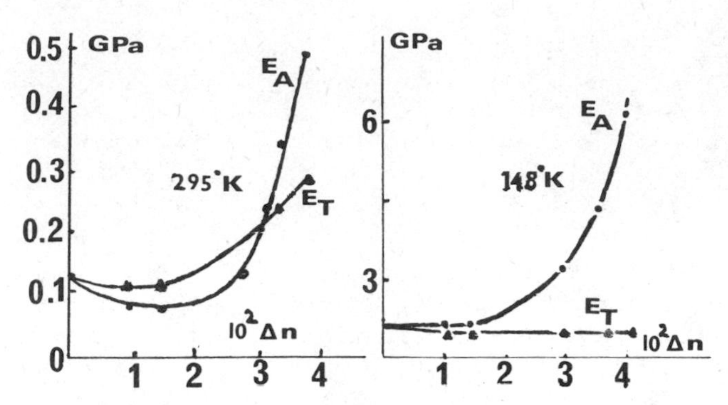

Fig.9: Axial and transverse Young's moduli of uniaxially drawn low density polyethylene above and below the glass transition temperature. Adapted from Ref. 29.

transverse modulus decreases with increasing uniaxial orientation[29].

The predicted Reuss-averages for S_{33} and S_{11} based on values for s_{11}, s_{33} and $(2s_{13} + s_{44})$ which Ward and his co-workers assumed to be equal to the measured macroscopic values for the sample having the highest degree of uniaxial orientation, were in fair agreement with the experimental results if the orientation distribution was calculated from X-ray diffraction or N.M.R. measurements[38]. Both these methods provide orientation distribution functions of the crystalline regions only, but the aggregate model as used by Ward et al. remains a single-phase model in which the noncrystalline regions are considered to form an integral part of the structural elements and not to support the stress as a separate phase.

The unusual anisotropy pattern observed above the glass transition temperature, probably does not critically depend on a particular structural model but is essentially due to the large value of the shear compliance S_{44} compared with S_{33} and S_{11}. This conclusion is confirmed by the observed angular dependence of the Young's modulus in the plane of the highly uniaxially oriented polyethylene films, showing a pronunced minimum at 45° with respect to the draw direction. The elastic compliance in any direction ϕ from the draw direction is given by:

$$S(\phi) = S_{33}\cos^2\phi + S_{22}\sin^2\phi + (2S_{13} + S_{44})\sin^2\phi\cos^2\phi \quad (22)$$

If the last term in Eq. (22) is the dominant one, the elastic compliance passes indeed through a maximum for $\phi = 45°$ and the modulus passes through a corresponding minimum.

The single-phase aggregate model has also been applied[39,40] with success to the analysis of the tensile deformation of medium and highly oriented poly(p-phenylene terephthalamide) fibers which may be regarded as being built up of paracrystalline microfibrils that contain crystallites with a narrow orientation distribution and are devoid of the distinct noncrystalline domain which are always present in conventional semicrystalline synthetic polymer fibers. The mechanical behaviour in uniaxial extension, investigated by means of dynamic techniques, was found to be in good agreement with the predicted behaviour based on a Reuss averaging procedure of compliances. Moreover, the analysis yields reasonable values for the compliance constants of the structural element which compare well with those estimated from the known crystal structure.

For most polymers in either the crystalline or glassy, amorphous state, the shear compliance, s_{44}, is expected to be of the same order of magnitude as the transverse compliances s_{11} and s_{22}.

If shear and transverse compliances of the structural element are exactly identical, Eqs.(19) or (21) may be simplified and reduced to[24]:

$$S_{33} = S°(1 - f) + \frac{1}{5}(f + 4g)\, s_{33} \quad (23)$$

where S° is the tensile compliance of the isotropic material. Equations (23) is to be compared with Eq. (14) which may be written as follows:

$$S_{33} = S°(1 - f) + f\, s_{33} \quad (14')$$

Since S° in always larger than s_{33}, the second term on the right-hand side of Eqs. (23) and (14') will be smaller than the first term and may even become negligible if s_{33} becomes very small which is often the case.

If the latter approximation is valid Eqs. (23) and (14') are identical and reduce to Eq. (16):

$$S_{33} = S°(1 - f) \quad (16')$$

It has already been shown above (Fig. 8) that Eq. (16') describes fairly well the tensile modulus measurements on PVC. Similar results have been reported for sonic modulus measurements on various amorphous and even semicrystalline polymers[41]. In the latter case, however, it is more appropriate to apply a generalized form of Eq. (16') which takes account of the presence of two distinct phases, as first proposed by Samuels[42] for the analysis of uniaxially oriented isotactic polypropylene fibers and films.

Properties of Anisotropic Semicrystalline Polymers

In a semicrystalline polymer with a volume fraction of crystallinity v_X, bulk properties can be predicted with the aid of the two-phase aggregate model theory by combining the average properties of each phase in an appropriate way:

$$P_{ij} = v_X \langle P_{ij} \rangle_X + (1 - v_X) \langle P_{ij} \rangle_{am} + \Phi_{ij} \tag{24}$$

where $\langle P_{ij} \rangle_X$ and $\langle P_{ij} \rangle_{am}$ are the average aggregate properties in the crystalline and noncrystalline phases, respectively.

Because the average aggregate properties for each phase represent either lower or upper bound values, the bulk properties calculated from Eq. (24) without its last term on the right-hand side, will also represent corresponding lower or upper bound values if the applied combination rule has been the same in both stages of the calculation[23,25]. The mechanical behaviour at small strains of conventionally processed semicrystalline polymers, is found to be fairly well represented, in a first approximation, by the Reuss-model implying a uniform stress distribution and, hence, a certain kind of interfacial discontinuity. In order to account for deviations from the Reuss model, Seferis has proposed[25] to include the term Φ_{ij} in Eq. (24) which he calls, in analogy with the well known treatment of birefringence in heterogeneous systems, a "form factor".

The form factor which is related to the contiguity parameter introduced in Eq. (6), depends according to Seferis'analysis on the relative magnitudes of the mechanical properties of both phases, on the aspect ratio of the aggregates and on the orientation of the structural elements. Its maximum contribution would occur in an unoriented material with crystalline aggregates of infinite aspect ratio.

The introduction of a form factor for mechanical properties is, of course, purely formal and not founded on a theoretical treatment comparable to the classical analysis of form birefringence[43]. It provides, nevertheless, a means for incorporating certain morphological features of a semicrystalline polymer into the final expression for the bulk properties. In many cases, however, we may neglect the form factor and it follows then from Eq. (24) that the number of parameter required to describe either lower or upper bound of property P_{ij} becomes equal to 2N + 1, if N parameters were needed to describe this property in a single-phase material (N parameters for each phase plus v_X).

Moreover, the orientation distribution of the structural

elements in both phases has to be known in order to calculate $\langle P_{ij} \rangle_X$ and $\langle P_{ij} \rangle_{am}$; the combination of several experimental techniques (X-ray diffraction, optical birefringence, infra-red dichroism, etc.), generally required to obtain this information, has been systematically applied in a few cases only, until now.

The most detailed investigations, at present, concern the structure-properties relationship in uniaxially[42] and biaxially[18] oriented isotactic polypropylene fibers and films. For this polymer both storage and loss tensile compliances are in fair agreement with the predicted Reuss bounds calculated from the simple two-parameter equation obtained by generalizing Eq. (16'):

$$S_{33}(\phi) = v_X S^{\circ}_X \left[1 - f_X F(\phi) \right] + (1 - v_X) S^{\circ}_{am} \left[1 - f_{am} F(\phi) \right] \tag{25}$$

where S (ϕ) is the tensile compliance measured at an angle from the 3-direction,

S°_X and S°_{am} are the tensile compliances of the randomly oriented, isotropic crystalline and noncrystalline phases, respectively,

f_X and f_{am} are the Hermans'orientation functions in both phases, and $F(\phi)$ is the "loading function" introduced by Seferis:

$$F(\phi) = \frac{1}{2} \left(3 \cos^2 \phi - 1 \right) \tag{26}$$

The intrinsic parameters S°_X and S°_{am} representing either storage or loss compliances, are temperature-and frequency dependent, and must be determined experimentally by measuring the compliances of a series of unoriented polymer samples with varying degrees of crystallinity[42] or by adjusting Eq. (25) to the experimental results obtained at different angles ϕ on a series of anisotropic samples[44].

The ratio of the intrinsic storage compliances $S^{\circ}_{am}/S^{\circ}_X$ in the glassy state of isotactic polypropylene, varies slowly with temperature from about 2.5 to 4.0; beyond the glass transition temperature this ratio increases rapidly. On the other hand, the loss compliance of the randomly oriented noncrystalline phase is always much larger than in the crystalline phase, even at temperatures below the glass transition. Since, moreover, orientation in the noncrystalline phase was always found to be significantly lower than crystalline orientation in the same sample (see Figure 10), it is obvious from Eq.(25) that the second term is, in general, much larger than the first one and that, consequently, the viscoelastic behaviour at small strains is mainly controlled by the molecular orientation in the noncrystalline phase.

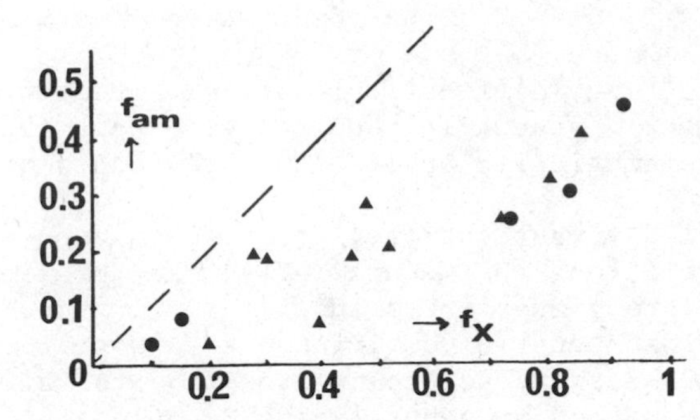

Fig.10: Noncrystalline versus crystalline orientation in uni - and biaxially drawn polypropylene films. Adapted from Ref. 18.

Non-linear mechanical behaviour associated with finite strains and technologically important characteristics such as tenacity,and impact resistance of polypropylene films was also observed to be strongly affected by the degree of noncrystalline orientation. In general, however, it is rather difficult to evaluate with exactitude the relative importance of orientation effects in crystalline and noncrystalline phases, respectively.

At very high degrees of orientation, corresponding to f_{am} - values larger than about 0.7, the two-parameter model for the tensile behaviour breaks down; however, the two-phase generalization of Eq. (23) was found sufficient to represent the experimental results even if it was further assumed that $s_{33,X} = s_{33,am}$. This means that only one term needs to be added to Eq. (25), which for $\phi = 0$ (the only case tested until now for these ultra-high oriented polypropylene films)[24,18] reduces to:

$$S_{33} = v_X S^\circ_X (1 - f_X) + (1 - v_X) S^\circ_{am} (1 - f_{am}) + \frac{1}{5} \langle f + 4g \rangle s_{33} \tag{27}$$

where

$$\langle f + 4g \rangle = v_X (f_X + 4g_X) + (1 - v_X) (f_{am} + 4g_{am})$$

The value for s_{33} to be substituted in Eq. (27) was found to be of the order of 0.025 to 0.028 GPa^{-1}, in good agreement with theoretical and experimental estimates of the extensional compliance of the isotactic polypropylene crystal.

Above the glass transition temperature the visco-elastic behaviour becomes more complex: the transverse compliance S_{11} or S_{22}

starts to decrease with increasing degree of uniaxial orientation, as in the case of polyethylene (Fig.9): apparently the shear compliance s_{44} is growing now more rapidly than the transverse compliances s_{11} and s_{22} with increasing temperature.

As a result the approximation which led to Eq. (23) is no longer valid and the usual anisotropy effects observed in low density polyethylene, shown in Fig. 9, may be expected to occur in polypropylene too. However, since the glass transition temperatures for both polymers differ by at least 100°K, similar effects in polypropylene can only be expected at temperatures much higher than room temperature.

The simple two-parameter model described by Eq. (25) has also been applied with some success to sonic modulus measurements on polyethylene terephthalate fibers[45]; modulus determinations on one way drawn films of the same polymer were found to be in semiquantitative agreement only with the model predictions, but fully confirmed the predominant effect of molecular orientation in the noncrystalline phase on the mechanical behaviour[10].

Choy et al. applied the same simple two-phase model to both modulus and dielectric loss measurements on uniaxially drawn samples of poly (chlorotrifluoroethylene)[46]. Orientation functions in the noncrystalline phase, f_{am}, calculated from both types of measurements with the aid of the model, were in reasonable agreement, values deduced from the dielectric data being 10 to 30% higher than those calculated from the modulus measurements.

Further Development of Anisotropic Two-phase Models

The relatively simple two-phase aggregate model defined by Eq. (24), combined with appropriate expressions for the average aggregate properties in the crystalline and noncrystalline phases contains, in its most general form, a sufficient number of parameters to describe many properties of anisotropic semicrystalline polymers. Upper and lower bounds for the properties can be predicted exactly from a limited amount of structural information, viz. degree of crystallinity plus orientation distribution functions for both phases. It is often found for various properties that the experimental data lie relatively close to one of the bounds predicted on the basis of physically acceptable model parameters.

Although various experimental techniques for the determination of orientation distributions in both phases of an anisotropic semicrystalline polymer exist, their systematic use in the study of structure-properties relationship implies a considerable effort and has remained, until now, rather exceptional. Nevertheless, it seems obvious that further development and, in particular, the introduction of additional information on specific structural details in existing models is only justified if the latter, in their present form, have been definitely demonstrated to be inadequate.

Serious candidates which may be expected to provide useful additional structural information, are the geometrical parameters (size and shape) of the aggregated anisotropic phases as well as the degree of continuity of each phase. We have already mentioned the probable importance of the aspect ratio of the crystalline aggregates for the definition of an appropriate phase average.

Another point of interest, abundantly discussed in the literature, concerns the structure of the material confined between two more or less voluminous crystalline domains (see Figure 11). These intercrystalline regions are supposed to contain various kinds of noncrystalline material: loose chain-ends, loops and also more or less taut tie-molecules of restricted mobility which interconnect two adjacent crystalline domains. Zachmann has been able to estimate the fraction of taut tie-molecules and their degree of tautness from an analysis of the N.M.R. absorption signal above the glass transition temperature in the case of uniaxially drawn polyethylene terephthalate films[47]. His results show that an important fraction of the noncrystalline material may be composed of taut tie-molecules characterized by high values of the Hermans' orientation function (f_{am} 0.9), much larger than the average noncrystalline orientation as determined by birefringence e.g.

Such a collection of highly extended chains will, obviously, exhibit properties along the chain direction very similar to those of a crystal; in fact, the main difference with respect to a crystal is the lack of lateral order.

The existence of so-called "crystalline bridges", first proposed by Fischer et al.[48] has been further explored by Gibson et al.[49] in order to account for the continuity of the crystalline phase in ultra-high oriented high density polyethylene. The relative degree of continuity of the crystalline phase, expressed as a fraction of crystalline bridges, was estimated from X-ray dif-

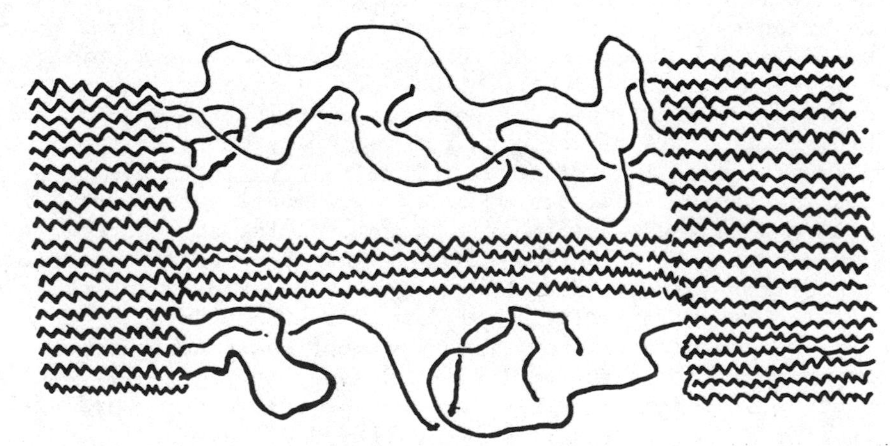

Fig.11: Schematic representation of the structure of intercrystalline regions in an oriented polymer, containing loops, taut tie-molecules, crystalline bridges, etc.

fraction measurements and found to increase with increasing degree of uniaxial orientation up to a maximum value of about 0.4. Not unexpectedly, this new structural parameter was also found to be strongly correlated with Young's modulus but this observation does not exclude the possibility that the measured increase in stiffness remains essentially due to molecular orientation in both crystalline and noncrystalline phases and is only indirectly related to the degree of continuity of the crystalline phase.

On the other hand, Choy and co-workers[50] have argued that the presence of crystalline bridges and/or taut tie-molecules must have a significant effect on the thermal expansivity along the direction of uniaxial orientation. A similar argument has been advanced for the thermal conductivity of oriented polymers[51].

The effect of crystalline bridges can be formally accounted for with the aid of a particular version of the Takayanagi two-phase composite model [52], in which the crystalline bridges are assumed to constitute a continuous phase connected in parallel with the bulk of the material composed of crystalline and noncrystalline regions coupled in series (see Fig. 12). In this model the noncrystalline phase is assumed to be isotropic ($f_{am} = 0$), and the crystalline phase in a state of perfect uniaxial orientation ($f_X = 1$). At a given total degree of crystallinity, the only remaining variable is the fraction of crystalline bridges, b, equal in a first approximation to:

$$b \simeq s_{33,X} / S_{33}$$

where S_{33} is $1/E_A$ and $s_{33,X}$ the crystal tensile compliance.

The axial Young's modulus E_A is proportional to b according to this simple model and the same proportionality holds for the thermal conductivity along the orientation axis. Thermal expansivity along the same direction, on the contrary, will rapidly decrease with increasing b, because of the constraining effect of the crystalline bridges and, consequently, rapidly approach the axial

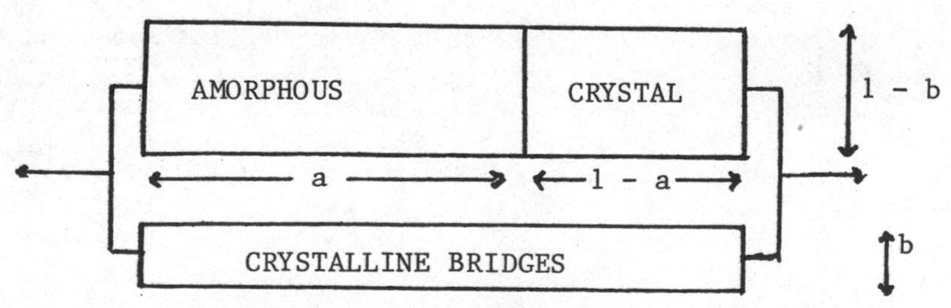

Fig.12: Schematic representation of Takayanagi model with intercrystalline bridges (Refs. 50 - 52).

thermal expansivity of the crystal, which for all polymers investigated until now, appears to be negative and to range from - 1 to about $- 4 \cdot 10^{-5}\ K^{-1}$, in good agreement with theoretical estimates[53].

The experimental data on ultra-high oriented linear polyethylene confirm that a small amount of crystalline bridges (as calculated from the observed Young's modulus) is very effective in constraining the thermal expansivity along the orientation axis and bringing it close to the negative value for the polyethylene crystal.

Still larger negative thermal expansivities have been observed in uniaxially oriented semycristalline polymers at temperatures above the glass transition temperature; Choy et al.[54] have reported values for low density polyethylene which are about 30 times more negative than the thermal expansivity of the polyethylene crystal. Similar behaviour has been reported for several other polymers of relatively low crystallinity (v_X 0.5).

These large negative thermal expansions are attributed to a rubber elastic effect comparable to the effects observed by Struik[8] in oriented glassy polymers and which we have already mentioned above. The retractive stresses in the noncrystalline taut tie-molecules are supposed to be counteracted by the rigidity of the crystalline bridges (see Fig. 11) and the resulting reversible thermal contraction is assumed to depend on the relative magnitude of these opposing forces. A quantitative analysis by Choy et al.[54] based on a simple model in which these forces are coupled in parallel, leads to the conclusion that the contribution of the rubber elastic effect to the thermal expansivity must be proportional to:

$$— \quad \frac{1}{T} \quad \frac{tE^t}{bE_c}$$

where t is the fraction of tie-molecules, E^t the Young's modulus of the tie-molecules and E_c the modulus of the crystal: $E_c = 1/s_{33,X}$

It follows then, that the crystalline bridge/tie-molecules ratio: b/t, should be the controlling parameter. From an analysis of the experimental data in terms of this model, it was calculated that the ratio b/t would be of the order of 0.01 for a low density polyethylene sample with a large negative expansivity ($-40. \ 10^{-5} K^{-1}$) at 320° K.

For a sample of high density polyethylene with an expansivity of only $-1. \ 10^{-5}\ K^{-1}$, the ratio b/t was calculated to be of the order of unity, i.e. about 100 times larger than in low density polyethylene. Since small b/t ratios will be associated, in general, with low degrees of crystallinity, important rubber elastic effects of the kind discussed here, are not expected to occur in oriented highly crystalline polymers.

In conclusion there seems to exist some evidence that a more or less detailed characterization of the intercrystalline domains by means of a limited number of well defined and measurable variables, may be useful and even indispensable for an improved description and understanding of the properties of anisotropic, two-phase macromolecular materials.

CONCLUSION

Considerable progress has been made in recent years with regard to a comprehensive description of polymer properties with the aid of more or less elaborate structural models. In the general case of anisotropic semicrystalline materials the fundamental importance of molecular orientation in both crystalline and noncrystalline phases, is clearly recognized and has been firmly established for various physical and technologically important properties. A number of different experimental techniques is available for the quantitative characterization of molecular orientation distributions in crystalline and noncrystalline phases but a systematic exploration of these methods combined with an adequately conceived program of property determinations has been achieved in a few cases only, at present.

The extensive experimental studies on the preparation and properties of ultra-high uniaxially oriented polymers, in particular linear high molecular weight polyethylene, seem to indicate the possible need for a more detailed characterization with respect to certain specific morphological features which may affect various physical properties to a significant extent.

However, for the large majority of anisotropic macromolecular materials, processed by means of existing industrial techniques, the presently available structural models, in spite of their obvious limitations, seem to satisfy the basic requirements with regard to a comprehensive description of various important properties in terms of a reduced number of physically relevant model parameters.

REFERENCES

1. L.R.G. Treloar, "The Physics of Rubber Elasticity", Oxford Univ. Press, London (1949).
2. I.V. Yannas and R.R. Luise, J. Macromol. Sci.-Phys. B21 (3), 443 (1982)
3. J. Meissner, Pure & Appl. Chem. 42, 553 (1975)
4. H.M. Laun, Colloid and Polymer Sci. 259, 97 (1981)
5. A.S. Lodge, "Elastic Liquids", Academic Press,London-NewYork (1964).
6. P.G. de Gennes, J. Chem. Phys. 55, 572 (1971); "Scaling Concepts in Polymer Physics", Cornell Univ. Press, Ithaca (1979)
7. M.Doi and S.F.Edwards, J.C.S. Faraday Trans.II 74, 1789,1802, 1818 (1978); 75, 38 (1979).
8. L.C.E. Struik, Polym.Eng.Sci. 18, 799 (1978)
9. A.J. de Vries and C.Bonnebat, Polym.Eng.Sci. 16, 93 (1976)

10. A.J. de Vries, C.Bonnebat and J. Beautemps, J. Polym.Sci.,Polymer Symp.58, 109 (1977).
11. P.J. Flory, Proc.Roy.Soc.Lond. A351 (1976)
12. P.R. Pinnock and I.M.Ward, Trans. Faraday Soc. 62, 1308 (1966)
13. W. Kuhn and F. Grün, Kolloid-Z. 101, 248 (1942)
14. J.L.S. Wales, "The Application of Flow Birefringence to Rheological Studies of Polymer Melts",Delft Univ.Press (1976)
15. P.J.Flory and Y. Abbe, Macromolecules 2, 335 (1969)
16. K. Nagai, J. Chem. Phys. 40, 2818 (1964)
17. R.S. Stein and G.L. Wilkes, "Physico-chemical Approaches to the Measurement of Anisotropy", Chapter 3 in: "Structure and Properties of Oriented Polymers", I.M.Ward, Ed., Applied Science Publish. London (1975).
18. A.J. de Vries, Pure & Appl.Chem. 53, 1011 (1981); 54,647(1982)
19. W.Voigt, "Lehrbuch der Kristallphysik", Teubner, Leipzin(1910)
20. A. Reuss, Z.Angew.Math.Mech. 9, 49 (1929)
21. J. Hennig, Kolloid-Z, 196, 136(1964); 200,46(1964); 202,127 (1965) J.Polym.Sci. C16, 2751(1967), Kunststoffe 57,385(1967)
22. I.M.Ward, Proc.Phys.Soc. 80, 1176 (1962)
23. R.L.McCullough, C.T.Wu, J.C.Seferis and P.H.Lindenmeyer, Polym.Eng.Sci. 16, 371 (1976)
24. J.C.Seferis and R.J. Samuels, Polym.Eng.Sci. 19,975 (1979)
25. J.C.Seferis, Ph.D.Thesis,Univ.of Delaware,Dept.Chem.Eng.(1977)
26. W.Ruland and W.Wiegand, J.Polym.Sci. Polym.Symp.58,43(1977)
27. H.C. Biangardi, J.Polym.Sci.Polym.Phys.Ed. 18,903 (1980)
28. D.W.Hadley and I.M.Ward, "The Macroscopic Model Approach to Low Strain Properties", Chapter 8 in "Structure and properties of Oriented Polymers", I.M.Ward,Ed.,Appl.Sci.Publ.,London (1975)
29. V.B.Gupta and I.M.Ward, J.Macromol.Sci.-Phys.B 1, 373 (1967)
V.B.Gupta, A.Keller and I.M.Ward, J.Macromol.Sci.-Phys.B2, 139 (1968)
Z.H.Stachurski and I.M.Ward,J.Macromol.Sci.-Phys. B3,427(1969)
30. Z.Haskin and S.S.Shtrikman, J.Mech.Phys.Solids 10,343 (1962)
31. C.T. Wu,Ph D.Thesis, Univ.of Delaware,Dept.of Chem.Eng.(1976)
32. J.C.Halpin and S.W.Tsai, AFML-TR 67, 423 (1969)
33. J.C.Halpin and J.L.Kardos, J.Appl.Phys. 43,2235 (1972)
34. R.S.Stein, J. Polym.Sci. A2,7, 1021 (1969)
35. S.Nomura, S.Kawabata, H.Kawai, Y.Yamaguchi, A.Fukushima and H. Takahara, J. Polym.Sci.A2, 7,325 (1969)
36. C.L.Choy, "Thermal Expansivity of Oriented Polymers", Chapter 4 in: "Developments in Oriented Polymers", I.M.Ward,Ed.,Appl. Sci.Publish.,London (1982)
37. H.H. Kausch, J.Appl.Phys. 38, 4213(1967); Kolloid-Z.237, 251 (1969)
38. V.J.McBrierty and I.M.Ward,Brit.J.Appl.Phys.(2) 1, 1529(1968)
39. M.G.Northolt and J.J. van Aartsen, J.Polym.Sci.Polymer Symp. 58, 283 (1977)
40. M.G.Northolt, Polymer 21, 1199 (1980)
41. H.M.Morgan, Textile Res.J. 32, 866 (1962)

42. R.J. Samuels, J.Polym.Sci. A3, 1741 (1965)
43. O. Wiener, Abh.Sächs.Ges.D.Wiss.Math.-Phys.Kl. 32,503(1912)
44. J.C. Seferis, R.L. McCullough and R.J.Samuels, Polym.Eng. Sci. 16, 334 (1976)
45. J.H.Dumbleton, J. Polym.Sci.A2, 6, 795 (1968)
46. C.L.Choy, K.H. Cheng and Bay-Sung Hsu, J.Polym.Sci.Polym. Phys. Ed. 19, 991 (1981)
47. H.G. Zachmann, Polym.Eng.Sci.19, 966 (1979)
48. E.W.Fischer, H.Goddar and W. Peisczeck, J.Polym.Sci.C32, 149 (1971)
49. A.G.Gibson, G.R.Davies and I.M.Ward, Polymer 19, 683 (1978)
50. C.L. Choy, F.C.Chen and E.L.Ong, Polymer 20, 1191 (1979)
51. A.G.Gibson, D.Greig, M.Sahota, I.M.Ward and C.L.Choy, J. Polym.Sci. Polym.Lett.Ed. 15, 183 (1977)
52. M. Takayanagi, K. Imada and T. Kajiyama, J. Polym. Sci. C15, 263 (1966)
53. F.C.Chen, C.L. Choy, S.P.Wong and K.Young, J Polym.Sci. Phys. Ed. 19, 971 (1981)
54. C.L. Choy, F.C. Chen and K.Young, J. Polym.Sci. Phys.Ed. 19, 335 (1981).

THE ROLE OF THE BINDER IN THE AGEING OF A FILLED ELASTOMER

Z. Laufer, Y. Diamant, S. Gonen and D. Katz

Department of Materials Engineering
Israel Institute of Technology
Technion City, Haifa, Israel

ABSTRACT

Accelerated ageing experiments were conducted on the elastomer used as binder of a composite material which had been investigated earlier. Specimens of the binder were kept at various temperatures with and without prestrain. After their ageing the physical and chemical properties of the specimens were tested in order to find out if any internal changes in the elastomer happened during the ageing. Testing methods which included swelling, X-ray, scanning electron microscopy and dynamic testing were applied and all these tests showed that no significant changes in the microstructure of the binder occurred.

In our previous study of the filled composite elastomer we assumed that the observed changes in the mechanical properties which occurred during ageing were due to internal changes in the binder and/or to changes on the interphase between the binder and the filler particles. On the basis of the present findings it can be concluded that the changes in the properties of the aged filled elastomer are not due to the internal changes in the binder, but they are probably related to some processes which occurred in the interphase during ageing.

INTRODUCTION

In an earlier paper (Diamant et al.,[1]) we reported the experimental results of strain endurance experiments with a filled elastomer. We observed changes in its mechanical properties during the accelerated ageing, which were enhanced by superimposing a prestrain on the samples. It was assumed that during the ageing process of the filled elastomer, certain chemical and/or physical changes occurred in its microstructure. Some of the possible ageing processes that might be involved are:

a) Orientation of the elastomeric binder chains due to the applied strain;
b) Changes in the network structure of the binder;
c) Effect of the microheterogeneity in the cured binder system;
d) Changes in the interphase between the binder and the filled particles.

The first three processes refer to changes that may occur in the binder. Therefore in order to understand better the ageing process in the composite material we decided to study the behaviour of the binder alone in this process.

We investigated the mechanical properties of the aged samples of the binder elastomer and by various techniques tried to find if any changes occurred in the elastomer during ageing.

MATERIALS AND EXPERIMENTAL

The Elastomer

The investigated elastomer is similar to the binder of the composite material tested earlier (Diamant et al.,[1]); it is a hydroxyl-terminated polybutadiene cross linked by a diisocynate. A minor change had to be introduced in the composition of the elastomer as compared with the binder and the equivalent ratio between the crosslinking agent and the polymer was increased. This change was necessary in order to increase the strength of the specimens to a value which makes it possible to perform the experimental work.

The Specimens

The specimens (see figure 1) were rectangular with dimensions 10 × 10 × 100 mm. and they were prepared by vacuum casting in molds. Perforated steel cubes were attached to the ends of each specimen. The elastomer adhered to the metal cubes during the curing process. The cubes were necessary for holding the soft specimens in the testing fixure.

Prestraining

The prestraining of the specimens was done on a special fixure shown in figure 2, in which three specimens were strained; the strain was applied by turning a screw. In order to age the specimens we kept these test devices with the strained specimens in an oven at different temperatures for several periods of time; the accelerated aging was executed at 50°, 65° and 80°C for 1, 4 and 7 days. At each temperature reference unstrained specimens were kept for the same period of time. The prestraining was done at two levels: 10% and 20% of the measured ultimate strain of the respective temperature. If we consider the high strain capability of the elastomer, then these strains are very high and we could not exceed them because the specimens failed after a very short time.

Tension Experiments

The tension experiments were conducted at a constant tension rate of 10 cm/min by use of an Instron testing machine. The first step was determination of the ultimate properties of the material at the ageing temperatures, in order to evaluate the prestrain level (see table 1). After ageing of the specimens in the above mentioned conditions the mechanical properties in tension were determined.

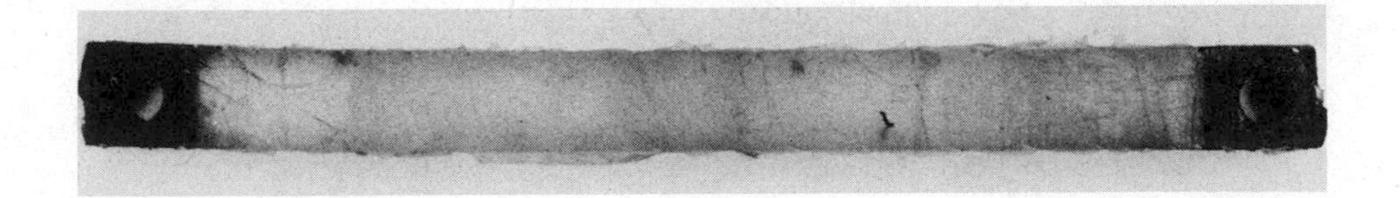

Figure No. 1: A specimen for tension experiments.

Figure No. 2: A fixure for keeping specimens under constant strain.

Table 1: Mechanical properties of the reference specimens. Tension at a constant strain rate 10 cm/min.

Tension temp. (°C)	Max. strain (%)	Max. true stress (kg/cm^2)	Initial modulus (kg/cm^2)
25	255	19.4	7.7
50	150	10.5	7.3
65	120	7.7	6.7
80	105	7.0	6.7

Swelling Experiments

The prestrained and reference specimens were let to swell until equilibrium in toluene for seven days and compressed under different loads in order to determine the average molecular weight between crosslinks $\bar{M}c$ according to the method of Cluff, Glading and Priser[2].

The solvent used for swelling experiments was tested by an I.R. spectrometer (Perkin-Elmer type 403) in order to evaluate its content.

Dynamic Tests

The dynamic properties of the specimens were determined by use of a torsion pendulum. The description of the apparatus and the theoretical background for calculating the dynamic moduli and loss-tangent are given in reference 3.

Electron Microscopy

In order to find out if any orientation occurred in the strained specimens while ageing, scanning electron microscopy (S.E.M.) was applied.

X-ray Radiography

Another method for observing orientation in polymers is X-ray radiography. We used a X-ray generator with a Tungsten tube and each specimen was exposed for radiation for two hours.

EXPERIMENTAL RESULTS AND DISCUSSION

Tension Experiments

Table No. 2 shows the results of tension experiments performed at a constant strain rate at ambient temperature. The results are tabulated in four groups according to the accelerated ageing temperatures. It can be seen that no significant changes in the mechanical properties occurred during ageing even in specimens kept prestrained at rather high levels of strain. The small deviations in the tabulated results are within the range of the usual experimental scattering which often happens in tension experiments of this type of elastomer.

Swelling Experiments

The results of the swelling experiments indicate that the swelling was almost constant in all cases (about 700%) while the percentage of extract was about 10 (see table 3). These results as well as, the calculated average molecular weight between crosslings (Mc), show that the ageing had no effect on the network structure of the elastomer. In the I.R. spectroscopy of the extracts no changes were found in the spectrum due to the ageing of specimens.

Dynamic Tests

From the dynamic tests conducted on aged specimens the loss tangent was calculated. The results presented in table 4 show that ageing has no effect on the dynamic properties of the elastomer.

Electron Microscopy

No orientation was observed in the unstrained and strained aged specimens, neither on their external surface nor on the failure surfaces obtained by tension testing.

X-ray Radiography

In the radiographic examination no signs of orientation could be observed.

Table No. 2: Mechanical properties of aged specimens, Tension at ambient temp. at a constant strain rate 10 cm/min.

Ageing Temp. (°C)	Ageing time (days)	Prestrain at Ageing (% of ε_{max})	max. true stress (kg/cm^2)	max. strain (%)	Initial Modulus (kg/cm^2)
25	REFERENCE	0	19	250	7.7
25	1	10	18	250	6.6
	1	20	17	240	6.8
	4	10	14	190	8.0
	7	10	14	200	7.7
	7	20	16	220	7.2
50	1	0	18	220	8.3
	1	20	18	230	8.1
	4	0	19	230	9.1
	4	10	18	230	8.9
	4	20	18	230	8.0
	7	0	18	230	8.2
	7	10	17	210	7.9
	7	20	22	260	8.6
65	1	0	20	250	8.2
	1	10	20	250	7.8
	1	20	21	260	8.3
	4	0	20	260	8.4
	4	10	15	200	8.3
	4	20	15	210	8.1
	7	0	22	270	8.4
	7	10	19	240	8.1
	7	20	18	250	7.2
80	1	0	22	270	7.8
	1	10	21	250	8.6
	1	20	14	180	8.3
	4	0	21	250	8.5
	4	10	20	250	8.1
	4	20	22	280	8.1
	7	0	13	190	7.9
	7	10	22	260	8.8
	7	20	17	230	8.2

Table No. 3: Swelling and extraction measurements.

	Ageing temp °C	REF	1 day 10% ε_{max}	1 day 20% ε_b	4 days REF	4 days 10% ε_{max}	4 days 20% ε_{max}	7 days REF	7 days 10% ε_{max}	7 days 20% ε_{max}
Swelling (%)	25	700	710	710		740			720	710
Extraction (%)		9	12	12		10			10	10
Mc		12650	12450			12300			13400	13450
Swelling (%)	50	720	700	710	710	700	700	700	700	710
extraction (%)		9	9	9	8	9	9	9	9.5	9
Mc		13600	13000	13500	13700	13800	13150	14000	12250	12150
Swelling (%)	65	700	700	700	690	700	695	720	725	710
extraction (%)		10	10	10	10	10	10	9	10	10.5
Mc		12760	13150	13000	12100	12450	14300	15200	12550	14000
Swelling (%)	80	710	710	715	700	720	740	735	725	725
extraction (%)		10	10	10	10	10	9.5	10	10	9.5
Mc		13350	13650	13550	12400	13800	15200	13400	13150	14000

Table No. 4: Loss tangent of specimens aged for 7 days at various temperatures without and with a prestrain of 20% of the ultimate strain. Measurements at ambient temperature.

Aging temp. °C	25	50	65	80
without prestrain	0.143	0.146	0.133	0.16
with prestrain	0.19	0.15	0.153	0.168

DISCUSSION

The observed changes of the mechanical properties of the filled elastomer due to ageing were assumed to result either from physical and/or chemical changes in microstructure of the binder and/or from changes in the interphase between the binder and the filler particles.

The experimental results presented in this study show that the ageing had almost no effect on the mechanical properties of the binder. The various tests in which physical and chemical properties were investigated indicate that there are no significant changes in the binder material due to ageing with or without high prestrain.

Therefore it can be assumed, that the observed changes in the mechanical properties of the aged filled elastomer originate mainly in changes occurring in the interphase. For better understanding of ageing process which takes place in the filled elastomer, further investigation of the structure of the interphase and its changes during ageing is necessary.

REFERENCES

1. Y. Diamant, Z. Laufer and D. Katz, J. of Materials Sci., 17, 2107 (1982).

2. E.F. Cluff, E.K. Gladding and R. Priser, J. of Polymer Sci., 45 341 (1960).

3. D. Katz, Y. Smooha and A. Ysoyeo, J. of Material Sci., 15 1167 (1980).

MOLECULAR FIELD THEORY FOR THE RHEOLOGICAL PROPERTIES OF MODERATELY CONCENTRATED POLYMER SOLUTIONS (Abstract)

Walter Hess

Fakultat fur Physik, Universitat Konstanz

Postfach 5560,7750 Konstanz,West Germany

The conformational distribution function of a weakly concentrated polymer solution in an external stationary flow field is calculated in the framework of a harmonic dumbbell model. Interactions between the polymers are treated by a multiple expansion up to the Quadrupole-Quadrupole term. The resulting equations are simplified by a mean field approximation. In this approximation a linear Langevin equation for the orientation vector od the dumbbells is obtained, which can be solved exactly and yields a Gaussian distribution function.

Using this distribution function all quantities characterizing the conformational and rheological properties can be calculated. The shear viscosity shows a typical non-Newtonian behaviour, with a power-law dependence on shear rate. The point of onset of shear dependence and the power law exponent depend on the two interaction parameters of the model. They can be easily adjusted to get quantitative agreement with experimental results.

A TREATISE ON THE ELASTIC AND HYGROEXPANSIONAL PROPERTIES OF PAPER BY A COMPOSITE LAMINATE APPROACH

L. Salmén and A. de Ruvo*

*Swedish Forest Products
Research Laboratory
Box 5604
S-114 86 Stockholm, Sweden

L. Carlsson

The Aeronautical Research
Institute of Sweden
Box 11021
S-161 11 Bromma, Sweden

SUMMARY

In this study paper is modelled as a homogeneous structure where the cell wall layer of the fiber constitutes the main building unit. By this approach the macroscopic properties of paper are related to the properties of the constituent polymers cellulose and hemicelluloses. The effects of fiber orientation and the influence of moisture content on the elastic moduli and hygroexpansivity of paper is examined and compared to experimental data. It is advocated that the mechanical properties of paper are dependent on a mixture of transverse and longitudinal cell wall layer properties and not only on the proportion of fibers in any direction.

INTRODUCTION

The relation of paper properties to the properties of its constituents has been the subject of a great number of investigations. Most theories have focused on the network structure of paper in relating the paper properties to the properties of an average fiber of those constituting the network[1-3]. Nissan[4] on the other hand regarded paper as a hydrogen bonded solid and estimated paper properties from a Morse function for the hydrogen bond. In fact a well bonded normal sheet of paper may more appear as a homogeneous structure containing holes than a regular array of fibers. The individual fibers are bonded to the extent that failure not only occurs in the fiber to fiber bonding sites but across or along fibers. Thus in describing the properties of paper it may be more relevant to describe the composite structure of a laminate of cell walls and relate the properties to the constituent polymer components, cellulose, hemicellulose and lignin.

In this report a model is discussed where relations are given between the wood polymers and the macroscopical properties of paper by modelling the paper structure as a laminate with a cell wall layer of the fiber wall as the main building unit. Thus the paper is modelled as a macroscopically homogeneous material. By the use of this model the effect of fiber or fibrillar orientation and the influence of moisture content on the paper properties are examined and compared to experimental data.

COMPOSITE MODEL FORMULATION

The paper structure is modelled as a laminate built up of plies consisting of cell wall layer material. The fiber cell wall material is considered as a unidirectional fiber reinforced composite where the reinforcing material is the cellulose microfibrils with the matrix consisting of the hemicellulose.

A useful micromechanics approach to obtain structure-property relationships for fiber-reinforced and filled composites has been suggested by Halpin and Tsai[5]. Salmén[6] has shown that these relations also give a good prediction of the moduli for individual wood fibers. These relations are here applied in estimating the moduli of the cell wall layer material.

For the prediction of the hygroexpansion of the composite material Shapery's analysis[7] is employed.

Table 1 Properties of the constituents of a cell wall layer.

Cellulose (Dry)	Hemicellulose (Dry)
$E_{x,f} = 134$ GPa [8,9]	$E_m = 8$ GPa [10]
$E_{y,f} = 27.2$ GPa [8]	$\nu_m = 0.33$
$\nu_{xy,f} = 0.1$ [8]	$V_m = 0.45$
$G_{xy,f} = 4.4$ GPa [8]	$\rho_m = 1.5$ g/cm^3
$V_f = 0.55$	

Cell wall layer properties

The properties of a cell wall layer may suitably be calculated using the micromechanical equations as indicated by the analysis of Salmén[6]. In the present study the fibers are considered delignified. The properties of the fiber wall constituents under dry conditions used in the calculations are given in Table 1.

The hemicellulose constituting the matrix material is an hygroscopic amorphous polymer. Any amorphous cellulose present is here regarded as part of the hemicellulose matrix. The water acts as a softener for this polymer and its modulus will therefore change drastically with increasing amounts of moisture content. The modulus of hemicellulose as a function of moisture ratio is here based on the experimental data obtained by Cousins[10]. Due to the high crystallinity of the cellulose microfibril it is assumed to be unaffected by water in the RH range here considered i.e. up to 85 % RH[6].

To obtain the expansional strain in the hemicellulose the following relation is employed, based on the assumption of volume additivity and isotropic swelling.

$$e_m = (1/V_w)^{1/3} - 1 \qquad (1)$$

in which

$$V_w = 1/(1+m_m\rho_m/\rho_w) \qquad (2)$$

where e_m is the expansional strain of hemicellulose, V_w is the volume fraction of hemicellulose in the mixture of water and hemicellulose, m_m is the moisture ratio of the hemicellulose, ρ_m is the density of hemicellulose and ρ_w is the density of water.

Relation: cell wall - paper properties

In the model of paper here presented it is assumed that a well bonded paper sheet may be regarded as a multilayer composite laminate. Even if a small sectional layer in a paper is composed of fibers with a wide orientation distribution each layer of the model laminate is considered as a unidirectional composite with the same properties in the principal directions as that of a cell wall layer.

The paper sheet may then be represented by a suitable laminate. The actual angular distribution of the plies in the laminate then has to match the actual angular distribution of microfibrils in the paper sheet since the basic building block is a cell wall layer. The model also has to display a symmetrical orientation which means that for each angle-ply above the mid-plane of the laminate there has to be an identical ply at the same position below the mid-plane with the same orientation.

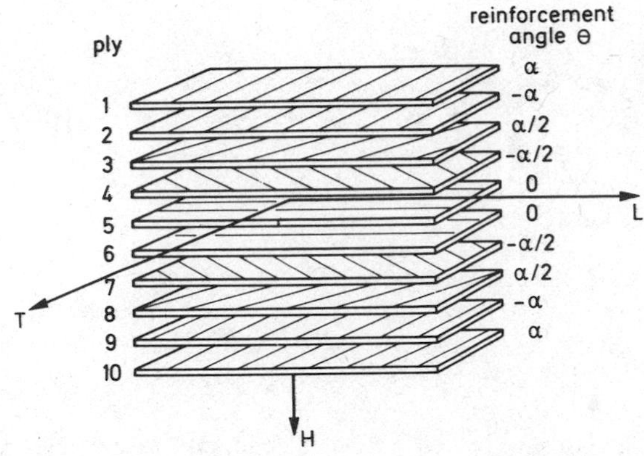

Fig. 1
Laminate model of papersheet. Oriented paper is obtained by the $[0, \pm\alpha/2, \pm\alpha]_s$ tacking sequence of the angle plies.

Fig. 1 shows the symmetrical laminate model used in this study. It is composed of 10 layers where the different anisotropy ratios are obtained by the stacking sequence $[\pm\alpha, \pm\alpha/2, 0]_s$. The thicknesses of ply 1, 2, 9 and 10 are half the thickness of the other plies in order to obtain a quasi-isotropic laminate corresponding to isotropic paper by a $[\pm 90, \pm 45^o, 0]_s$ stacking sequence. Lamination theory[11] is then used to obtain the laminate properties from the geometry, orientation and material properties of the constituent plies. The hygroexpansional strains of the laminate may be calculated by using a Duhamel-Newman form of Hooke's law incorporated into classical lamination theory[12].

ORIENTATION DEPENDENCE OF MECHANICAL PROPERTIES

The merit of a model to describe the properties of paper must be judged by its ability to predict the influence of climatic changes and structural factors on properties of paper. In the present case of estimating the paper moduli and hygroexpansivity the key functions are the dependence of fiber orientation and the influence of moisture content. These are here discussed in more detail.

With increasing fiber orientation in the longitudinal direction the modulus in this direction will increase while the modulus in the transverse direction decreases as a consequence of the change in orientation distribution of the cellulose microfibrils. The resulting anisotropy of paper can be defined in several ways. Here the anisotropy is defined as the ratio of the elastic moduli i.e. E_L/E_T of the paper. To compare papers with different fiber orientations the geometric mean value of the elastic modulus in the L- and T-directions, $\sqrt{E_L \times E_T}$ has been shown to be a useful measure of the isotropic modulus[13].

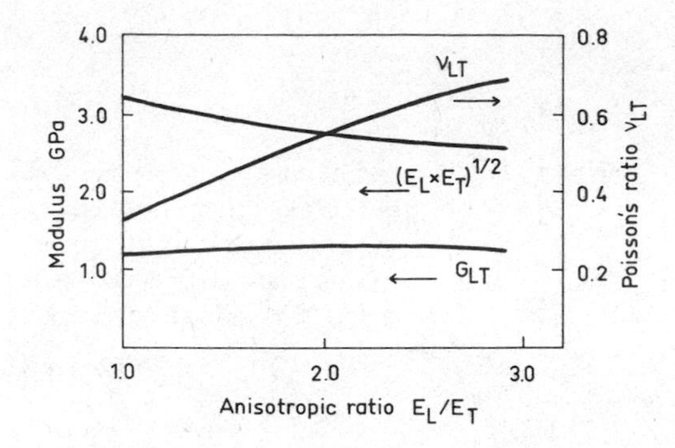

Fig. 2
Shear modulus, G_{LT}, geometric mean value of the elastic modulus, $\sqrt{E_L \times E_T}$ and Poisson's ratio ν_{LT} of the laminate model as a function of the anisotropic ratio E_L/E_T.

In Fig. 2 the effect of an increased lamellar orientation is shown for the shear modulus, the Poisson's ratio and the geometric mean value of the elastic modulus for the conditions at 60 % RH calculated from the laminate model shown in Fig. 1. Obviously the laminate model predicts a slightly decreasing geometric mean value with increasing anisotropy which is also reported by experimental findings[13].

From Fig. 2 it is also concluded that the shear modulus is approximately constant with increasing anisotropy, a behaviour which may be concluded from the experimental data of Baum et al[14]. They have shown that the in-plane shear modulus G_{LT} is a function only of the geometric mean value of the elastic moduli of the paper, $G_{LT} = 0.387 \sqrt{E_L \times E_T}$. Thus, due to the fact that the geometric mean value of the elastic moduli is nearly constant as a function of anisotropy, the shear modulus should be approximately constant with increasing anisotropy. Perkins and Mark[3] on the other hand predict a small decrease of the shear modulus G_{LT}.

The properties in any two perpendicular directions in the plane of a laminate can be obtained from appropriate transformation relations[11]. Such calculations shows that irrespective of the angle with respect to the L-T directions the laminate model gives geometric mean values and shear moduli which are approximately constant. These results agree with the experimental findings of Schrier and Verseput for the geometric mean value of anisotropic paperboar[15]. Craver and Taylor[16] have also found the shear modulus to be invariant with the direction in the plane of the sheet.

INFLUENCE OF MOISTURE

With increasing moisture content the elastic modulus of the paper decreases quite markedly. In the model this is achieved by a softening of the amorphous carbohydrates.

In order to compare the effect of moisture on the elastic modulus of different papers with different absolute moduli it is convenient to compare the effect on the relative rigidity, i.e. to compare E/E_o, where E_o in this case is the modulus of the paper when dry or at low moisture content.

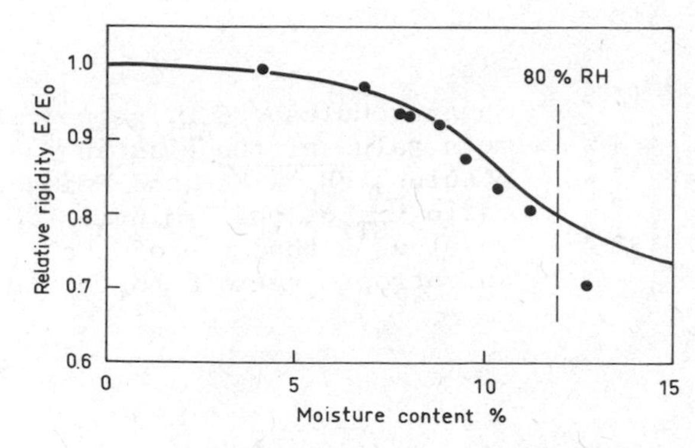

Fig. 3

The relative rigidity E/E_o, where E_o is the modulus at 0 % moisture content, as a function of moisture content, for isotropic sheets.

From the comparison of an isotropic sheet in Fig. 3, it is obvious that the agreement between the predicted and measured decrease of the modulus with increasing moisture content is quite good up to a moisture content of about 12 % corresponding to about 80 % RH. The discrepancy above this RH-level may be due either to the onset of softening of the disordered zones in the cellulose microfibrils as suggested by Salmén[6] or to a softening of the bonding areas between fibers, factors which are not included in the proposed model of the paper structure.

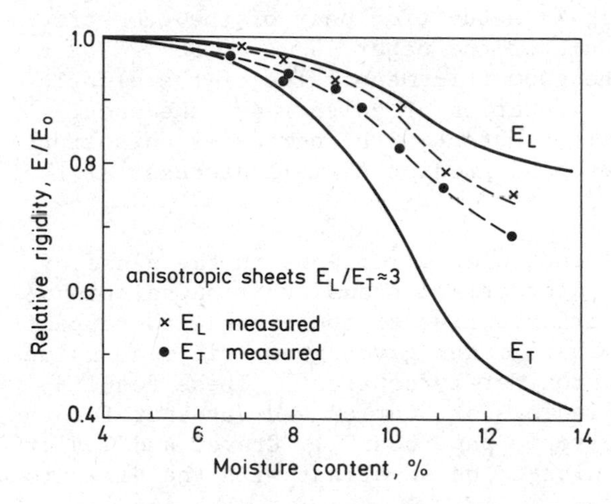

Fig. 4

The relative rigidity E/E_o as a function of moisture content for an oriented sheet with anisotropic ratio of about 3 at 9 % moisture content. E_o represents the modulus at 4 % moisture content.

It is obvious that the laminate model is successful to predict changes in elastic modulus due to moisture uptake for an isotropic sheet. However the same degree of predictive power is not obtained when the laminate is given anisotropic characteristics by changing α to 60°. In Fig. 4 it is seen that the prediction from the model is showing a difference to real sheets already at low moisture contents. Thus the change in elastic modulus in the cross-direction is less than prescribed by the laminate model, while the change in the longitudinal direction is predicted to be less than in reality. It is notable that the change in geometric mean value would be in agreement with the behaviour of a real sheet. Evidently the laminate model is not capable to properly predict the magnitude of the changes in the directional properties. However, paper does display a difference in the change of modulus in the two directions. The tendency is the same as a laminate but the effect is smaller.

HYGROEXPANSIVITY

It is generally recognised that with increasing fiber orientation the hygroexpansion decreases in the longitudinal (machine) direction, L, while it increases in the transverse (cross) direction, T. This is in line with the predictions given by the laminate model. In Fig. 5 the hygroexpansional ratio is shown versus anisotropy, E_L/E_T, for the conditions at 60 % RH.

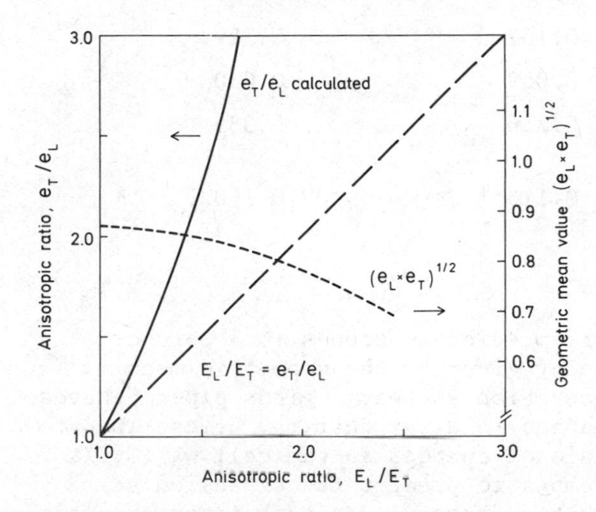

Fig. 5

The hygroexpansional ratio e_T/e_L and the geometric mean value of the hygroexpansional strain $\sqrt{e_L x e_T}$ for the laminate model as a function of the anisotropic ratio E_L/E_T.

The hygroexpansional anisotropy increases substantially more than the stiffness anisotropy. This is in contrary to experimental findings by Green[17] who has found that the T/L expansivity ratio is linearly correlated to the L/T-stiffness ratio. The same result can also be concluded from the findings of de Ruvo et al[18] in that the elastic modulus times the hygroexpansivity is a constant.

For the hygroexpansional strains the geometric mean value also decreases more markedly with the increased anisotropy than is the case for the elastic modulus.

In Table 2 the hygroexpansional coefficients, are compared for measured and calculated data for freely dried sheets. As seen the geometric mean value for an isotropic laminate calculated from the model is close to the measured properties. The discrepancies are again large when the calculations are made on the anisotropic laminate. As for elastic properties there are limitations with regard to the laminate model to describe the influence of matrix changes. These deficiencies are exposed when the model is applied on an oriented structure.

Table 2 Hygroexpansional coefficients for isotropic and anisotropic sheets, freely dried. For the anisotropic sheets the anisotropy defined as E_L/E_T is in both cases about 3.

sheets		β (%/% moisture content) measured	calculated
isotropic		0.107	0.129
anisotropic	β_L	0.059	0.030
	β_T	0.216	0.335
geometric mean value	$(\beta_L\beta_T)^{1/2}$	0.113	0.100

The laminate model is, however, predicting trends in a correct manner. The expansional tendency given by the geometric mean is properly calculated from the sorption isotherm. Thus paper behaves with regard to dimensional changes in a way that may be ascribed to the moisture uptake and dimensional changes in the cell wall. It should also be added that attempts to predict the expansion at higher moisture contents are not successful. A much lower expansion than measured is predicted. This may be partly ascribed to the inadequacy of the laminate model. However, it should also be taken into account that the applicability of the Shapery equation becomes limited when large changes in matrix properties take place.

DISCUSSION

The laminate model presented here is clearly a simplification of the complex fiber network structure of paper. Indeed, as was stated in the introduction, most theories developed to describe the mechanical properties of paper mathematically describe the stress or strain situation in a network in relation to the properties of the individual fibers. Particularly these theories relate the elastic modulus to the longitudinal fiber modulus. The laminate model shows however that the effects of moisture on both the elastic properties and the hygroexpansion are rather well predicted for moderate increases of the moisture content. It is also apparent that for anisotropic sheets the effects of moisture on the moduli are different in the two directions. Thus as exemplified by the laminate model calculations in Fig. 4 it is apparent that the mechanical properties of paper are dependent on a mixture of transverse and longitudinal cell wall layer properties and not only on the proportion of fibers in any direction. However, the laminate model gives a too simplistic view of the directional dependency of the paper properties. In the longitudinal direction the modulus is overestimated while it is underestimated in the transverse direction. In the longitudinal direction the calculated properties will approach those of the longitudinal fiber properties. It may be noticed that for single wood fibers the longitudinal fiber modulus reduces less than 20 % up to about 15 % moisture content[19]; well in agreement with similar theoretical calculations for fibers[6] as those here presented for paper. Thus in both the longitudinal and transverse directions it seems reasonable that the elastic properties will depend both on the longitudinal and transverse fiber properties. However, as the experimental data on moduli show a less anisotropy with regard to the effect of moisture sensitivity than actually predicted, this implies that the directional properties are also dependent on the quantity of longitudinal fiber elements in the various directions. These features may be taken into account in a mosaic model which joins together the structural elements of longitudinal fibers with laminated bonding zones[20].

Another point where the laminate model does not work satisfactory is the behaviour at higher moisture levels. The physical reason has been attributed to a softening of either bond sites or disordered microfibril zones. To describe these phenomena it is necessary to either incorporate a relaxion of the bond stiffness or to assume a reduction of the length of the cellulose reinforcing element in the cell wall due to a softening of disordered regions[6]. The latter mechanism could be incorporated into a laminate model and allow this to describe also the behaviour at these higher moisture levels.

EXPERIMENTAL

The measurements were performed on laboratory sheets made from a commercial bleached sulphate pulp. The pulp was beaten to 20°SR. Sheets of 100 g/m^2 were made either as isotropic on a Finnish sheet machine or anisotropic on a sheet former, Formette Dynamique[21].

The sheets were pressed to a density of 770 to 850 g/m^2. The drying was made at 50 % RH, 20°C, either freely dried or restrained dried in drying frames[13]. All measurements were performed at equilibrium conditions at 20°C with samples approaching the condition while adsorbing moisture. In order to avoid hysterisis effects all samples were predried at about 15 % RH, 20°C.

REFERENCES

1. Cox, H.L., Brit. J. Appl. Phys. 3, (1952) 72.

2. Page, D.H., Seth, R.S. and de Grace, J.H., Tappi 62(1979), 9:99.

3. Perkins, R.W. and Mark, R.E., Preprints of the Conference on "The role of fundamental research in paper-making", Cambridge, Sept. 1981 session 5, paper 3.

4. Nissan, A.H., Trans. Faraday Soc. 53, (1957) 700.

5. Halpin, J.C. and Kardos, J.L., Polym. Eng. Sci. 16(1976), 5:344

6. Salmén, N.L., Dr. dissertation, Stockholm, 1982.

7. Shapery, R.A., J. Composite Mat. 2(1968), 3:380.

8. Mark, R.E., "Cell wall mechanics of tracheids", Yale Univ. Press, 1967.

9. Sakurada, I., Nukushina, Y. and Ito, T., J. Polym. Sci. 57(1962), 651

10. Cousins, W.J., Wood Sci. Techn. 12(1978)161.

11. Tsai, S.W. and Hahn, H.T., "Introduction to composite materials", Technomic, Westpoint, Conn. 1980.

12. Whitney, J.M. and Ashton, J.E., AIAA Journal 9(1971)1708.

13. Htun, M. and Fellers, C., Tappi 65(1982), 4:113.

14. Baum, G.A., Brennan, D.C. and Habeber, C.C., Tappi 64(1981), 8:97.

15. Schrier, B.H. and Verseput, H.W., Tappi 50 (1967), 3:114.

16. Craver, J.K. and Taylor, D.L., Tappi 48(1965), 3:142.

17. Green, C., Ind. Eng. Chem. Prod. Res. Dev. 20(1981), 1:151.

18. de Ruvo, A., Lundberg, R., Martin-Löf, S. and Söremark, C., In "The fundamental properties of paper related to its uses", B.P. & B.I.F., (1976), p. 785.

19. Kersavage, P.C., Wood Fiber 5 (1973), 2:105.

20. Rigdahl, M., Andersson, H., Westerlind, B. and Hollmark, H., Fibre Sci. Techn., in print.

21. Sauret, G., Trinh, H.J., and Lefebvre, G., Das Papier 23(1969), 1:8.

VISCOELASTIC PROPERTIES OF COPOLYMERS:

RELATIONS WITH STRUCTURE

J.P. Montfort, J. Lebez, G. Marin, and Ph. Monge

Laboratoire de Physique des Matériaux Industriels
I.U.R.S. - Avenue Louis Sallenave
64000 Pau (France)

INTRODUCTION

Thermoplastic copolymers are of great importance in industry because of the large range of their physical properties. This comes from the nature of the components, their ratio, the sequence lengths (randoms, blocks, grafts copolymers), their solid state morphology (crystalline or amorphous) and the multiphasic or monophasic structure.

The viscoelastic properties in the melt have not been widely studied[1,2] excepted for blocks copolymers. We investigated commercial samples, with various structural properties and discussed their effect on the rheological behaviour. The peculiarity of copolymers with respect to homopolymers and blends will be emphasized.

STRUCTURAL ANALYSIS

The various samples we used have the following caracteristics:

(i) ethylene-vinylacetate copolymers (EVA) : random distribution of monomeric sequences, crystalline and monophase morphology, 5 to 45 % vinylacetate content.

(ii) ethylene-propylene copolymers (EP) : at low ethylene content (<5%) we have random (EPS) or block (EPB) copolymers with respectively one and two crystalline phases and a dispersed microstructure ; at high ethylene content (∿50%), the copolymer is a monophasic amorphous system, with random distribution (EPR).

(iii) triblock styrene-butadiene copolymers (SBS) , with 30 to 40 % styrene content, have an amorphous and well-ordered domain

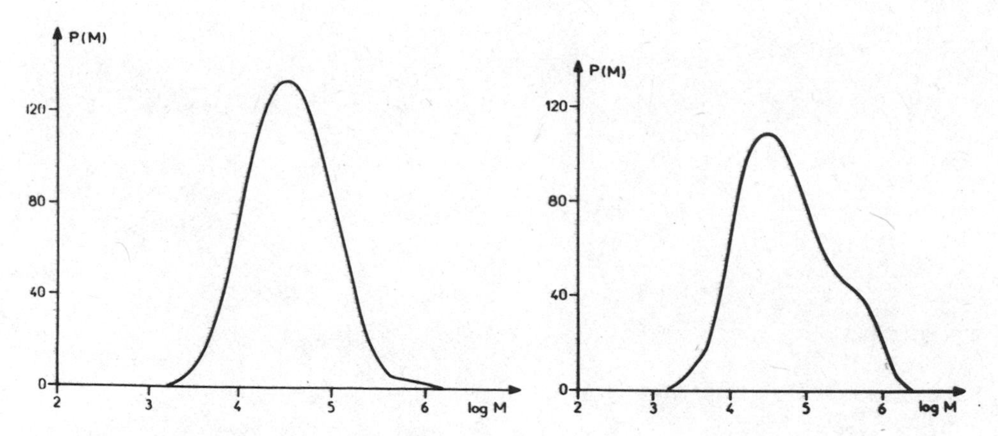

Fig. 1. Different chain length distributions obtained from Gel. Permeation Chromatography for two EVA copolymers (% VA = 9)

structure with a morphology depending on the processing conditions.

Various technical analysis have been used to get the copolymer structure :

(i) with Gel Permeation Chromatography, the chain length distribution has been specified. G.P.C. has shown that EVA copolymers obtained from various processes have either a symetrical distribution or an important tail of long chains as illustrated in Fig. 1.

(ii) with Differential scanning Calorimetry, glass transitions and melting temperatures have been measured and compared with those of the pure components. For amorphous random copolymers (EPR) the glass transition temperature is between the Tg's of the pure components and for an amorphous block copolymer (SBS), there are two Tg corresponding to the homopolymer components. For crystalline copolymers, the melting temperature of the ethylene phase in EVA decreases with increasing vinylacetate content as indicated in Fig. 2 . For EPB, the two melting peaks are broader and brought together because of the effect of one component on the ordered structure of the other.

(iii) Infrared Spectroscopy and Nuclear Magnetic Resonance allows to titrate the short and long branchings and have shown that the increase of vinylacetate component decreases the number of short branchings and increases the number of long branchings.

(iiii) From electron micrography, we see the effect of chemical composition on the domain structure of SBS[3]. The polybutadiene matrix contains cylindrical polystyrene domains, and other authors have shown that this structure has a big effect on the mechanical properties [4,5,6,7].

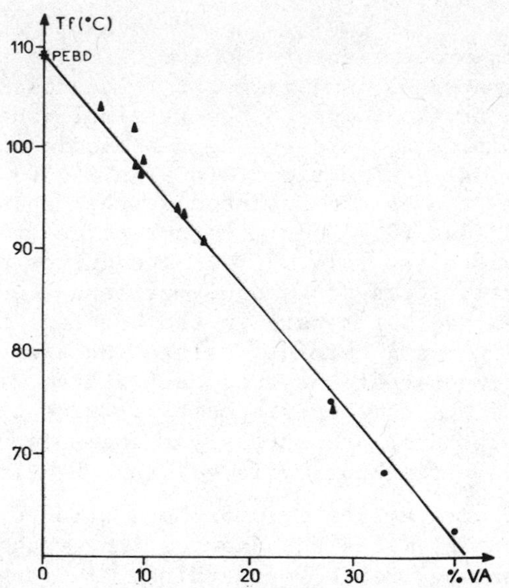

Fig. 2. Decrease of melting temperature for EVA copolymers versus vinylacetate content.

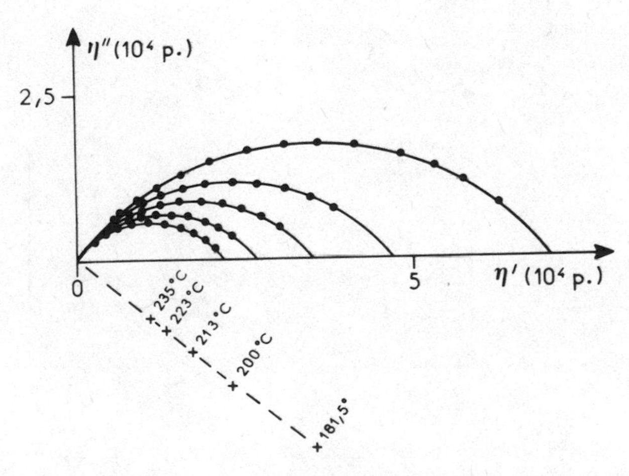

Fig. 3. Representation of complex viscosity η* at different temperatures for a random copolymer EPS (% PE = 3.4)

VISCOELASTIC PROPERTIES

The dynamics properties of copolymers melts are investigated at various temperatures, in a range of frequencies varying from 10^{-3} to 10^{2} Hz, corresponding to the terminal zone of relaxation spectrum. The samples are undergoing a periodic shearing at small amplitude sinusoïdal deformations. (CONTRAVES and INSTRON 3250 Rheometers). The results are reported as the dependence of the storage (G') and loss (G") moduli versus circular frequency (ω), or the complex viscosity ($\eta^* = G^*/j\omega$). The curves presented in Fig. 3 are caracteristics of homopolymers and also random copolymers. We get then two parameters in the terminal zone : the zero-shear viscosity η_o and a terminal relaxation time τ_o , ($\tau_o = \omega_m^{-1}$ where ω_m is the inverse of the frequency at the maximum of the imaginary part η'' of the complex viscosity). These two parameters depend on the structural properties (weight-molecular weight M_w , polydispersity P , branching ratio g) and the temperature.

In general, crystalline or amorphous structures don't influence the behaviour of melts because the long-range effects are dominated by the entanglements, cross-links or long branchings. For polydisperse materials, τ_o is increased by the long-chain tail and we have observed large variations of τ_o for ethylene-vinylacetate copolymers obtained from various processes and offering different weight distributions.

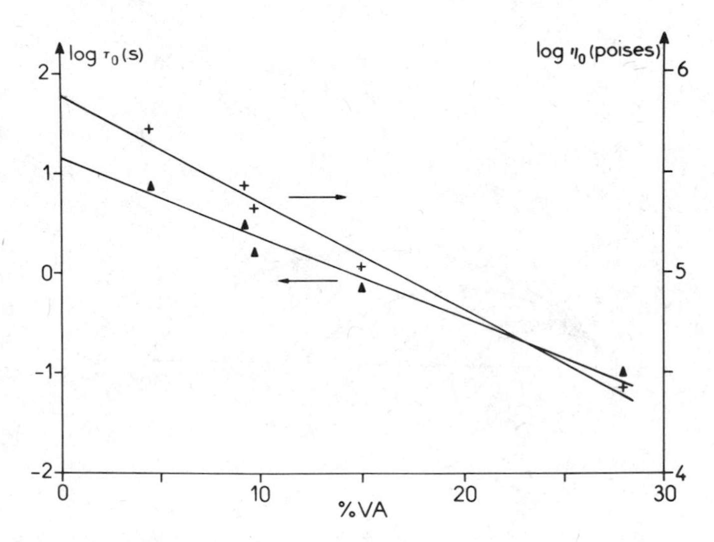

Fig. 4. Variations of the terminal relaxation time τ_o (▲) and zero-shear viscosity (+) versus vinylacetate content, for $gM_w \simeq 47.000$ and T = 135°C

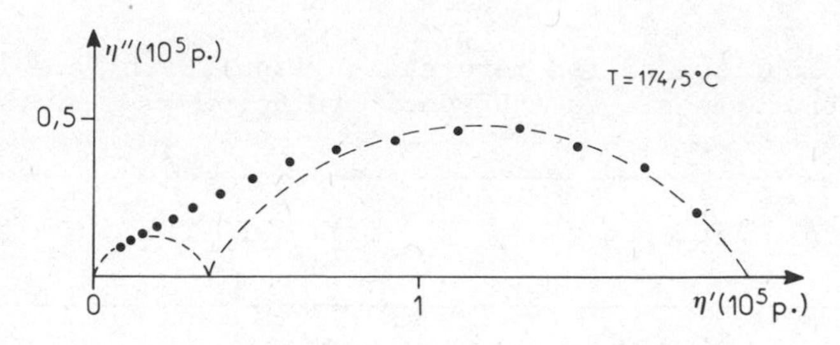

Fig. 5. Example of double terminal relaxation for a SBS triblock copolymer (40 % styrene content)

The increase of long branching for EVA, when the amount of vinylacetate content increased, implies a decrease of η_o and τ_o , as shown in Fig. 4 , because, at same weight, the free-volume fraction of a branched chain is higher than for a linear chain. Short branchings have no effect on the terminal relaxation.

As homopolymers, random copolymers (EVA , EPS and EPR) exhibit a single relaxation region in the terminal zone, with values of η_o and τ_o between these of the pure components with same length. The sequencies of every species aren't long enough to have their own relaxation. The triblocks copolymers SBS exhibit, for temperatures higher than the range 150 - 170 °C , two relaxation domains, as we can see in Fig. 5 . We can assign the high frequency part of the spectrum to the relaxation of the central sequence of polybutadiene and the low frequency part to the whole chain. The polystyrene domains remain in the melt, acting as physical crosslinks and leading to high values of viscosity and compliance.

We have studied the effect of the multiphase structure comparing, for two given species (ethylene and propylene) at same composition (10 % ethylene content), on the viscoelastic properties of three blends : E-P , EPS-E and EPB-E . We have related the rheological behaviour to their microstructure. The results listed in the Table 1 below show in particular that the viscosity of the EPB-E blend is higher than that of every component. This seems to indicate a better interfacial adherence induced by an interpenetration of the added polyethylene chains and large sequences of polyethylene of the EPB copolymer.

Table 1. Comparison of the zero-shear viscosity η_o at T = 200 °C for three blends A-B specified below

Component A	Component B	η_o (A)	η_o (B)	η_o (blend AB)
Polypropylene	PEHD	25.000	8.900	20.000
EPS (% PE = 3.4)	PEHD	47.500	8.900	45.000
EPB (% PE = 4)	PEHD	44.000	8.900	65.000

So, the viscoelastic properties of copolymers melts, in the terminal zone of relaxation spectrum are very dependent on structure. These various properties allow to specify materials fitted with a wide range of industrial needs.

REFERENCES

1 . S. K. Ahuja, Structure and rheology of styrene methyl methacrylate copolymers, Rheologica Acta , 18 : 374 (1979)
2 . D. S. Pearson and W. W. Graessley , Elastic properties of well-characterized ethylene-propylene copolymers networks , Macromolecules , 13 : 1001 (1980)
3 . S. L. Aggarwal , Structure and Properties of blok polymers and multiphase polymer systems : an overview of present stature and future potential, Polymer , 17 : 938 (1976)
4 . G. Kraus and J. T. Gruver , Properties of random and block copolymers of butadiene and styrene, II : melt flow , Journal of Applied Polymer Science, 11 : 2121 (1967)
5 . K. R. Arnold and D. J. Meier , A rheological characterization of SBS block copolymers , Journal of Applied Polymer Science , 14 : 427 (1970)
6 . G. V. Vinogradov , Viscoelastic properties of butadiene-styrene block copolymers, Rheologica Acta , 17 : 258 (1978)
7 . G. Kraus and K. W. Rollman , Effects of domain and molecular orientations on the mechanical properties of a styrene-butadiene block polymer , Journal of Macromolecular Science, Physics , B 17 : 407 (1980)

RELATIONSHIP BETWEEN PTITSYN-EIZNER λ PARAMETER AND CONFORMATIONAL TRANSITION

I. Katime*, P. Gutierrez Cabañas** and C. Ramiro**

Dpto. Química Física
*Universidad País Vasco, Bilbao
and **Universidad Complutense, Madrid-3

INTRODUCTION

Eizner and Ptitsyn[1] have treated the intrinsic viscosity of semi-rigid macromolecules in the light of the intrinsic viscosity equation of Peterlin[2,3], and the "wormlike" chain of Kratky and Porod[4]. In this communication we study the behavior of the Ptitsyn-Eizner's λ parameter for several PMMA fractions, of very narrow distributions, in several solvents in a wide range of temperature.

EXPERIMENTAL PROCEDURES AND RESULTS

The syndiotactic PMMA samples were obtained by anionic polymerization in THF solution initiated with butyllithium[5]. The polymer was reprecipitated from benzene solution with isopropylalcohol. The fractions were obtained by gradual precipitation of the polymer from the benzene solution with ethanol.

The heterotactic PMMA sample was prepared by free radical polymerization in benzene solution, initiated with 1,2-azobisisobutyronitrile. It was divided into 12 fractions by fractional precipitation.

Molecular weights of the fractions were determined by both light scattering and osmometry. Polydispersities of the fractions used in this work lie between 1.1 to 1.3.

Viscometry measurements were carried out in a Ubbelhode suspended level viscometer adapted for dilution in situ.

DISCUSSION

According to the theory of Eizner and Ptitsyn, the intrinsic viscosity of the semi-rigid molecules is given by Equation (1):

$$(\eta)=\frac{2^{3/2}\Phi_o(b^3/M_o)\quad N.\chi(N,\lambda)}{(45(2\pi/3)^{1/2}/32(3-\sqrt{2}))(b/\lambda r_o)+(1/\lambda^{3/2})\Phi_o(\lambda,N)N^{1/2}} \tag{1}$$

where (η) is the intrinsic viscosity in dl. per gram, ϕ is the Flory coefficient whose limiting value is 2.87×10^{21} mol^{-1} at high molecular weight, N is the degree of polymerization, r_o is the hydrodynamic radius of the monomer unit, M_o is molecular weight, and $\lambda=a/b$ (a is the persistence length introduced by Kratky-Porod for wormlike chain[4] and b is the length of the monomer unit), is the length of the monomer unit along the direction of the main chain. The function $\chi(N,\lambda)$ is a geometric factor defined as

$$\chi(N,\lambda) = \frac{3<s>^2}{b^2\lambda N} = 1 - \frac{3}{(N/\lambda)^3}\left(\frac{N}{\lambda}\right)^2 - 2\left(\frac{N}{\lambda} - 1+e^{N/\lambda}\right) \tag{2}$$

or

$$\chi(z) = 1 - (3/z^3)\left[z^2 - 2(Z - 1 + e^{-z})\right]$$

where $z = N/\lambda$. This factor is easy to calculate.

The hydrodynamic function, $\Phi_0(N,\lambda)$ is given by:

$$\Phi(N,\lambda) = \frac{15(\pi/3)^{1/2}}{4(3-\sqrt{2})}\frac{1}{\sqrt{\lambda}N^{5/2}}\frac{(k^2+K-Nk-2N)\Psi(x)}{(x-1+\exp(-x))^{1/2}} +$$

$$+\frac{((N^2/2)-2k^2+N)\Psi(x)}{(x-1+\exp(-x))^{1/2}} \tag{3}$$

where $\Psi(x)=0.427+0.573\left[(45x^2+156x+214-54(x+4)e^{-x}+2e^{-3x})/27(x-1+e^{-x})^2\right]$ and $x=k/\lambda$. The hydrodynamic function can be calculated with a computer. We have written a program in Basic for a CBM Commodore computer for the calculation of function $\Phi_0(N,\lambda)$. In Table 1 are some values of the hydrodynamic function.

The intrinsic viscosity may be written in the form

$$(\eta) = \Phi\frac{<R^2>_o^{3/2}}{M} = \Phi\frac{6^{3/2}<s^2>_o^{3/2}}{M} \tag{4}$$

Table 1. Some values of hydrodynamic function $\Phi(N,\lambda)$ from the Eizner-Ptitsyn theory

	λ					
N	1	5	10	15	20	100
10	-	-	0.349	0.460	0.550	1.324
40	0.508	1.067	1.566	1.944	2.261	5.125
100	0.700	1.095	1.525	1.878	2.181	4.954
250	0.814	1.056	1.349	1.611	1.846	4.164
400	0.854	1.039	1.270	1.480	1.674	3.697
1.000	-	1.018	1.159	-	1.413	-
2.500	-	1.008	1.093	-	1.248	-
4.000	-	1.005	1.071	-	1.192	-
10.000	-	1.002	1.043	-	1.117	-

where $\langle R^2\rangle_o$ and $\langle s^2\rangle_o$ are the mean-square and end-to-end distance and mean-square radius of gyration, respectively, of a chain in the theta state. According to the Eizner-Ptitsyn theory, the Flory coefficient is defined by

$$\Phi = \frac{\Phi_o}{\left[\Phi(\lambda,N)+(45(2\pi/3)^{1/2}/32(3-\sqrt{2}))(b/r_o)(\lambda/N)^{1/2}\right]\chi^{1/2}(N,\lambda)} \quad (5)$$

Combination of Equations (1), (2) and (5) gives

$$2^{3/2}\Phi_o\frac{b^3}{M_o}\frac{N}{(\eta)}(N,\lambda)=\left(\frac{2\pi}{3}\right)^{1/2}\frac{45}{32(3-\sqrt{2})}\frac{b}{\lambda r_o}+\frac{1}{\lambda^{3/2}}\Phi(\lambda,N)N^{1/2} \quad (6)$$

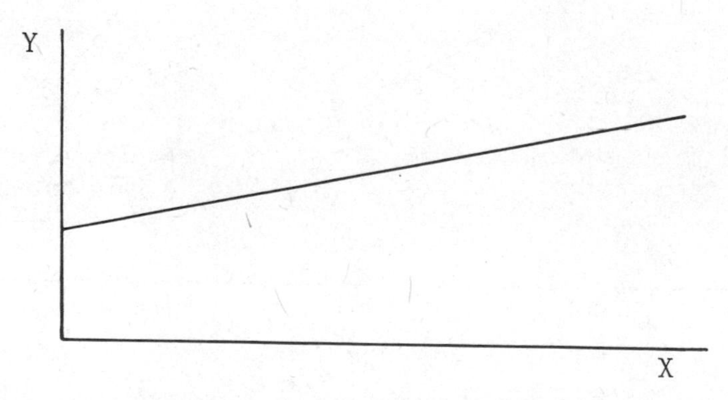

Fig. 1. Determination of Ptitsyn-Eizner parameter from intrinsic viscosity data for the system PMMA/Ethyl acetate.

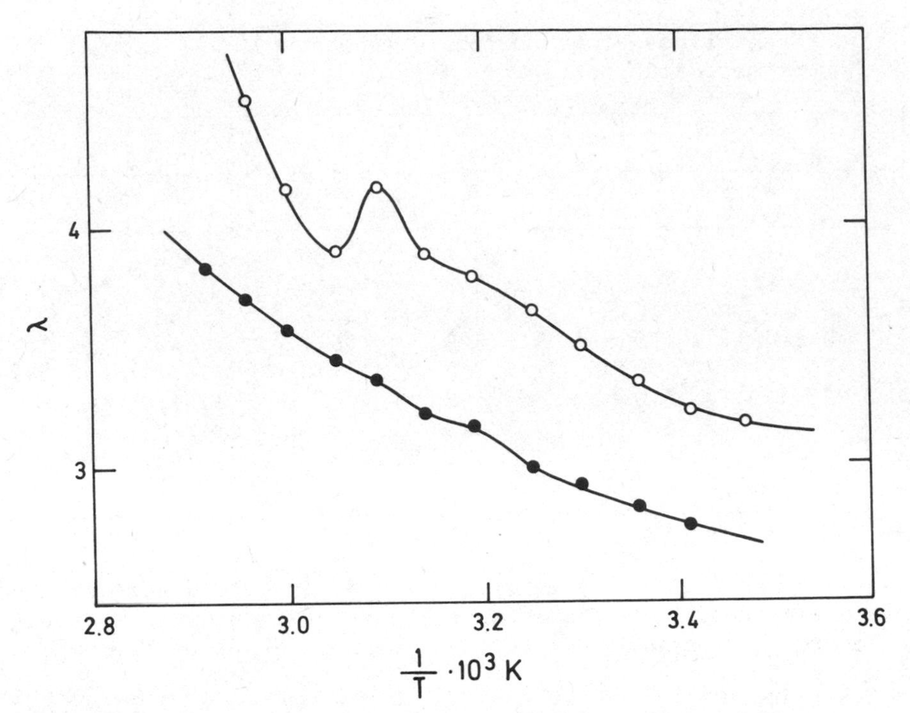

Fig. 2. Plot of λ parameter as a function of 1/T for ethyl acetate (o) and butyl acetate (●).

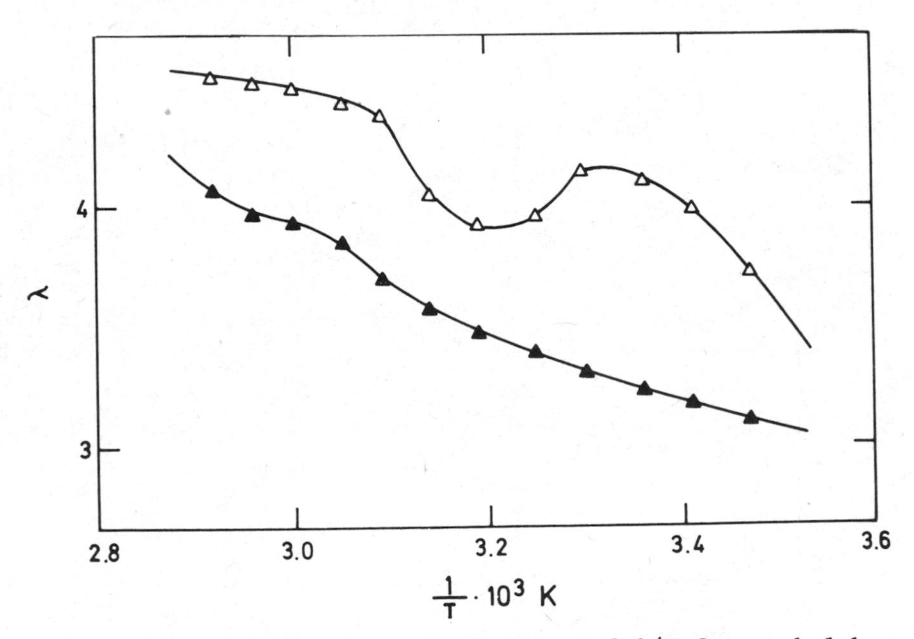

Fig. 3. Plot λ parameter as a function of 1/T for methyl benzoate (▲) and ethyl benzoate (△).

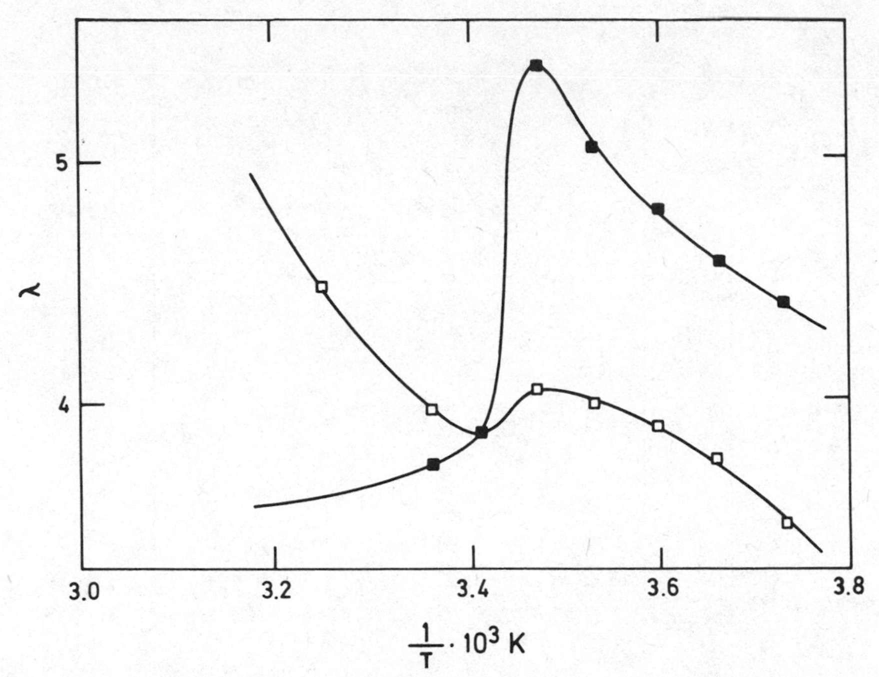

Fig. 4. Variation of λ with temperature in acetone (□) and cyclohexanone (■).

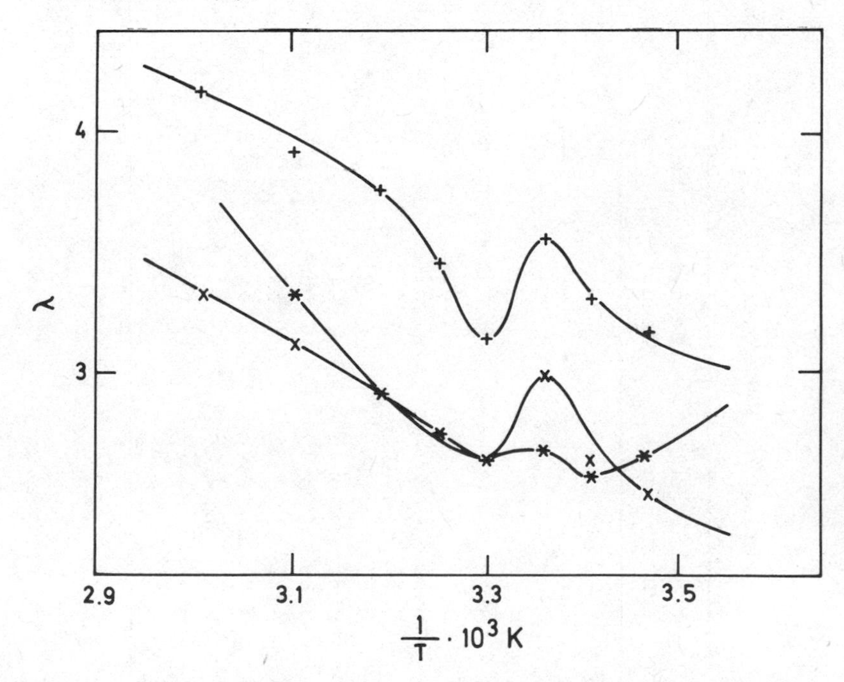

Fig. 5. Plot of λ parameter as a function of 1/T for chlorobenzene (+), o-dichlorobenzene (*) and m-dichlorobenzene (x).

Table 2. λ Parameter for PMMA as a Function of Solvent and Temperature

T°C	Ethyl acet.	Butyl acet.	Methyl benz.	Ethyl benz.	Acetone	Cyclo- hexan.	Chlorob.	1,2-Di- chlorob.	1,3-Di- chlorob.
-5	-	-	-	-	3,29	4,86	-	-	-
0	-	-	-	-	3,74	4,61	-	-	-
5	-	-	-	-	3,88	4,85	-	-	-
10	-	-	-	-	3,98	5,11	-	-	-
15	3,14	-	1,74	0,26	4,06	5,50	3,16	2,48	2,64
20	3,19	2,67	1,80	0,30	3,86	3,87	3,32	2,62	2,54
25	3,33	2,78	1,80	0,40	3,96	3,73	3,57	2,98	2,65
30	3,56	2,88	1,96	0,54	-	-	3,13	2,61	2,61
35	3,65	2,94	2,16	0,69	4,61	-	3,46	2,74	2,74
40	3,82	3,15	2,24	0,75	-	-	3,77	2,91	2,91
45	3,89	3,19	2,50	1,16	-	-	3,94	3,11	3,11
50	4,22	3,35	2,56	1,33	-	-	4,21	3,34	3,33
55	3,92	3,46	2,59	1,57	-	-	-	-	-
60	4,20	3,58	2,75	1,63	-	-	-	-	-
65	4,62	3,73	2,86	1,80	-	-	-	-	-
70	-	3,85	3,06	1,91	-	-	-	-	-

For PMMA, this equation becomes:

$$1.302.10^{-6}\chi(N,\lambda)\,\frac{M}{(\eta)} = 3.237.10^{-9}(\lambda,r_o)^{1+\lambda^{-3/2}}\Phi(\lambda,M)M^{1/2} \qquad (7)$$

A plot of $Y=(1.302.10^{-6}\chi(N,\lambda)M(\eta)^{-1}$ against $\Phi(\lambda,M)M^{1/2}$ should be a straight line.

In Figure 1 we can see an example for this system PMMA/ethyl acetate.

Figures 2-5 and Table 2 show the variation of the parameter as a function of temperature for several solvents (ethyl and butyl acetate, methyl and ethyl benzoate, acetone and cyclohexanone, mono- and dichlorobenzenes (ortho and meta isomers). It can be seen that a sharp decrease of λ is always observed in a determined temperature range. This sudden change corresponds to a conformational transition in PMMA[7-9].

A comparative study of the conformational transition for PMMA in different solvents has been made. It has been found that a dependence exists between the magnitude of the transition and the solvent nature, particularly solvent polarity.

Finally, it is interesting to point out that the polymer tacticity is a very important factor in the temperature as transition occurs.

REFERENCES

1. Yu. E. Eizner and O. B. Ptitsyn, Visokomol.Soedin., 4:1725 (1962).
2. A. Peterlin, J.Polymer Sci., 5:473 (1950).
3. A. Peterlin, J.Chem.Phys., 33:1979 (1960).
4. O. Kratky and G. Porod, Rec.Trav.Chim., 68:1106 (1949).
5. I. Katime and C. Ramiro Vera, Anal.Qui.(Madrid), 68:9 (1972).
6. P. J. Flory and T. G. Fox, Jr., J.Am.Chem.Soc., 73:1904 (1951).
7. I. Katime and J. R. Quintana, Polymer Bull., 6:455 (1982).
8. I. Katime, P. Gutierrez Cabañas, J. R. Ochoa, M. Garay, and C. Ramiro Vera, J.Polymer, (in press).
9. I. Katime, X. Ibarra, M. Garay, and R. Valenciano, European Polymer J., 17:509 (1981).

PROPERTIES OF PMMA STEREOCOMPLEX IN SOLID STATE AND SOLUTION

I. Katime and J. R. Quintana

Departamento de Química Física
Universidad del País Vasco
Apartado 644, Bilbao, España

INTRODUCTION

On mixing solutions of isotactic and syndiotactic poly(methyl methacrylate) (PMMA) a stereocomplex is formed by intermolecular association. This phenomenon occurs in most solvents but to different degrees[1-5].

Stereocomplexes are produced both in bulk and in solution. In this sense, the solvents can be classified into three groups: the strongly complexing type A, the weakly complexing type B, and the non-complexing type C[1]. It has been found that the stereocomplex has a crystalline structure as x-ray analysis shows[2].

The optimum ratio of iso- and syndiotactic PMMA in the stereocomplex is for most investigators 1/2, but ratios of 1/1 and 1/1.5 are also reported.

The influence of solvent on the stereocomplex formation has also been studied by viscometry, ultracentrifugation and NMR, and in very few cases by DTA. The stereocomplex formation heat in several solvents has been studied by Biros et al.,[7]. Boer et al.,[8] have studied the melting behavior of the stereocomplex obtained in acetone and DMF.

In this communication we have studied the influence of solvent quality on the stereocomplex formation. For this we have chosen the three binary cosolvent mixtures: acetonitrile/carbon tetrachloride, acetonitrile/butyl chloride and butyl chloride/carbon tetrachloride.

EXPERIMENTAL

Isotactic PMMA was prepared by polymerization of MMA (Fluka purum) in toluene at low temperature with phenyl magnesium bromide as initiator. Syndiotactic PMMA was obtained by anionic polymerization in THF solution initiated with diphenyl sodium.

The weight average molecular weight, $\bar{M}_w$, of both samples were determined by light scattering. The values obtained for it- and st-PMMA were $6.6 \cdot 10^5$ and $1.7 \cdot 10^5$, respectively.

The stereocomplex was prepared by mixing, in a ratio 1/2, 0.2 g/dl it- and st-PMMA solution in the binary mixture. The solutions mixed were kept in thermostatically controlled bath at 30.0°C for one day. The precipitate was separated from the solution in a centrifuge with an acceleration of 22,000 g and then dried in a vacuum at room temperature.

The calorimetric measurements were carried out in a differential thermal analyzer Mettler TA 2000 system with a heating rate of 5°C/min. The data were calibrated in absolute units by comparison with the specific heat of a sample of indium.

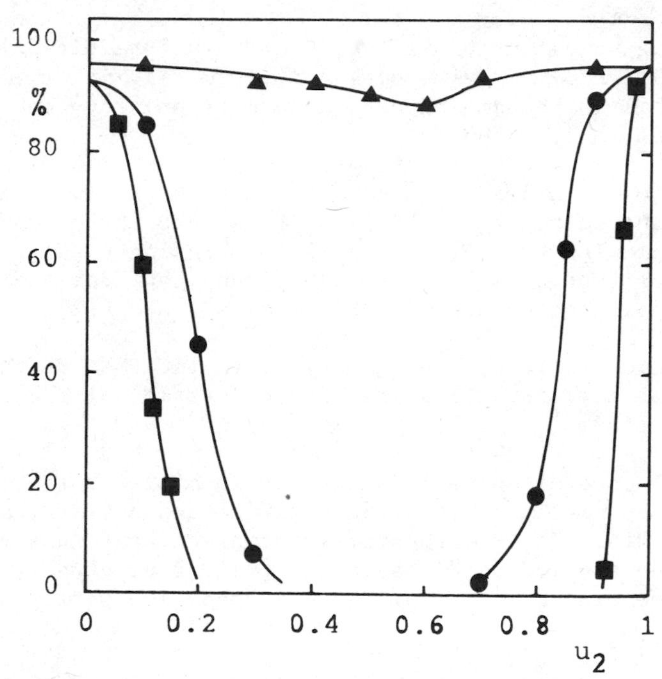

Fig. 1. Stereocomplex percentage as a function of the composition of the binary mixtures acetonitrile (1)/carbon tetrachloride (2) (■), acetonitrile (1)/butyl chloride (2) (●) and butyl chloride (1)/carbon tetrachloride (2) (▲). (u_2, volume fraction of component (2)).

Laser light scattering measurements (λ=632 nm) were made at 298 K using a modified FICA photometer. The instrument was calibrated using benzene as standard.

RESULTS AND DISCUSSION

The thermogrammes of the stereocomplex cannot be performed at all solvent binary mixture compositions due to the absence of precipitate at certain compositions, as can be seen in Figure 1.

This can be explained by taking into account the loss of the "strongly complexing" character of the cosolvent binary mixture.

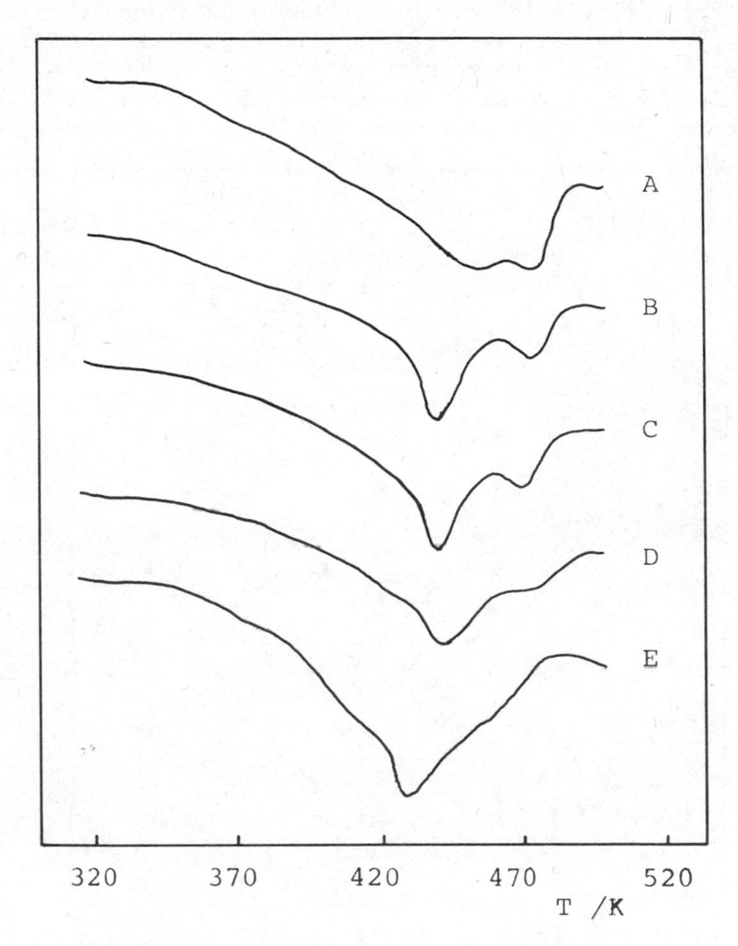

Fig. 2. Thermogrammes of different samples of stereocomplex obtained at different compositions of the binary mixture acetonitrile (1)/carbon tetrachloride (2) (u_2=0.95 (A), 0.96 (B), 0.97 (C), 0.99 (D), 1 (E)).

This behavior prevents the precipitation of the stereocomplex by centrifugation, due to the absence of clusters of high molecular weight.

In the thermogrammes we have found two endothermic peaks in the temperature range 155-207°C. These peaks are characteristic of two different melting temperatures (Tm_1 and Tm_2). Challa et al.,[9] have shown that both peaks are due to two different crystallizations of the stereocomplex, and not to the melting of crystalline it- or st-PMMA.

In Figures 2 and 3 can be seen the solvent influence on both melting temperatures and the ratio of the peak areas, which are related to the variation of the fusion heat, ΔH_f, and therefore to crystallinity. On going away from both pure liquids the first endothermic peak decreases with respect to the second one. This effect is accompanied by a decrease of stereocomplex precipitate. We think that the second peak (Tm_2) corresponds to the stereocomplex formation

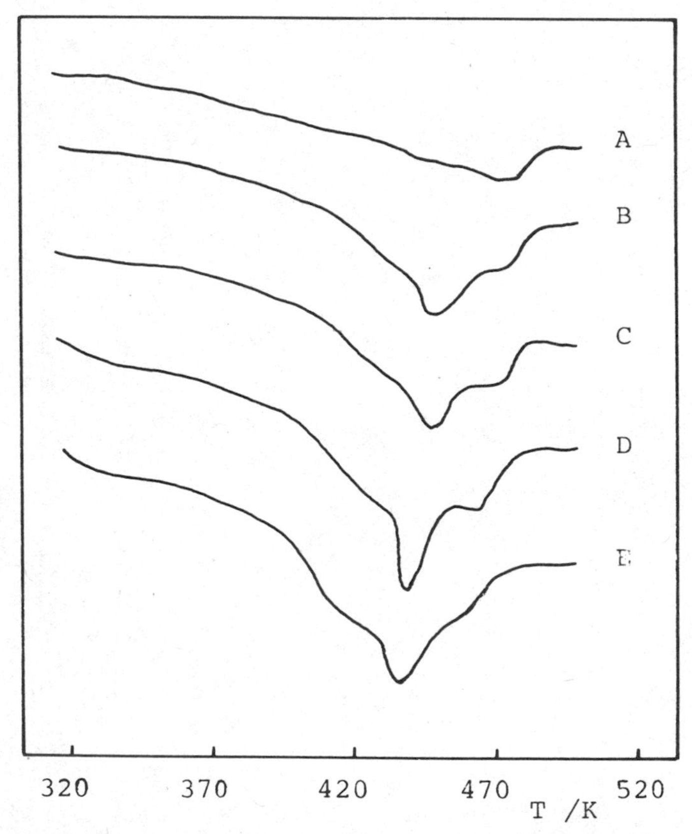

Fig. 3. Thermogrammes of different samples of stereocomplex obtained at different compositions of the binary mixture acetonitrile (1)/butyl chloride (2) (u_2=0.80 (A), 0.85 (B), 0.90 (C), 0.95 (D), 1 (E)).

and the first one (Tm_1) to aggregates of stereocomplex ("clusters"). As these are stereocomplex interactions (intermolecular forces of first class), it is reasonable that the melting temperature of the former is greater than the melting temperature of the latter. On the other hand, the second peak is almost independent of cosolvent power of the binary mixture. This means that, at least, at binary mixture compositions in which stereocomplex precipitate is obtained, this crystallization is not strongly affected by solvent power.

This behavior can be explained taking into account the excess Gibbs free energy, G^E. In fact, we have found that G^E (butyl chloride/carbon tetrachloride) < G^E (acetonitrile/butyl chloride) < G^E (acetonitrile/carbon tetrachloride). This implies that solvent-solvent interactions increase as G^E decreases, and therefore polymer-solvent interactions decrease. This interaction balance between the four components is the the controlling force of the process.

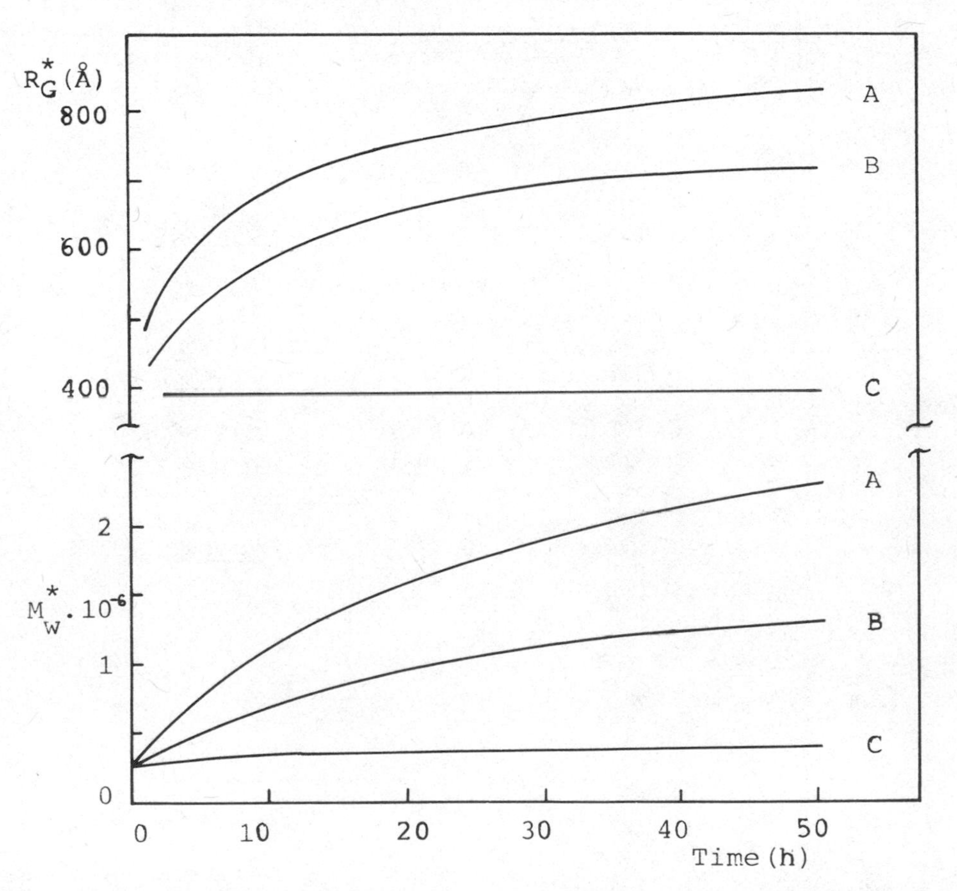

Fig. 4. Variation of M_w^* and R_G^* as a function of time at different compositions of the binary mixture acetonitrile (1)/butyl chloride (2) (u_2=0.90 (A), 0.80 (B) and 0.50 (C)).

The melting temperatures, Tm_1 and Tm_2, are also influenced by solvent power of the binary mixtures. We have found that Tm_1 ranges between the Tm_1 values obtained in pure solvents. Whereas Tm_2 increases as cosolvent power increases. This can be explained taking into account that as the macromolecular coil increases, the ester groups of it-PMMA and the α-methyl groups of st-PMMA can interact more easily and to a greater extent. On the other hand, as cosolvent power increases the cluster formation, decreases. The effect of this is that the quantity of stereocomplex obtained decreases, as the stereocomplex-solvent interaction increases.

We have measured by light scattering the formation of PMMA stereocomplex as a function of time and solvent mixture composition for the mixture acetonitrile/butyl chloride.

In Figure 4 can be seen the variation of apparent molecular weight and the apparent radius of gyration of PMMA stereocomplex as a function of time at u_2=90, 80 and 50% butyl chloride. The obtained results confirm that as the cosolvent binary mixture power increases the formation of PMMA stereocomplex decreases, that is to say, the same results that we have obtained by DTA.

REFERENCES

1. G. Challa, A. de Boer, and Y. Y. Tan, Int.J.Polym.Mat., 4:239 (1976).
2. A. M. Liquori, G. Anzuino, V. M. Coiro, M. D'Alagni, P. de Santis and M. Savino, Nature, 206:358 (1965).
3. H. Z. Liu and K. J. Liu, Macromolecules, 1:157 (1968).
4. W. Borchard, M. Pyrlik, and G. Rehage, Die Makromol.Chem., 145: 169 (1971).
5. E. J. Vorenkamp and G. Challa, Polymer, 22:1705 (1981).
6. F. Bosscher, D. Keekstra, and G. Challa, Polymer, 22:124 (1981).
7. J. Biros, Z. Masa, and J. Pouchly, Eur.Polym.J., 10:629 (1974).
8. A. de Boer and G. Challa, Polymer, 17:633 (1976).
9. E. J. Vorenkamp, F. Bosscher, and G. Challa, Polymer, 20:59 (1979).

THE EFFECT OF PLASTICIZATION ON THE STRUCTURE OF HIGHLY CROSSLINKED MATERIALS, AND THE FORMATION OF THE INTERPHASE IN TWO-PHASE SYSTEMS (Abstract)

Basil Kefalas

Chair of Mechanics A

Athens National Technical University

Bright-field transmission electron micrographs of epoxy resins reveal a non-uniform structure. High crosslink density regions are loosely connected with lower molecular weight interstitial regions during the latter stages of the curing process. This two-phase system is affected by the amount of plasticizer during the network formation in the curing process.

Stress-strain dynamic measurements on epoxy resins specimens, with various amounts of plasticizer, reveal the effect of plasticization on the structure. Also the relation to the adhesion developed at the interface in two-phase polymer-polymer composites is examined by stress-strain dynamic measurements.

GENERALIZED ANISOTROPIC DESCRIPTIONS FOR THE DYNAMIC MECHANICAL PROPERTIES OF CRYSTALLINE POLYMERS

James C. Seferis, Andri E. Elia, and Alan R. Wedgewood

Polymeric Composites Laboratory
Department of Chemical Engineering
University of Washington
Seattle, Washington, 98195

INTRODUCTION

In earlier work, a two-phase aggregate model has been developed, which successfully described the anisotropic properties of uniaxially oriented semi-crystalline polymer films of isotactic polypropylene[1-7]. In this methodology, the polymer was considered to be composed of two phases, crystalline and non-crystalline. Each phase has orientation-independent intrinsic properties, which are simply a manifestation of anisotropic properties of structural elements that make up as an aggregate the particular phase of the polymer. The two-phase model provided fundamental relations for the anisotropic compliance properties of polymers in terms of such fundamental parameters as the unoriented phase compliance intrinsics, the orientation distribution of each of the phases, and the volume fraction crystallinity.

The significance of this modelling methodology is twofold. First, the model can be utilized to describe the anisotropic properties of a variety of uniaxially oriented polymer films with a unified treatment. Second, for the particular polymer under consideration compliances of the pure unoriented phases can be extracted from corresponding experimental measurements on oriented films. It is important to note here that the compliances of the pure unoriented phases can also be extracted from measurements on unoriented films as a function of crystallinity. The values of the unoriented phase compliances extracted from measurements on unoriented films should be equal to the values extracted from measurements on oriented films, if the fundamental assumptions of orientation independent structural element and uniform distribution of stress are valid[8]. This is found to be true when the model is applied to the case of dynamic mechanical behavior of isotactic polypropylene[8]. Also, from a number of model reductions, a two-parameter form is applicable for polypropylene[1,6,7]. For the case of dynamic mechanical behavior of linear polyethylene however, the compliances of the pure unoriented non-crystalline phase extracted from measurements on oriented films and those extracted from measurements on unoriented films are equal below -70°C, while they diverge above -70°C[8]. To account for this, a conformational modification is made to the model, by incorporating conformation dependent real and imaginary compliances for the unoriented non-crystalline phase[8].

In the present study, the two-phase model is applied in describing dynamic mechanical compliance properties of isotactic polypropylene as a function of frequency. The unoriented crystalline and non-crystalline phase real and imaginary compliances are extracted, thus providing the frequency dependence of the real and imaginary compliances of the pure unoriented non-crystalline phase to our earlier reported results at 11 Hz[6,7]. To these results a time-temperature superposition is applied and master curves are constructed for the unoriented non-crystalline compliance properties. From these curves, our corresponding earlier sonic results are also predicted [1]. Finally, the conformationally modified two-phase model is successfully applied in predicting anisotropic dynamic mechanical properties of uniaxially oriented polyethylene films in a wide temperature range[8].

THEORETICAL BACKGROUND

The schematic representation of our modelling methodology on microscopic and macroscopic levels is depicted in Figure(1). The structural element shown may be representative of the crystalline or non-crystalline phases, and the angle θ between the chain direction of the structural element and the machine draw direction defines the microscopic orientation. For mechanical properties, quantitative description of the orientation distribution may be achieved by the second and fourth moment orientation functions f_p and g_p defined as[1-7,9]:

$$f_p = 1/2\ [3\langle\cos^2\theta\rangle-1] \qquad ; \qquad g_p = 1/4\ [5\langle\cos^4\theta\rangle-1]$$

Here, the subscript p stands for crystalline (p=c) or non-crystalline phase (p=a). These functions f_p, g_p assume the value 0 for random orientation of the structural elements, 1 for perfect orientation, and -1/2 and -1/4 respectively when the chain direction is perpendicular to the draw direction. The angle Θ of Figure(1) defines the location of experimentation or bulk property

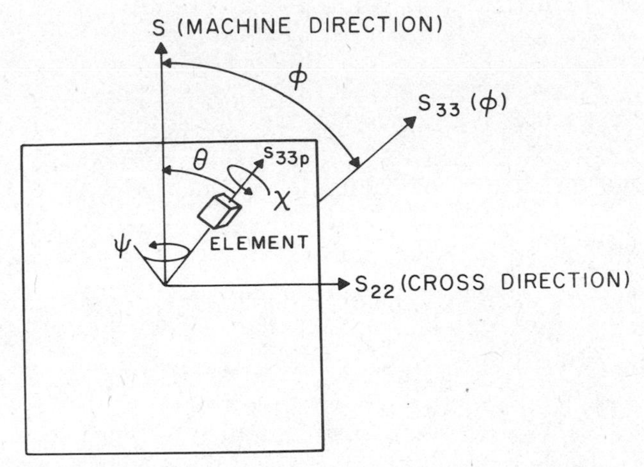

Figure(1): Definition of orientation and intrinsic compliance properties of the structural elements. The structural element shown may be crystalline (p=c) or non-crystalline (p=a). The angle θ defines the orientation of the chain direction with respect to the machine draw direction. The angle ϕ defines the location of the structural element on the film with respect to the draw direction. The compliance s_{33} is along the chain direction.

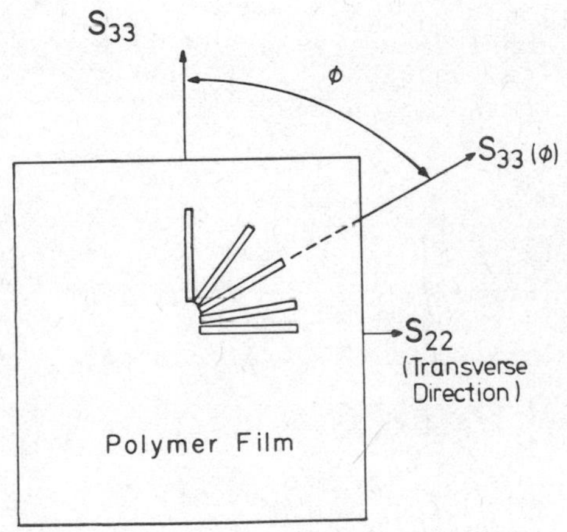

Figure(2): Macroscopic anuglar description. The angle ϕ defines the angle at which strips of film are cut with respect to the draw direction.

measurement with respect to the film draw direction. This is also indicated in Figure(2).

For a uniaxially oriented (transversely isotropic) film, the aggregate averaged compliances along the principal directions, $\langle S_{33} \rangle$ (chain direction), $\langle S_{11} \rangle$ and $\langle S_{22} \rangle$ (transverse directions), are given in terms of the structural elements compliance constants[1-8]:

$$\langle S_{11} \rangle_p = \langle S_{22} \rangle_p \tag{1}$$

$$\langle S_{22} \rangle_p = A_p - 1/2(B_p + 5/4C_p)f_p + 3/8C_p g_p \tag{2}$$

$$\langle S_{33} \rangle_p = A_p + B_p f_p + C_p g_p \tag{3}$$

with $$S_{ij} = V_c \langle S_{ij} \rangle_c + V_a \langle S_{ij} \rangle_a$$

Here, the parameters A_p, B_p, and C_p, are defined as linear combinations of the compliance descriptors of the structural elements denoted by s_{ijp}[1-8]:

$$A_p = 2/30[3(s_{11}+s_{22}+s_{33})+2(s_{13}+s_{23}+s_{12})+(s_{44}+s_{55}+s_{66})]_p \tag{4}$$

$$B_p = 5/30[-3(s_{11}+s_{22})+2(2s_{13}+s_{23}-s_{12})+(2s_{44}+s_{55}-s_{66})]_p \tag{5}$$

$$C_p = 3/30[3(s_{11}+s_{22})+8s_{33}-2(4s_{13}+4s_{23}-s_{12})-(2s_{44}+2s_{55}-s_{66})]_p \tag{6}$$

while V_c and V_a are the volume fractions of the crystalline and non-crystalline phases.

On a macroscopic level, compliance measurements are performed on strips of film cut at various angles ϕ to the draw direction, as shown in Figure(2). Quantitatively, the angular dependence is introduced by loading functions, $F(\Phi)$, and $G(\Phi)$, defined by analogy to the microscopic orientation functions[1,7]:

$$F(\Phi) = 1/2\ [3\cos^2\Phi - 1] \quad ; \quad G(\Phi) = 1/4\ [5\cos^4\Phi - 1]$$

These functions $F(\Phi)$, $G(\Phi)$ also assume the value 1 for a cut at 0° to the draw direction, and -1/2 and -1/4 respectively for a cut at 90° to the draw direction. Thus, the bulk film compliance $S_{33}(\phi)$, measured at an angle ϕ to the draw direction, is given by:

$$S_{33}(\Phi) = A + DF(\Phi) + EG(\Phi) \tag{7}$$

where D and E are linear combinations of the compliance descriptors of the structural elements, and equation(7) is analogous to the microscopic description of equation(3) [1-7].

Combining the microscopic and macroscopic descriptions, a set of equivalent expressions are obtained for the macroscopic constants, in terms of orientation functions, volume fraction crystallinities, and intrinsic compliance properties of the crystalline and non-crystalline phases:

$$A = A_c V_c + A_a V_a \tag{8}$$

$$E + (7/5)D = -[Q_c A_c V_c f_c + Q_a A_a V_a f_a] \tag{9}$$

$$E + D = [2.5(1-Q_c)A_c f_c + 1/2(5Q_c-7)A_c g_c]V_c + [2.5(1-Q_a)A_a f_a + 1/2(5Q_a-7)A_a g_a]V_a \tag{10}$$

In the above expressions the parameter Q_p, the phase material index, has been incorporated, which is a combination of the unoriented phase intrinsic, A_p, and the phase intrinsics B_p and C_p, if s_{33p} is assumed to be negligible:

$$Q_p = -[((7/5)B_p + C_p)/A_p] \tag{11}$$

The values of Q_p fall in the range $-1 < Q_p < 2$ [8]. Two advantages of adapting the material index formulation are its weak dependence on temperature and that the crystalline and non-crystalline material indices were found to be approximately equal, $Q_c = Q_a$ [8].

As can be seen using equations (7 and 8) the compliance descriptors of the unoriented crystalline (A_c) and the unoriented non-crystalline phase (A_a) may be extracted from measurements on oriented film samples. However, when equations (7-10) are also applied to unoriented film samples the orientation functions f_p and g_p are zero, thus making the parameters D and E vanish. This results in the following expression:

$$S^{un} = A^{un} = A_c^{un} V_c + A_a^{un} V_a \tag{12}$$

which should be identical to equation(8) describing unoriented properties extracted from oriented films. Thus, using equation(12) or equations (7 and 8), the compliances of the pure unoriented phases can be extracted from measurements on oriented and unoriented films, and compared for consistency of the model.

When the two-phase model is applied to isotactic polypropylene, it has been found experimentally that the model is indeed consistent and $Q_c = Q_a = 1.4$. Thus, a simple two-parameter form is sufficient for adequate description of the compliance properties, as had also been shown by earlier empirical results[6,7,10]. The expression for polypropylene simply reduces to[6,7]:

$$S_{33}(\Phi) = A_c[1-f_cF(\Phi)]V_c + A_a[1-f_aF(\Phi)]V_a \quad (13)$$

To analyze the measured dynamic mechanical properties, the static two parameter model equation is extended through the application of the correspondence principle, as has been previously demonstrated[4,6,7]. Thus, equation(13) was extended to provide the two-phase reduced model constitutive equations for dynamic mechanical compliance properties, both storage (′) and loss (″)[12].

$$S'_{33}(\Phi,\omega,T) = A'_c(T)[1-f_cF(\Phi)]V_c + A'_a(\omega,T)[1-f_aF(\Phi)]V_a \quad (14a)$$

$$S''_{33}(\Phi,\omega,T) = A''_c(T)[1-f_cF(\Phi)]V_c + A''_a(\omega,T)[1-f_aF(\Phi)]V_a \quad (14b)$$

In the above expressions, the frequency and temperature dependence is illustrated. It should be noted that application of these expressions to experimental data showed that the unoriented crystalline phase compliance was dependent only on temperature, while the unoriented non-crystalline phase compliance was dependent on both temperature and frequency[12,14].

The experimentally verified form of the model has been successfully used to predict compliance properties of isotactic polypropylene in stress relaxation as well[12]. However, it should be recalled that polypropylene is simply a special case of the generalized treatment, $Q_c = Q_a = 1.4$[8]. Also, as mentioned earlier, the unoriented phase compliances extracted from measurements on oriented films are equal to those extracted from measurements on unoriented films[8]. This point indicates that no conformational considerations are necessary for this polymer, and orientation independent intrinsic properties may be associated with the structural elements[8]. However, in generalizing the application of the modelling methodology to other polymeric systems, the material index value and conformational considerations must always be examined, and specific forms of the model investigated[8].

For linear polyethylene, the material index is found to be 0.5 (i.e. $Q_c = Q_a = 0.5$) [8]. This makes the angular dependence of the compliance properties more complex than for polypropylene. But along the draw direction ($\Phi=0^o$) the loading functions $F(\Phi)$ and $G(\Phi)$ become equal to 1, and a simple form of the model results, which is quite comparable to the two-parameter model (equation(13))[8]:

$$S_{33}(0^o) = A_cV_c[1-h_c] + A_aV_a[1-h_a]$$

where h_p replaces f_p, and has the following dependence on Q_p:

$$h_p = -(1/2)[5(1-Q_p)f_p + (5Q_p-7)g_p$$

Equation(15) can be expanded through the correspondence principle into real and imaginary components, as was done for equation(13). Note that the parameter h_p is equal to f_p only when $Q_p = 1.4$. Again, this is the case for isotactic polypropylene, hence the validity of equation(13). But for all other values of Q_p, the parameter h_p is equal to a linear combination of f_p and g_p, according to equation(16).

Finally, a modification is made to the two-phase model, to account for the fact that the compliance of the unoriented non-crystalline phase of linear polyethylene is conformation dependent[8]. This was indicated by the fact that the unoriented non-crystalline phase compliance extracted from measurements on oriented films is not equal to that extracted from measurements on unoriented films. A conformation dependent compliance descriptor for the unoriented non-crystalline phase, Z_t, is defined by the following expression[8]:

$$A^*_a = A^{un}_a + Z_t(q_{tttt} - 1) \tag{17}$$

where A^*_a is the compliance of the unoriented non-crystalline phase extracted from measurements on oriented films, A^{un}_a is the compliance of the unoriented noncrystalline phase extracted from measurements on unoriented films, and q_{tttt} is the relative conformation ratio for sequence of four or more chain segments in the trans conformation with respect to the unoriented material[8]. Values of q_{tttt} are obtained from infrared measurements[8,15]. Values of Z_t can be obtained by equating equations (8 and 12), eliminating the crystalline contributions and consolidating the non-crystalline values obtained from oriented and unoriented film measurements.

EXPERIMENTAL

The experimental application of the model in describing dynamic mechanical properties of polyethylene and polypropylene is summarized in the next section.

The instrument used to generate the data was the Rheovibron model DDV-IIb, modified for improvement in reproducibility[13].

The structural features of the polypropylene samples used in this study are listed in Table(1). The polypropylene data were measured at the frequencies of 3.5, 11, and 110 Hz[7,14]. The structural features of the oriented polyethylene samples are listed in Table(2). Also, measurements on unoriented film samples of different crystallinities were performed. The polyethylene data were measured at only one frequency, 11Hz [8].

Table 1

Characterization Data for Isotactic Polypropylene Film

Film	Drawing Temp °C	Elongation %	Density g/cm^3	v_c	f_c	f_a
A-1	110	12	.9030	.604	.38	-.06
A-3	110	73	.9035	.610	.41	.11
A-5	110	246	.9050	.630	.77	.28
A-6	110	447	.9030	.604	.90	.51
B-2	135	188	.9110	.704	.45	.18

Table 2

Structural and Morphological Parameters for The Linear Polyethylene Films

Film	v_c	f_c	g_c	f_a	g_a	q_{tttt}
H1	0.630	0.63	0.50	0.17	0.19	1.27
H2	0.641	0.83	0.74	0.21	0.23	1.36
H3	0.656	0.87	0.80	0.24	0.27	1.48
H4	0.663	0.93	0.88	0.28	0.33	1.53
H5	0.671	0.94	0.90	0.28	0.30	1.51

RESULTS

A) Isotactic Polypropylene

Extraction of Intrinsic Parameters

The compliance descriptors A'_c, A''_c, A'_a, and A''_a were determined at each frequency from equations (14a, 14b) at various temperatures[12,14]:

$$A'_c = [0.12 + 0.0001T(^oC)] \times 10^{-10} \ cm^2/dyne \quad (18)$$

$$A''_c = [0.15 + 0.0003T(^oC)] \times 10^{-12} \ cm^2/dyne \quad (19)$$

The values A'_a and A''_a may be used to determine E'_a, E''_a, and $\tan\delta_a$, the storage and loss moduli and the tangent of the phase angle of the unoriented, non-crystalline phase:

$$E'_a = A'_a/(A_a'^2+A''^2_a) \quad ; \quad E''_a = A''_a/(A'^2_a+A''^2_a)$$

$$\tan\delta_a = A_a''/A'_a$$

The compliances of the unoriented crystalline and non-crystalline phases are shown in Figure(3) as functions of temperature. These results illustrate the frequency dependence of the non-crystalline phase. On the other hand, as expected, the crystalline phase is frequency independent (Figure(3)).

Time-Temperature Effects

Figure(4a) shows the real compliance of the unoriented non-crystalline phase plotted as a function of frequency at various temperatures[12,14]. A simple time-temperature superposition is used to shift these curves along the log frequency axis[12,14]. This shift results into the master curves of Figure(4b). No vertical shift is necessary for these data. The importance of the master curves of Figure(4b) is that they may be used to predict the storage compliance of any isotactic polypropylene sample. As an example of the predictive ability of the master curves, our earlier sonic results extracted for the non-crystalline phase compliance at 10 KHz and 20°C are successfully predicted, $A_a = 6.3 \times 10^{-11}$ cm^2/dyne, as indicated by the dashed lines in Figure(4b).

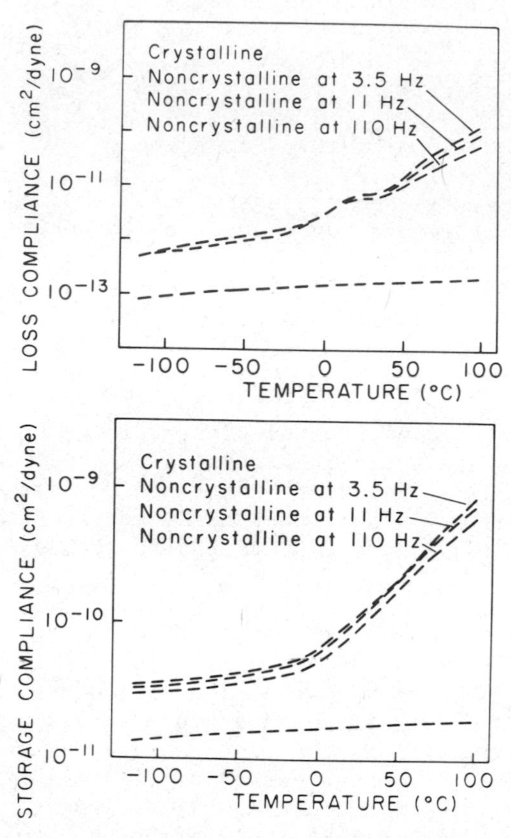

Figure(3): Dynamic mechanical properties of isotactic polypropylene. The top three curves are the properties of the pure unoriented non-crystalline phase at 3.5, 11, and 110 Hz respectively. The bottom curve represents the pure unoriented crystalline phase. a) Loss (imaginary) compliance as a function of temperature. b) Storage (real) compliance as a function of temperature.

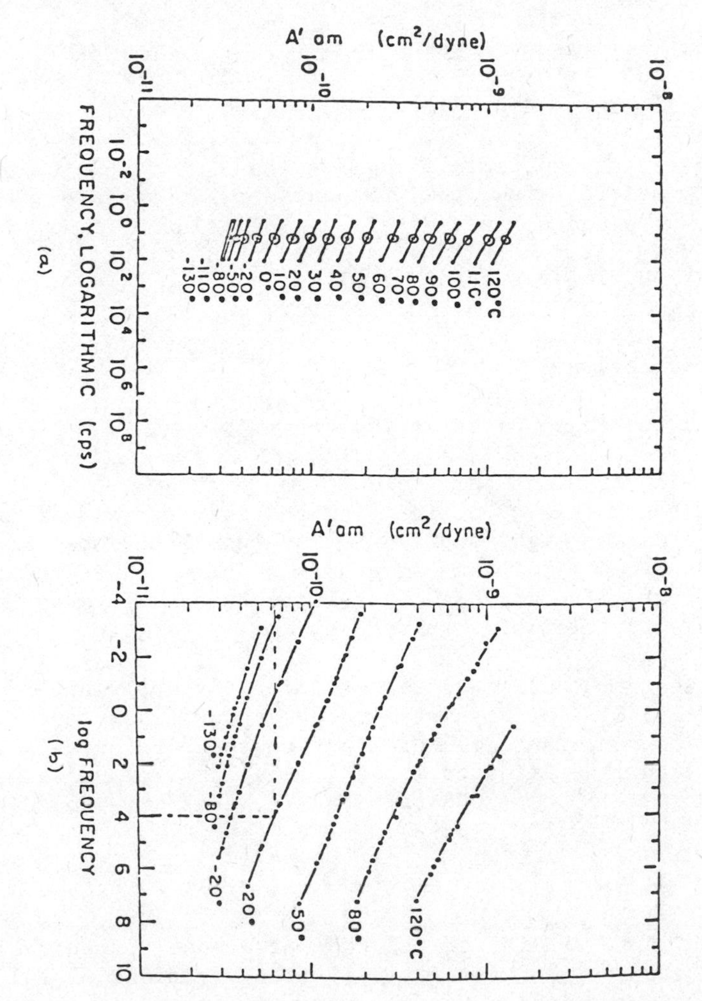

Figure (4): a) Loss compliance of the unoriented non-crystalline phase as a function of frequency, at various temperatures. b) Master curves obtained from shifting the curves of part a) along the log frequency axis, according to the time-temperature superposition principle.

Figure(5) shows a plot of the shift factor log a_T, used to shift the real compliance curves, to the reference temperature of 20^oC, obtained by the equation:

$$\log a_T = (\log f - \log f_o) = \log(f/f_o) .$$

The dashed line is the prediction from the WLF equation

$$\log a_T = -[a_1(T-T_r)]/[a_2 + (T-T_r)] ,$$

using the universal constants $a_1 = 17.4$ and $a_2 = 51.6$. The solid line is the prediction from the WLF equation, using data fitted values of $a_1=51.4$ and $a_2=237$. Either set of values, for a_1 and a_2 however, does not result into adequate predictions below the glass transition temperature of 12^oC [12,14].

B) Linear Polyethylene

Evaluation of the Phase Material Indices

The unoriented phase tensile compliance values, A'_c, A''_c, A'_a, and A''_a, were determined by a linear regression analysis of the compliance values of eight unoriented samples. Equations (8, 18, and 19) were used[8]. The real and imaginary compliances of the pure unoriented crystalline and non-crystalline phases are shown in Figure(6) as a function of temperature.

The values of Q_c and Q_a were determined by applying equations (7-10)[8]. The data used in these equations were the angular tensile compliance measurements on the five uniaxially oriented polyethylene films of Table(2)[8]. The average values of Q_c and Q_a in the -150 to -70^oC temperature range were found to be equal, $Q_c = Q_a = 0.50 \pm 0.15$ [8].

Predictions

The predictive ability of the two phase model is shown in Figure(7). The calculated (solid line) from equations (7-10) and experimentally determined (data points) storage and loss compliances are plotted as a function of temperature. The agreement in the -150 to -70^oC temperature range is excellent. Above -70^oC, however, deviations as large as 25% are observed. This can be partly attributed to the fact that, in this unmodified form, the model does not account for conformational changes in the non-crystalline phase of the sample caused by the drawing process[8].

With the incorporation of the conformation dependent

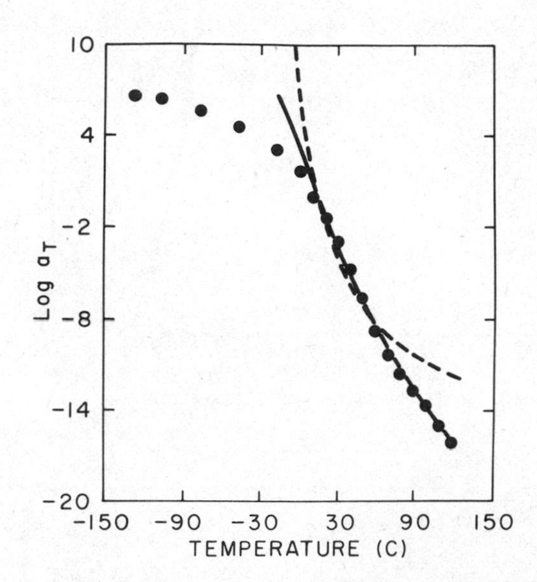

Figure(5): The shift factor $\log a_T$ as a function of temperature (dots). The dashed line is the prediction of the WLF equation when the universal constants $a_1=17.4$ and $a_2=51.6$ are used. The solid line is the prediction of the WLF equation when the empirical values $a_1=51.4$ and $a_2=237$ are used.

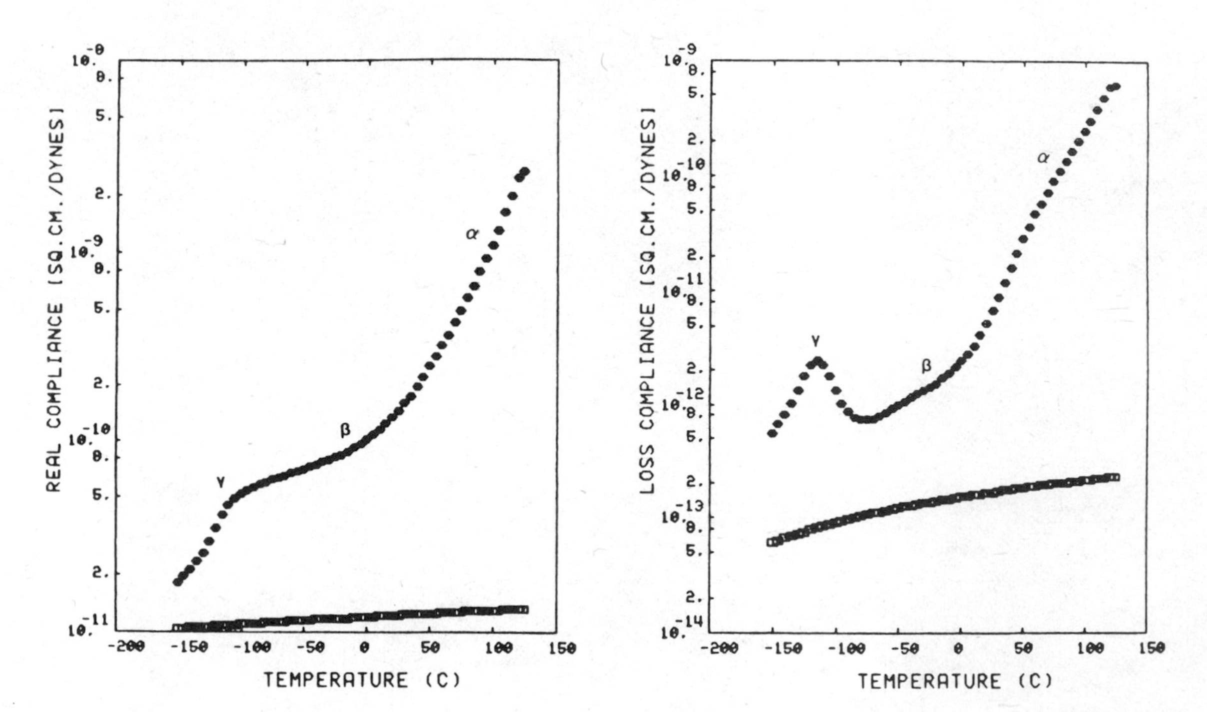

Figure (6): Dynamic mechanical properties of linear polyethylene. The top curve (hexagons) represents the pure unoriented non-crystalline phase and the bottom curve (squares) represents the pure unoriented crystalline phase. a) Storage (real) compliance as a function of temperature. b) Loss imaginary compliance as a function of temperature.

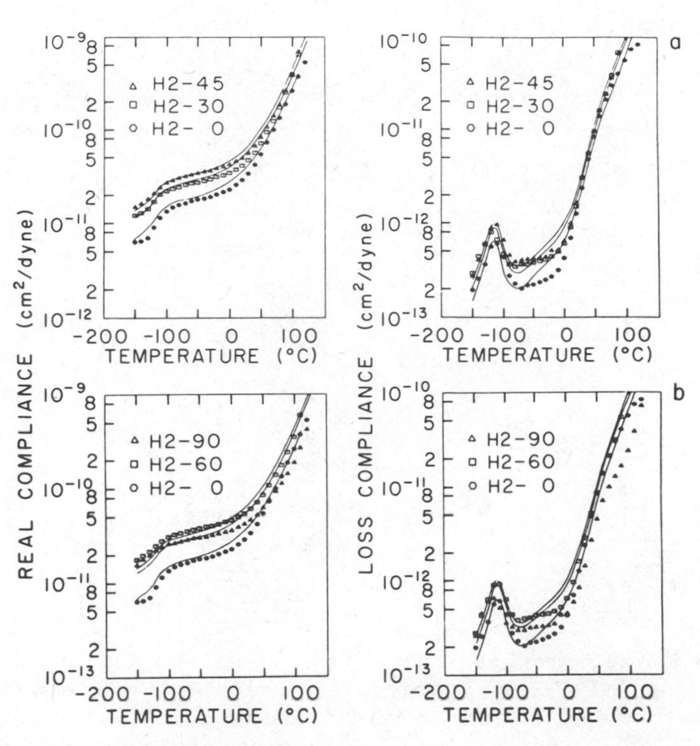

Figure(7): Storage and Loss compliance as a function of temperature, for film H2 of linear polyethylene. The solid lines are the predictions of the two phase model. a) Results for 0° (hexagons), 30° (squares), and 45° (triangles) to the draw direction. b) Results for 0° (hexagons), 60° (squares), and 90° (triangles) to the draw direction.

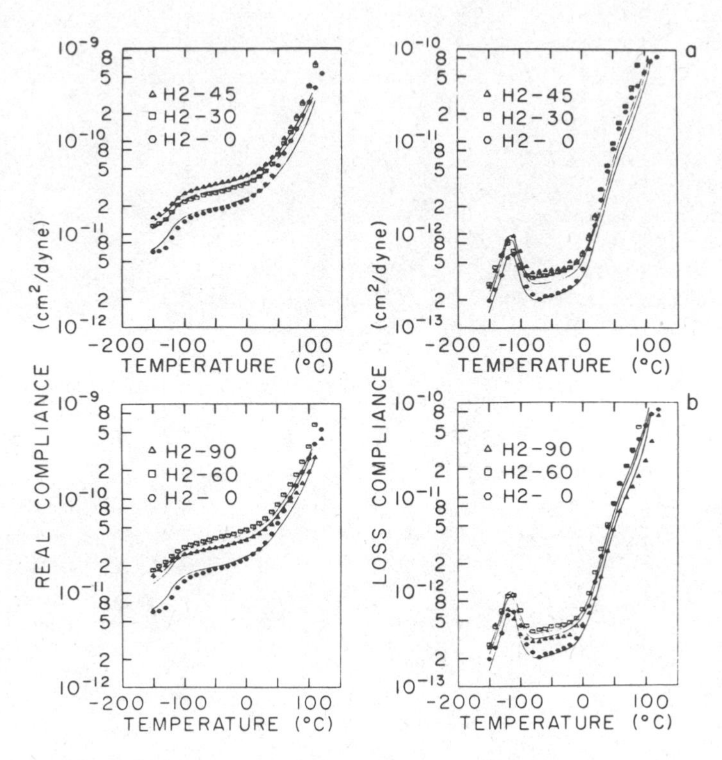

Figure(8): Storage and Loss compliance as a function of temperature, for film H2 of linear polyethylene. The solid lines are the predictions of the modified two phase model. a) Results for 0° (hexagons), 30° (squares), and 45° (triangles) to the draw direction. b) Results for 0° (hexagons), 60° (squares), and 90° (triangles) to the draw direction.

non-crystalline phase intrinsic compliance Z_t, the predictive ability of the modified two-phase model improves tremendously, as demonstrated in Figure(8). The real and imaginary values of Z_t were calculated from equation(17) and applied in the two phase model equations (7-10), in substitution of the A_a storage and loss values. The agreement of the experimental data (data points with the calculated curve (solid line) in Figure(8) remained excellent in the -150° to $-70^{\circ}C$ temperature range, and improved tremendously above $-70^{\circ}C$.

CONCLUSIONS

This study summarizes recent developments of our two phase reduced model which can be used successfully to predict anisotropic dynamic mechanical properties of uniaxially oriented film samples of isotactic polypropylene, as a function of temperature and frequency. To make such predictions, only the knowledge of fundamental parameters, like percent crystallinity, crystalline and non-crystalline orientation distributions, and the intrinsic compliance based dynamic mechanical properties of the unoriented crystalline and non-crystalline phases are needed. We have also shown that, for applications to polyethylene, the two phase model may be modified with the inclusion of a conformation dependent intrinsic compliance of the unoriented non-crystalline phase. The modified two-phase model can be used to successfully predict the anisotropic dynamic mechanical properties of linear polyethylene films in a wide temperature range.

ACKNOWLEDGEMENTS

Support for this work was provided by the National Science Foundation, Polymers Program and Industry University Cooperative Program, Grant No. DMR-8113239. The assistance of E. I. du Pont de Nemours & Company as the Cooperating Industry for this project is also greatly aknowledged.

REFERENCES

[1] J. C. Seferis and R. J. Samuels, Polym. Eng. and Sci., 19, 975 (1979).
[2] J. C. Seferis, R. L. McCullough, and R. J. Samuels, Appl. Polym. Symp., 27, 205 (1975).
[3] J. C. Seferis, R. L. McCullough, and R. J. Samuels, ACS Plast. and Coat. Prepr., 35, 210 (1975).
[4] J. C. Seferis, R. L. McCullough, and R. J. Samuels, Polym. Eng. and Sci., 16, 334 (1976).
[5] R. L. McCullough, C. T. Wu, J. C. Seferis, and P. H. Lindenmeyer, Polym. Eng. and Sci., 16, 371 (1976).
[6] J. C. Seferis, R. L. McCullough, and R. J. Samuels, J. Macromol. Sci.-Phys., B13(3), 357 (1977).
[7] J. C. Seferis, PhD Dissertation, University of Delaware; University Microfilm No. 77-22, 201, Ann Arbor (1977).
[8] A. R. Wedgewood, PhD Dissertation, Department of Chemical Engineering, University of Washington (1982).
[9] P. H. Hermans, "Contributions to the Physics of Cellulose Fibers", Elsevier N.Y. (1946).
[10] R. J. Samuels, "Structured Polymer Properties", Wiley, N.Y. (1974).
[11] J. C. Seferis and R. J. Samuels, 2nd Joint Meeting U.S.-Japan Societies of Rheology, Hawaii, April (1979).
[12] S. M. Dowis, Masters Thesis, Department of Chemical Engineering, University of Washington (1981).
[13] A. R. Wedgewood and J. C. Seferis, Polymer, 22, 967 (1981).
[14] P. A. Coulis, Masters Thesis, Department of Chemical Engineering, University of Washington (1979).
[15] A. R. Wedgewood and J. C. Seferis, Pure and Appl. Chem., 55, 873 (1983).

MECHANISM OF MOISTURE ABSORPTION IN METAL-FILLED EPOXIES

P.S. Theocaris, G.C. Papanicolaou and E.A. Kontou

Department of Theoretical and Applied Mechanics
The National Technical University of Athens
5, Heroes of Polytechnion Avenue
Zographou, Athens 624, Greece

ABSTRACT

Moisture absorption is one of the main parameters affecting the thermomechanical behavior of viscoelastic composites. Absorbed moisture has as a result the lowering of the glass-transition temperature, while, on the other hand, contributes to a change of the residual stresses due to matrix swelling. These effects of absorbed moisture, are closely related to the microstructure of the composite under investigation. More precisely, as it is well known, during the manufacture of a composite material around each one of the inclusions a third phase is developed due to physicochemical interaction between filler and matrix. It has been shown that this extra phase affects the overall behavior of the composite since it regulates the degree of adhesion between filler and matrix. In the case of moisture absorption the extent of this extra phase, which is called "boundary interphase", changes, while, at the same time, there is a change in the mechanical behavior of the composite. Thus, there is an interrelation between moisture absorption, mechanical behavior of the composite and extent of the boundary interphase. In the present paper, a quantitative interrelation of the abovementioned three parameters is derived, while, based on these results, the mechanism of moisture absorption in metal-filled composites is explained.

I. INTRODUCTION

Interfacial interactions between filler and matrix is one of the main parameters affecting the thermomechanical properties as well as the hygrothermal behavior of polymeric composites. Especially in the case where the dispersed phase inclusions are virtually incompressible, (mineral fillers, fiber-glass, etc.), as compared with the polymeric matrix, an interphase layer of varying thickness whose properties differ from properties within the bulk is created at the close vicinity of the filler surface due to the adsorption interaction of the polymer with the solid surface [1,2].

The existense of the interphase layer has been proved by i.r. spectroscopy, e.s.r., n.m.r., electron microscopy and other experimental methods [3]. Moreover, it has also been proved that the thickness of this layer depends on the polymer cohesion energy, free surface energy of the solid, and on the flexibility of the polymer chains [3].

There are several types of microheterogeneity appeared in the interphase layer. Namely, molecular heterogeneity determined by the macromolecular structure of the polymer chains, structural microheterogeneity, microheterogeneity on the supermolecular level, and finally, chemical microheterogeneity caused by the effect of the interface on the reaction of formation of polymeric molecules [4].

All these types of microheterogeneities characterizing the interphase layer, affect the overall behavior of the composite, while the effective thickness of this layer depends on the property under consideration [4]. In a series of previous papers [5-9], the authors have studied extensively, the effect of the interphase layer on the thermomechanical properties of epoxy composites, reinforced either with metallic powders, or with glass fibers. In these investigations some theoretical models have been developed and theoretical results were in good agreement with experimental findings.

In the present paper, the effect of the interphase layer on the moisture absorption mechanism of iron-epoxy particulate composites was investigated, and the extent of this layer during the moisture absorption process was evaluated.

II. EXPERIMENTAL

The matrix material used in the tests was a diglycidyl ether of bisphenol A epoxy polymer, with an epoxy equivalent 185-192, a molecular weight between 370 and 384, and a viscosity of 15×10^{3}cP at 25°C. Triethylene-tetramine was employed as curing agent with 8 percent by weight of epoxy. The epoxy matrix was filled with iron particles of an average diameter of 150μm. The composites were manufactured in the way presented in reference [10]. Five distinct filler concentrations (0%,5%,10%,16%,20%) were used, in order to study the effect of filler-volume fraction on the properties of the composites. The thermomechanical behaviour of the same composites was investigated by the authors in previous publications [10-15].

For the percent weight change, due to moisture absorption, specimens in the form of thin plates with dimensions 0.003m in thickness, 0.012m in width and 0.050m in length were used in order to have one-dimensional mode of diffusion. Before placing the test specimens into the water bath they were dried in an oven at 50°C, until their weight-loss was stabilized. The dried specimens were then placed in a high vacuum for 24h to create full-dried specimens and their weights were measured in an analytical balance. Then the specimens were immersed in a distilled-water bath at constant temperature 20°C, controlled to less than ±0.5°C. The specimens, conditioned in water, were then removed periodically, wiped, air dried for 5min, and then weighed.

The tensile properties of the materials were determined by using dogbone tension specimens with standard dimensions at their measuring area, as follows: 0.003m in thickness, 0.006m in width and 0.042m in length. During the tests, the elastic moduli and their ultimate strength were measured in an Instron tester with a crosshead speed of 0.5mm/min.

Finally the glass transition temperature of the composites was measured with a thermal analyser, combined with a DSC analyser. Large diameter aluminum hermetically sealed pans were used in all runs. The specimens used for the DSC tests had the following dimensions: 0.001m in thickness, 0.004m in width and 0.004m in length. Scans were made at 5°C/min heating rate.

III. RESULTS AND DISCUSSION

The incorporation of fillers into polymeric matrices has as a result, the production of composites with highly improved mechanical properties. However, the degree of improvement of the mechanical properties is a function of adhesion efficiency between filler and matrix. This is due to the fact that the existence of the impermeable filler surface into the bulk matrix result in restrictions of segmental mobility thus creating different conformations of the macromolecules which are at the close vicinity to the inclusions. The change in mobility concerns not only the polymer layers which are in direct contact with the filler-surface, but also the more remote layers [4].

On the other hand, absorbed water behaves as a classical plasticizer [16,17] by lowering the T_g of the polymeric composite as a function of the amount absorbed.

Thus, in studying the mechanism of moisture absorption in polymeric composites, we are dealing with the combined effect of filler and plasticizer. A common feature of both effects is the

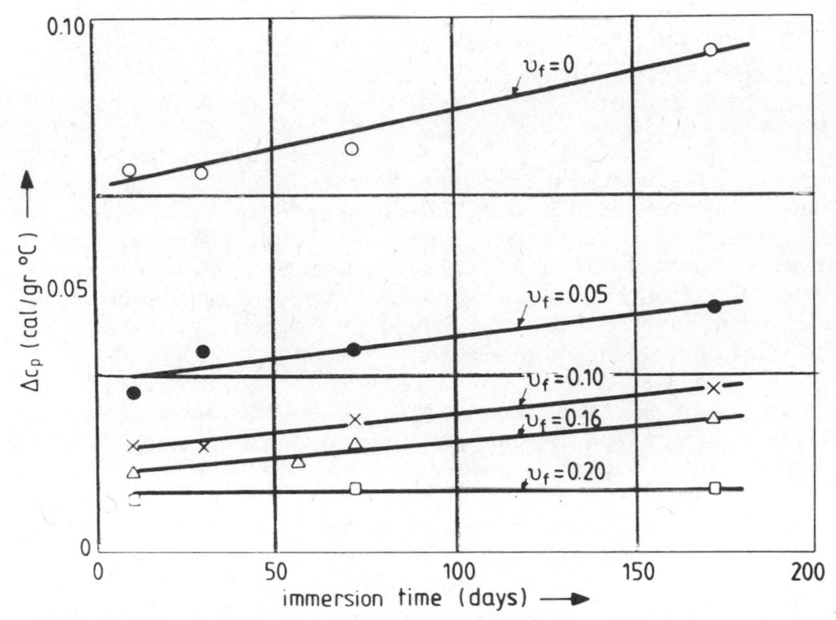

Fig. 1. Variation of Δc_p with time of immersion into distilled water for various filler-volume fractions.

variation of the interphase layer fraction with filler volume fraction as well as with time of immersion into water. In the latter case, the action of moisture results in the partial disruption of the bonds between filler and matrix and the formation of additional cavities, which would be filled with water. It has been established [18] that water enters at the interface at a rate of approximately 450times more reapidly than in the bulk matrix itself, and consequently any absorbed moisture is concentrated preferentially in the interphase.

According to Lipatov theory [4], the effective thickness of the interphase layer may be evaluated from the jump in heat capacity Δc_p observed in glass transition region. Fig.1 shows the variation of Δc_p in the glass transition region of iron-epoxy composites with time of immersion and for various filler-volume fractions. From this figure it becomes clear that the Δc_p magnitude is an increasing function of time of immersion while at the same time it is reduced with the increase of filler-volume fraction.

Following the theory presented in ref.[4], the decrease in magnitude of Δc_p is a definite indication that a certain part of macromolecules is excluded from the cooperative process of glass transition because of their interaction with the filler-surface.

This part of macromolecules forms the interphase layer proper of macromolecules close to the surface, where the mobility of macromolecules is considerably reduced. Thus, there is a relation between Δc_p and the interphase-volume fraction, which is expressed by:

$$\lambda_i = 1 - \frac{\Delta c_{p_f}}{\Delta c_{p_{unf.}}} \tag{1}$$

where Δc_{p_f}, and $\Delta c_{p_{unf.}}$ are the thermal capacity jump for the filled and the unfilled polymer respectively, while λ_i is the fraction of excluded macromolecules. Then the thickness of the interphase layer can be approximately evaluated from the equation

$$\left(\frac{\Delta r_i + r_f}{r_f}\right)^3 - 1 = \lambda_i \frac{\upsilon_f}{1-\upsilon_f} \tag{2}$$

where Δr_i is the thickness of the interphase layer, r_f is the average radius of the filler particle and υ_f is the filler-volume fraction.

From relations (1) and (2) it may be derived that the interphase-volume fraction υ_i is expressed by:

$$\upsilon_i = \frac{3\upsilon_f \Delta r_i}{r_f} \tag{3}$$

Relations (1) to (3) may now be used in order to study the combined effect of filling and plasticization in terms of the variation of interphase-volume fraction.

Figures 2 and 3, show the variation of both, the experimental values of the relative moduli E_c/E_m (where E_c=modulus of the composite and E_m=modulus of the matrix material) and that of the thickness Δr_i of the interphase layer, with time of immersion, as it is derived from Eq.(2). Moreover in Fig.4 the percentage weight-change of specimens for various filler-volume fractions is plotted against time of immersion.

If one considers the law of variation of the functions shown in Figs.2,3 and 4, then the following mechanism of moisture absorption may be established.

According to this mechanism, the moisture absorption process may be devided into three periods of time.

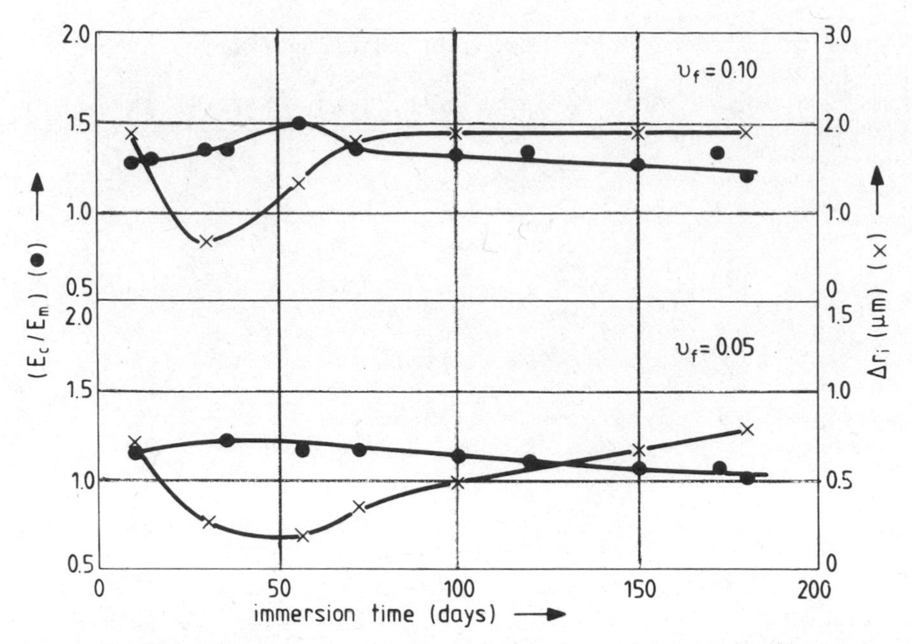

Fig. 2. Variation of both, the experimental values of the relative moduli E_c/E_m of the composites and that of the thickness Δr_i of the boundary interphase with time of immersion for various filler-volume fractions.

At the initial period of time, extended to about 30 to 60 days, the relative modulus E_c/E_m exhibit a slight increase while at the same time there is a simultaneous decrease of Δr_i. At the same initial period of time, as it is shown in Fig.4, the absorption rate is high and at the end of this period the specimen has attained the 70% of its equilibrium moisture content. This kind of behavior may be explained by the fact that since the interphase layer is characterized by voids, microcracks and other defects which create preferential diffusion path [19], it is obvious that at the initial time interval of immersion, absorbed moisture occupies the empty spaces available in the interphase layer, leading to a plasticization of the interphase material. Thus, at this period of time the interphase layer shrinks, because of the stresses exerted on it from the bulk matrix. However, the decreasing of the interphase thickness leads to a better adhesion between filler and matrix and consequently to an increase in the ratio E_c/E_m.

During the second period of time, the relative modulus E_c/E_m decreases with time of immersion while at the same time the thickness of the interphase layer increases. This period of time

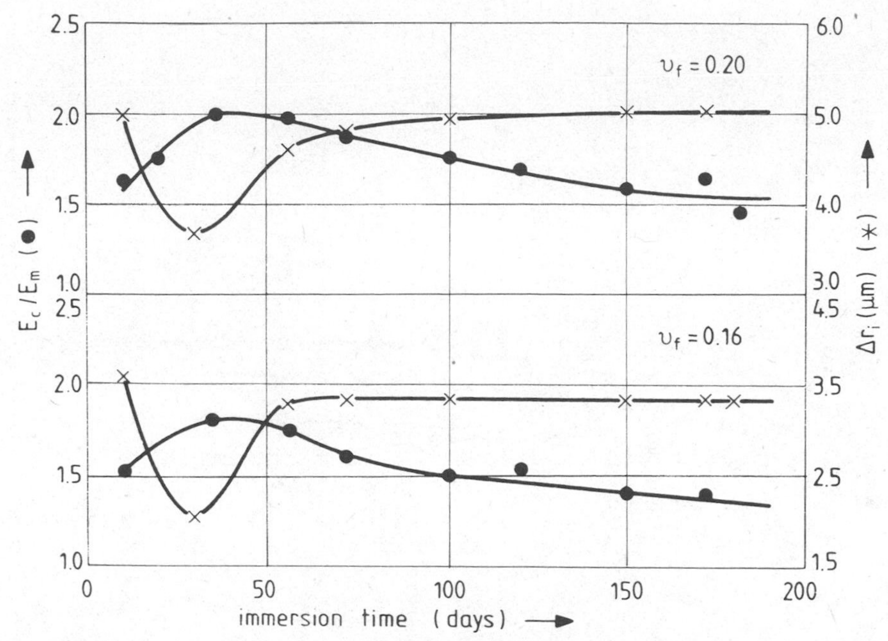

Fig. 3. Variation of the experimental values of the relative moduli E_c/E_m of the composites and the thickness Δr_i of the boundary interphase with time of immersion for various filler-volume fractions.

is also characterized by a low rate of absorption as it is shown in Fig.4. This kind of behavior shows that during this period, absorbed moisture plasticizes the bulk matrix and thus the already plasticized interphase layer is expanded. The increase of the interphase thickness leads to a weaker bonding of the two main phases and consequently to a respective decrease of the value of the ratio E_c/E_m. This period of time is extended up to the time where the degree of plasticization of both the bulk matrix and the interphase material become equal.

Then, a third period of time starts where the equilibrium is reached and both E_c/E_m and Δr_i attain their initial values which remain constant for the rest time of immersion.

CONCLUSIONS

In the present paper a mechanism of moisture absorption in particulate-epoxy composites was introduced. According to this mechanism the interphase layer regulates the moisture absorption process and the resulting variation of the mechanical behavior of the composite.

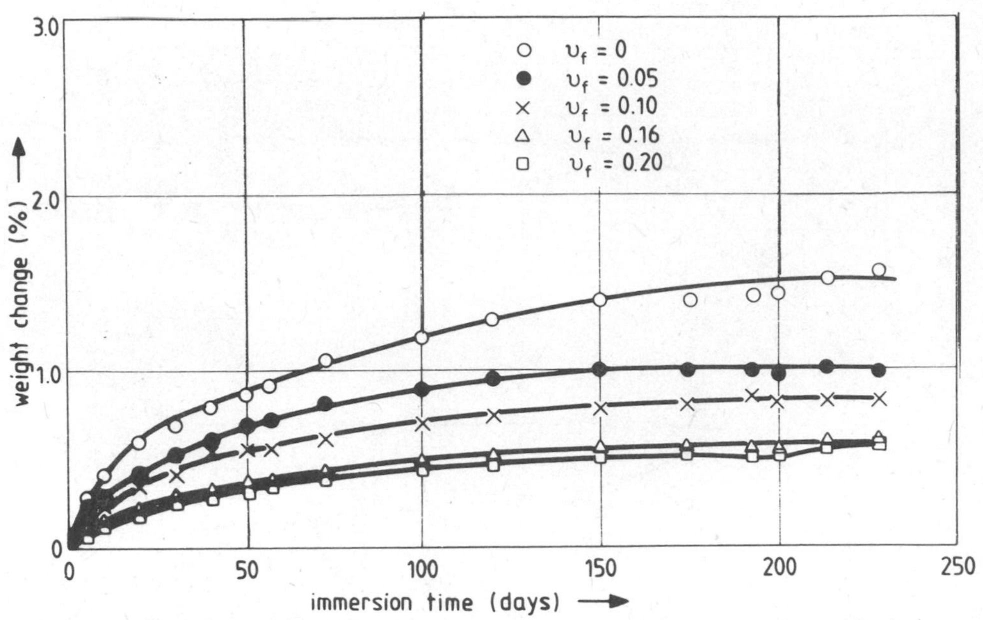

Fig. 4. Variation of the percent weight change due to moisture absorption versus time for different filler-volume fraction.

REFERENCES

1. Yu. S. Lipatov, "Physical Chemistry of Filled Polymers", (Russ). Kiev. (1967).
2. Yu. S Lipatov, L. Sergeeva, "Adsorption of Polymers", New York: J. Wiley (1974).
3. Yu. S. Lipatov, Jnl. Polymer Sci. C., Vol.42, p.855 (1973).
4. Yu. S. Lipatov, Advances in Polymer Science, Vol.22, pp.1-59, (1975).
5. G.C. Papanicolaou, S.A. Paipetis and P.S. Theocaris, Colloid and Polymer Science, Vol.256, 7, p.625 (1978).
6. G.C. Papanicolaou and P.S. Theocaris, Colloid and Polymer Science, Vol.257, 3, p.239 (1979).
7. P.S. Theocaris and G.C. Papanicolaou, Fibre Science and Technology, Vol.12, 6, p.421 (1979).
8. G.C. Papanicolaou, P.S. Theocaris and G.D. Spathis, Colloid and Polymer Science, Vol.258, 11, p.1231 (1980).
9. P.S. Theocaris, G.C. Papanicolaou and G.D. Spathis, Fibre Science and Technology, Vol.15, p.187 (1981).
10. G.C. Papanicolaou, S.A. Paipetis and P.S. Theocaris, Jnl. Applied Pol. Science, Vol.21, p.689 (1977).
11. P.S. Theocaris, G.C. Papanicolaou and E.P. Sideridis, Jnl. Reinforced Plastics and Composites, Vol.1, p.90 (1982).

12. S.A. Paipetis, G.C. Papanicolaou and P.S. Theocaris, Fibre Sci. and Technology Vol.8, p.221 (1975).
13. P.S. Theocaris, S.A. Paipetis and G.C. Papanicolaou, Jnl. Appl. Pol. Sci., Vol.22, p.2245 (1978).
14. P.S. Theocaris, G.C. Papanicolaou and G.A. Papadopoulos, Jnl. of Comp. Mat., Vol.15, p.41 (1981).
15. P.S. Theocaris, G.C. Papanicolaou and E.A. Kontou, Jnl. of Reinforced Plastics and Composites, Vol.1, p.206 (1982).
16. C.E Browning, Polymer Engng. and Science, Vol.18, 1, p.16 (1978).
17. C.E. Browning, G.E. Husman and J.M. Whitney, ASTM STP 617, p.481 (1977).
18. H. Lee and K. Neville "Handbook of Epoxy Resins", McGraw-Hill, New York, 1967, p.22-53.
19. AFML/AFOSR, Mechanics of Composites Review, Dayton, (1976).

INDEX